THE LAST SUPPER ON THE MOON

ALSO BY LEVI LUSKO

Through the Eyes of a Lion: Facing Impossible Pain, Finding Incredible Power

Swipe Right: The Life-and-Death Power of Sex and Romance

I Declare War: Four Keys to Winning the Battle with Yourself

Take Back Your Life: A Forty-Day Interactive Journey to Thinking Right So You Can Live Right

Roar Like a Lion: Ninety Devotions to a Courageous Faith

THE LAST SUPPER ON THE MOON

NASA'S 1969 LUNAR VOYAGE, JESUS CHRIST'S BLOODY DEATH, AND THE FANTASTIC QUEST TO CONQUER INNER SPACE

LEVI LUSKO

W Publishing Group
An Imprint of Thomas Nelson

The Last Supper on the Moon

Published in Nashville, Tennessee, by W Publishing Group, an imprint of Thomas Nelson.

Published in association with the literary agency of Wolgemuth & Associates, Inc.

Cover photo by Riley Connell

Front matter schematics, diagrams in Chapter 8, and small lunar module icon appearing throughout the body text (outlined image placed at text breaks) created by Marcin Szpak.

Thomas Nelson titles may be purchased in bulk for educational, business, fundraising, or sales promotional use. For information, please email SpecialMarkets@ThomasNelson.com.

All italics used in quoted Bible verses indicate emphasis added by the author.

ISBN 978-0-7852-5285-6 (hardcover)
ISBN 978-0-7852-5288-7 (audiobook)
ISBN 978-0-7852-5287-0 (eBook)
ISBN 978-0-7852-5286-3 (TP)

Library of Congress Control Number: 2021945428

Printed in the United States of America

23 24 25 26 27 LBC 5 4 3 2 1

TO LENNOX AND MY LUSKO LADIES.
I LOVE YOU TO THE MOON AND BACK. WAFFLES ON SATURDAY. ALWAYS.

Behold, we are going up to Jerusalem, and the Son of Man will be betrayed to the chief priests and to the scribes; and they will condemn Him to death, and deliver Him to the Gentiles to mock and to scourge and to crucify. And the third day He will rise again.

—JESUS CHRIST (MATT. 20:18–19)

No one takes it from Me, but I lay it down of Myself. I have power to lay it down, and I have power to take it again. This command I have received from My Father.

—JESUS CHRIST (JOHN 10:18)

Nevertheless I tell you the truth. It is to your advantage that I go away; for if I do not go away, the Helper will not come to you; but if I depart, I will send Him to you.

—JESUS CHRIST (JOHN 16:7)

Most assuredly, I say to you, he who believes in Me, the works that I do he will do also; and greater works than these he will do, because I go to My Father.

—JESUS CHRIST (JOHN 14:12)

I believe that this nation should commit itself to achieving the goal, before this decade is out, of landing a man on the moon and returning him safely to the Earth. No single space project in this period will be more impressive to mankind, or more important for the long-range exploration of space; and none will be so difficult or expensive to accomplish.

—PRESIDENT JOHN F. KENNEDY, ADDRESS TO JOINT SESSION OF CONGRESS IN WASHINGTON, DC, MAY 25, 1961

Those who came before us made certain that this country rode the first waves of the industrial revolutions, the first waves of modern invention, and the first wave of nuclear power, and this generation does not intend to founder in the backwash of the coming age of space. We mean to be a part of it. . . . We set sail on this new sea because there is new knowledge to be gained, and new rights to be won, and they must be won and used for the progress of all people. . . . But why, some say, the moon? Why choose this as our goal? . . . We choose to go to the moon in this decade and to do the other things not because they are easy, but because they are hard; because that goal will serve to organize and measure the best of our energies and skills; because that challenge is one that we're willing to accept, one we are unwilling to postpone, and one which we intend to win.

—PRESIDENT JOHN F. KENNEDY, ADDRESS AT RICE UNIVERSITY IN HOUSTON, TEXAS, SEPTEMBER 12, 1962

The problems of the world cannot possibly be solved by skeptics or cynics, whose horizons are limited by the obvious realities. We need men who can dream of things that never were, and ask why not.

—PRESIDENT JOHN F. KENNEDY, ADDRESS BEFORE THE IRISH PARLIAMENT IN DUBLIN, JUNE 28, 1963

CONTENTS

Before Liftoff . . . XI
Begin Here . . . XXVII
A Promise . . . XLIII

Chapter 1 Everyone Is a Moon . . . 1
Chapter 1.60769 What the Heck? . . . 9
Chapter 2 The Crowded Hour . . . 11
Chapter 3 So This Is What Elevators Look Like . . . 21
Chapter 4 There Is No Moon . . . 35
Chapter 4.5 On Crucifixion and Centrifuges . . . 53
Chapter 5 Let the Party Continue . . . 71
Chapter 6 Houston, We Have a Problem . . . 87
Chapter 7 Fifteen Seconds to Paradise . . . 109
Chapter 8 If Not Me, Another . . . 133
Chapter 9 Out of This World . . . 149
Chapter 10 Nine and a Half Fingers . . . 163
Chapter 11 Rite of Passage . . . 175
Chapter 12 Burn, Baby! Burn! . . . 193
Chapter 13 Go, No Go . . . 203
Chapter 14 Hakuna Matata . . . 221

Chapter 15 Good Luck and Godspeed 231
Chapter 16 Eight Comes Before Eleven 247
Chapter 17 Rock(s) of Ages 265
Chapter 18 On Shuttles and Roads 279
Chapter 19 Steak and Eggs 287
Chapter 20 Tattooed Soul 301
Chapter 21 NASA Meets Napa 315
Chapter 22 Radio Silence 327
Chapter 23 Masks and Thermometers 347
Chapter 24 Blood, Sweat, and Tears 359

End Here 373
Acknowledgments 377
Notes 383
About the Author 431

BEFORE LIFTOFF

GLOSSARY OF TERMS

Here is a list of terms that will be helpful for you on your journey.

Wherever a pronunciation is not given for an acronym, the term is sounded out by letter.

abort: the unscheduled termination of a mission prior to its completion

Apollo: the program that resulted in American astronauts walking on the moon; twelve missions using three-man spacecrafts between 1968 and 1972, after Mercury and Gemini (Apollo 1 ended in tragedy with three astronauts dying. Four of the twelve flights tested the equipment. Six of the other seven flights landed on the moon.)

Apollo spacecraft: the CSM (command and service module) and LM (lunar module) when docked together

CapCom: capsule communicator

CM: command module; the part of the Apollo spacecraft that contained the crew during takeoff from and reentry to earth

Columbia: the name of the CM (command module) on Apollo 11

cryo: hydrogen and oxygen fuel stored at extremely cold temps

CSM: command and service module; two distinct units, the CM (command module) and the SM (service module), when connected together

Eagle: the name of the LM (lunar module) on Apollo 11

EECOM [pronounced *eecom*]: electrical, environmental, and consumables manager; MCC (Mission Control Center) engineer responsible for electrical, environmental, and communications in the CSM (command and service module), including cryogenic, fuel cell, and structural systems

EVA: extravehicular activity; also called a space walk

FDO [pronounced *fido*]: flight dynamics officer; a specialist in launch and orbit trajectories

flight controller: NASA personnel who oversee various aspects of a spaceflight in real time, interpreting telemetry at their stations in the MOCR (Mission Operations Control Room); they are involved before, during, and after the mission

flight director: manager of flight controllers; role in the MOCR (Mission Operations Control Room) is like a conductor of a symphony

g-force: the force exerted upon an object by gravity or in reaction to acceleration or deceleration

Gemini or Project Gemini: America's second human space program (after Mercury, before Apollo), which tested movements and maneuvers necessary to attempt Apollo; ten missions in a two-man spacecraft between 1965 and 1966

go/no go: the decision to continue to the next event or abort an activity

GNC: guidance, navigation, and controls system engineer; MCC (Mission Control Center) engineer responsible for managing propulsion, altitude control, guidance and navigation systems in the CSM (command and service module)

Guido or Guidance: MCC (Mission Control Center) specialist in navigation and computer software systems

JSC: Johnson Space Center in Houston, Texas

KSC: John F. Kennedy Space Center on Merritt Island, Florida

LES: launch escape system; the part of the rocket that can propel the astronauts and their capsule away from the launch vehicle or launchpad in the event of an emergency during takeoff or ascent

LM [pronounced *lem*]: lunar module (originally called lunar excursion module); the part of the Apollo spacecraft that landed on the moon

MCC: Mission Control Center in Houston, Texas

Mercury or Project Mercury: America's first human space program involving a one-man spacecraft traveling to space and then into orbit (There were six missions on Redstone and Atlas rockets between 1961 and 1963.)

MOCR [pronounced *moker*]: Mission Operations Control Room at the MCC (Mission Control Center) in Houston, Texas; front row of this room was nicknamed "the trench"

NACA: National Advisory Committee for Aeronautics; formed in 1915 and absorbed by NASA in 1958

NASA [pronounced *nasa*]: National Aeronautics and Space Administration; created on October 1, 1958, to oversee US space exploration and aeronautics research; led by an administrator nominated by the president and confirmed by the Senate

pogo: a rapid up-and-down shaking of a rocket that, if not corrected, will cause failure

powered descent: a maneuver that involves firing thrusters to assist a spacecraft in landing on the surface of a planet or moon

reentry: when a spacecraft reenters the atmosphere after flying above it

retro: MCC (Mission Control Center) specialist in reentry trajectories

Saturn V: three-stage launch vehicle that transported the Apollo spacecrafts from the earth toward the moon and carried Skylab, the United States' first space station, into orbit (Thirteen were launched.)

SM: service module; the part of the Apollo spacecraft that contained the main engines and most of the consumables (oxygen, water, helium, fuel cells, and fuel) jettisoned before reentry

splashdown: the process of landing a spacecraft in the ocean using multiple sets of parachutes that slow it down before it hits the water

stage: a section of a rocket that contains an engine or group of engines; the stages usually separate from the rocket when they have used up their fuel

surgeon: directs medical activities during the flight and monitors the health of the astronauts via telemetry

TELMU [pronounced *telmu*]: telemetry, electrical, and EVA (extravehicular activity) mobility unit officer; monitors LM (lunar module) electrical and environmental control systems

thrust: the force produced by the engines of a rocket or plane directed forward or upward

TLI: translunar injection; a maneuver to leave a parking orbit around the earth toward the moon

Tranquility Base: the name of Apollo 11's landing site on the moon

CAST OF CHARACTERS

Though not exhaustive, this list includes some of the people you will meet in this book. Feel free to refer back to it to keep who's who straight in your mind.

Alan Shepard: first American in space on Mercury-Redstone 3; grounded due to inner-ear ailment; became chief of the astronaut office; returned to flight as commander of Apollo 14

Charlie Duke: air force test pilot; CapCom (capsule communicator) during Apollo 11 moon landing; lunar module pilot on Apollo 16

Christopher Columbus Kraft Jr.: the original flight director;

instrumental in developing the functionalities and procedures of Mission Control; at the helm of Mission Control from Mercury until the beginning of the Apollo program

Donald K. "Deke" Slayton: air force test pilot; Mercury astronaut grounded by a heart condition; became director of flight crew operations and was responsible for NASA crew assignments; returned to flight and was Apollo docking module pilot on the Apollo-Soyuz Test Project

Edward White II: air force test pilot; pilot on Gemini 4; made the first American EVA (extravehicular activity), or space walk, on Gemini 4 mission; command pilot on Apollo 1

Edwin "Buzz" Aldrin Jr.: air force pilot; first astronaut with a doctorate; nicknamed "Dr. Rendezvous"; pilot on Gemini 12; lunar module pilot on Apollo 11 (the first spaceflight to land humans on the moon)

Frank Borman: air force test pilot; command pilot on Gemini 7; commander on Apollo 8 (the Christmas flight)

Gene Cernan: pilot on Gemini 9; commander for Apollo 17; person who has walked on the moon most recently

Gene Kranz: air force pilot; long-standing NASA flight director throughout the space race; chief flight director for the Apollo 11 mission; wore iconic vests and flattop haircut

Jack Garman: computer engineer and specialist; part of Steve Bales's "back room" support; enabled the Apollo 11 mission not to abort during the 1202 program alarm crisis by giving knowledge of NASA computer codes in split-second timing; chief information officer at JSC (Johnson Space Center)

James [Jim] Lovell Jr.: navy test pilot; pilot on Gemini 7; command pilot on Gemini 12; command module pilot on Apollo 8 (the Christmas flight); commander on Apollo 13; first person to fly to the moon twice

James Webb: NASA administrator during Mercury and Gemini programs

John Glenn: navy test pilot; first American to orbit the earth on Mercury-Atlas 6; became a national hero; after a political career, returned to space at age 77 as a payload specialist on STS-95, becoming the oldest person to fly in space at that time

Katherine Johnson: American mathematician and human computer; one of the many women who were a vital part of the space program and without whom NASA would not have been able to get off the ground; provided calculations of orbital mechanics that were indispensable from Mercury all the way through Apollo; of her computations John Glenn said before his mission, "If she says they're good, then I'm ready to go."

Michael Collins: air force test pilot; pilot on Gemini 10; command module pilot on Apollo 11 (the first spaceflight to land humans on the moon)

Neil Armstrong: civilian; NACA test pilot; command pilot on Gemini 8; commander on Apollo 11; first person to walk on the moon

Peggy Whitson: biochemist; flight engineer on STS-111; first female commander of the ISS (International Space Station) on Expedition 16; commander of Expedition 51; current record holder for most cumulative days (665) spent in space; at age 57, oldest woman to go to space at the time of her mission; chief of the Astronaut Office

President John F. Kennedy: effortlessly cool, beloved visionary leader who threw down the gauntlet of the moonshot; was in office during Project Mercury and was the friend of several astronauts; was tragically assassinated in November 1963 before the dream of the moon landing was realized

President Lyndon B. Johnson: sworn in as president after President Kennedy was assassinated in Dallas, Texas; under his presidency Congress appropriated funds to fulfill President Kennedy's vision through Gemini and the beginning of the Apollo program

President Richard M. Nixon: president during the Apollo 11 moon landing; spoke to Neil Armstrong and Buzz Aldrin from the White House during their EVA and greeted them in person when they returned

Robert Goddard: father of modern rocketry; patented liquid-fueled and multi-stage rockets in 1914; postulated in 1920 that humans could reach the moon using rockets; was not recognized for his ideas or accomplishments in spaceflight until after he was dead

Roger Chaffee: navy pilot; pilot on Apollo 1

Shane Kimbrough: army helicopter pilot; mission specialist on STS-126; flight engineer on Expedition 49; commander of the ISS (International Space Station) on Expedition 50; commander of the NASA SpaceX Crew-2 mission (on which he flew with this book manuscript on a thumb drive)

Steve Bales: NASA engineer and flight controller; served as guido on Apollo 11

Virgil "Gus" Grissom: air force test pilot; pilot on Mercury-Redstone 4; command pilot on Gemini 4; commander on Apollo 1

Walter Cronkite: trusted CBS reporter whose narration was the soundtrack of the space race for the American people; known for his sign-off slogan, "And that's the way it is."

Wernher von Braun: German-born scientist who designed rockets in Germany during WWII; developed rockets for America's space program, most notably the Saturn V used in the Apollo moon landings

William (Bill) Anders: air force pilot; lunar module pilot on Apollo 8 (the Christmas flight)

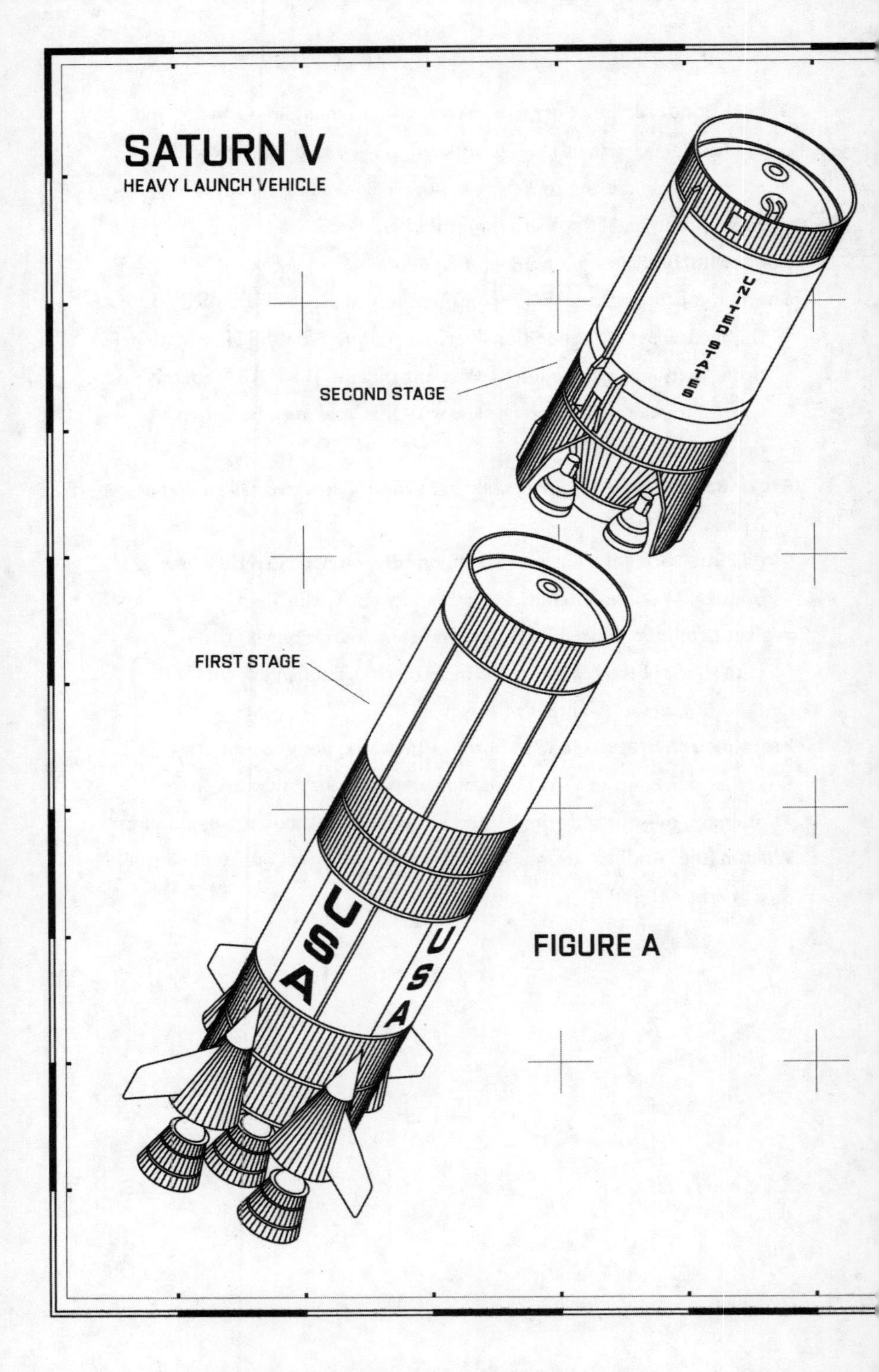
SATURN V
HEAVY LAUNCH VEHICLE
SECOND STAGE
UNITED STATES
FIRST STAGE
USA
USA
FIGURE A

Saturn V was a powerful three-stage rocket built by NASA to send people to the moon. Its three stages would burn until they ran out of fuel; then they would separate from the rocket and the next one would take over. The first stage used five powerful F-1 engines [thus the roman numeral V in the name]. Saturn Vs were flown between 1967 and 1973 for all Apollo moon missions and the last one launched Skylab, the first American space station.

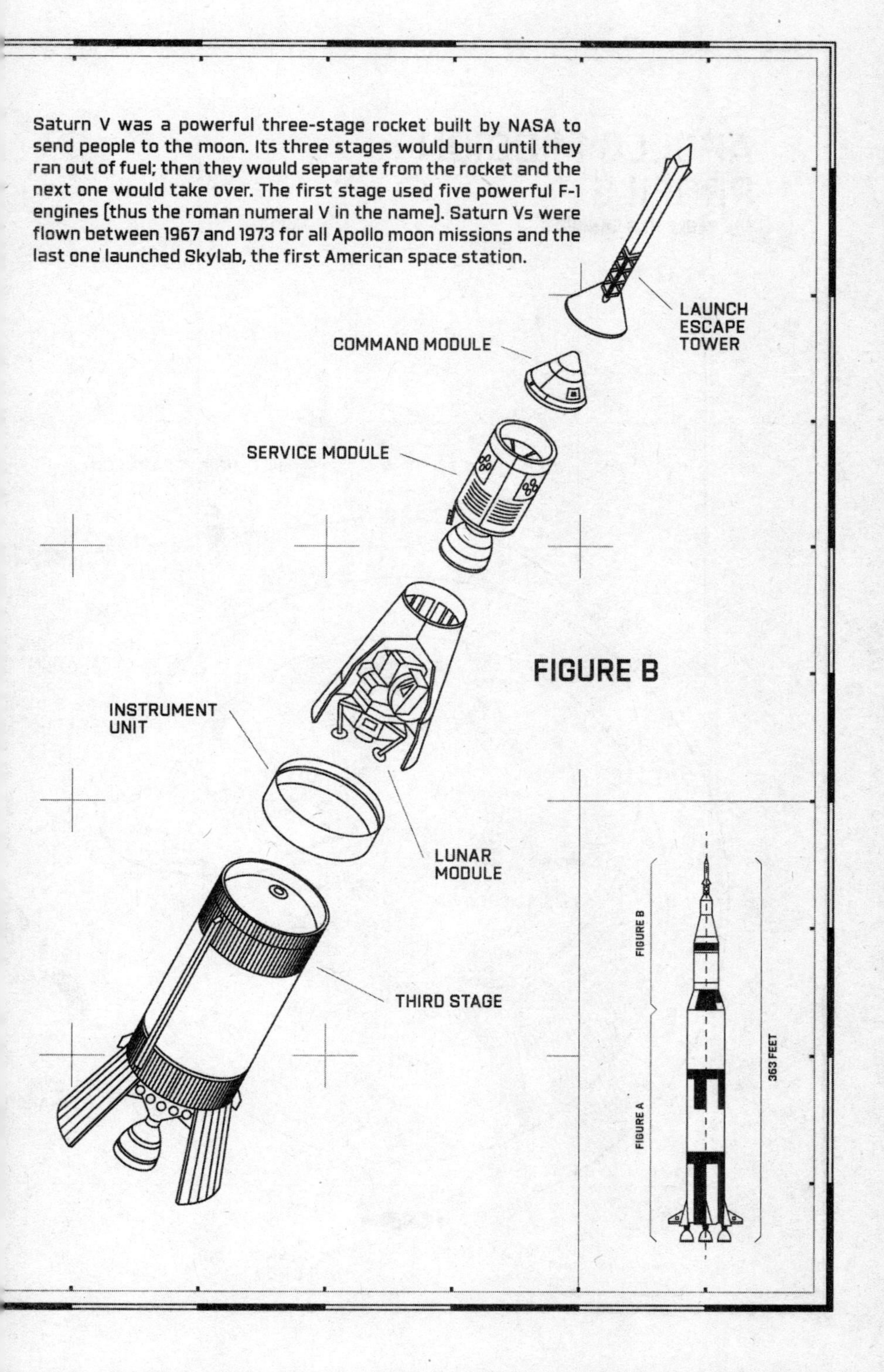

APOLLO MISSION PROFILE

THE EAGLE HAS LANDED...

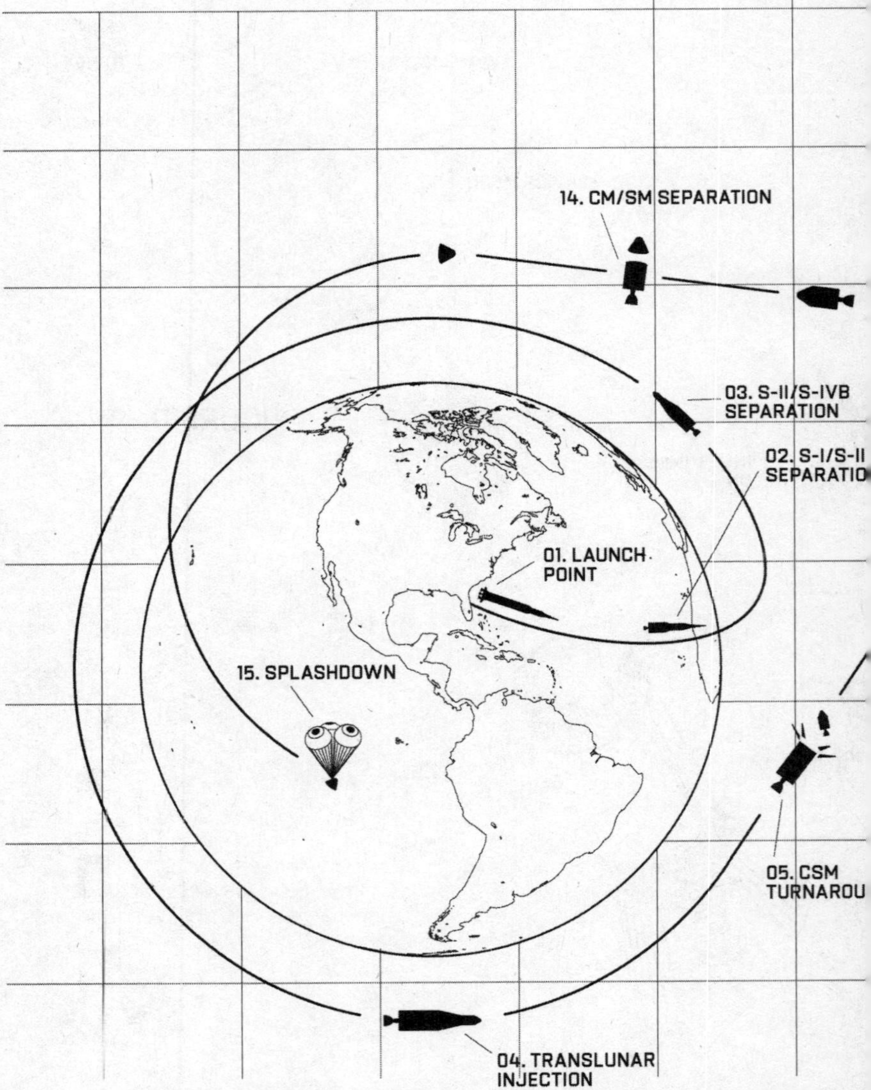

14. CM/SM SEPARATION

03. S-II/S-IVB SEPARATION

02. S-I/S-II SEPARATIO

01. LAUNCH POINT

15. SPLASHDOWN

05. CSM TURNAROU

04. TRANSLUNAR INJECTION

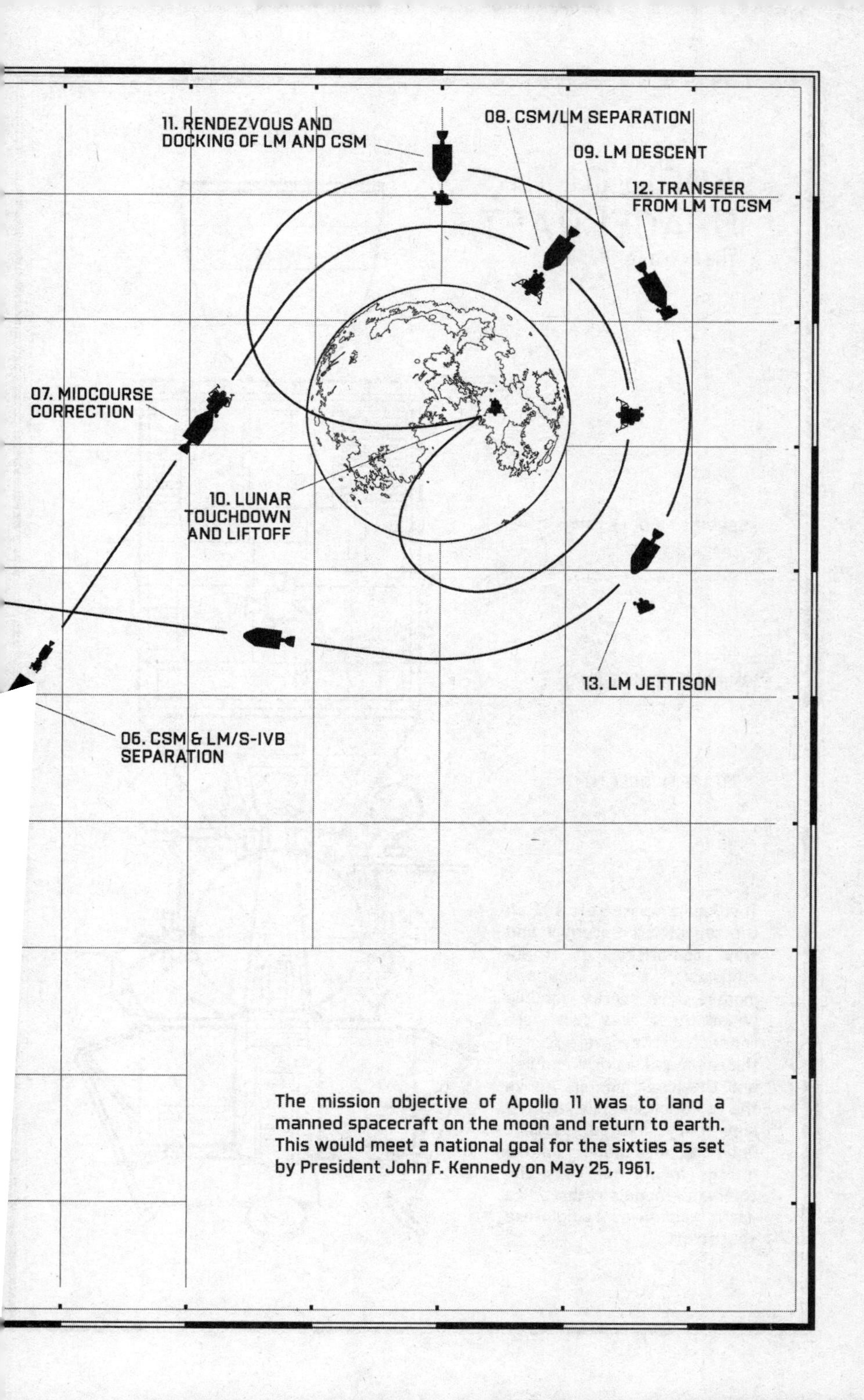

The mission objective of Apollo 11 was to land a manned spacecraft on the moon and return to earth. This would meet a national goal for the sixties as set by President John F. Kennedy on May 25, 1961.

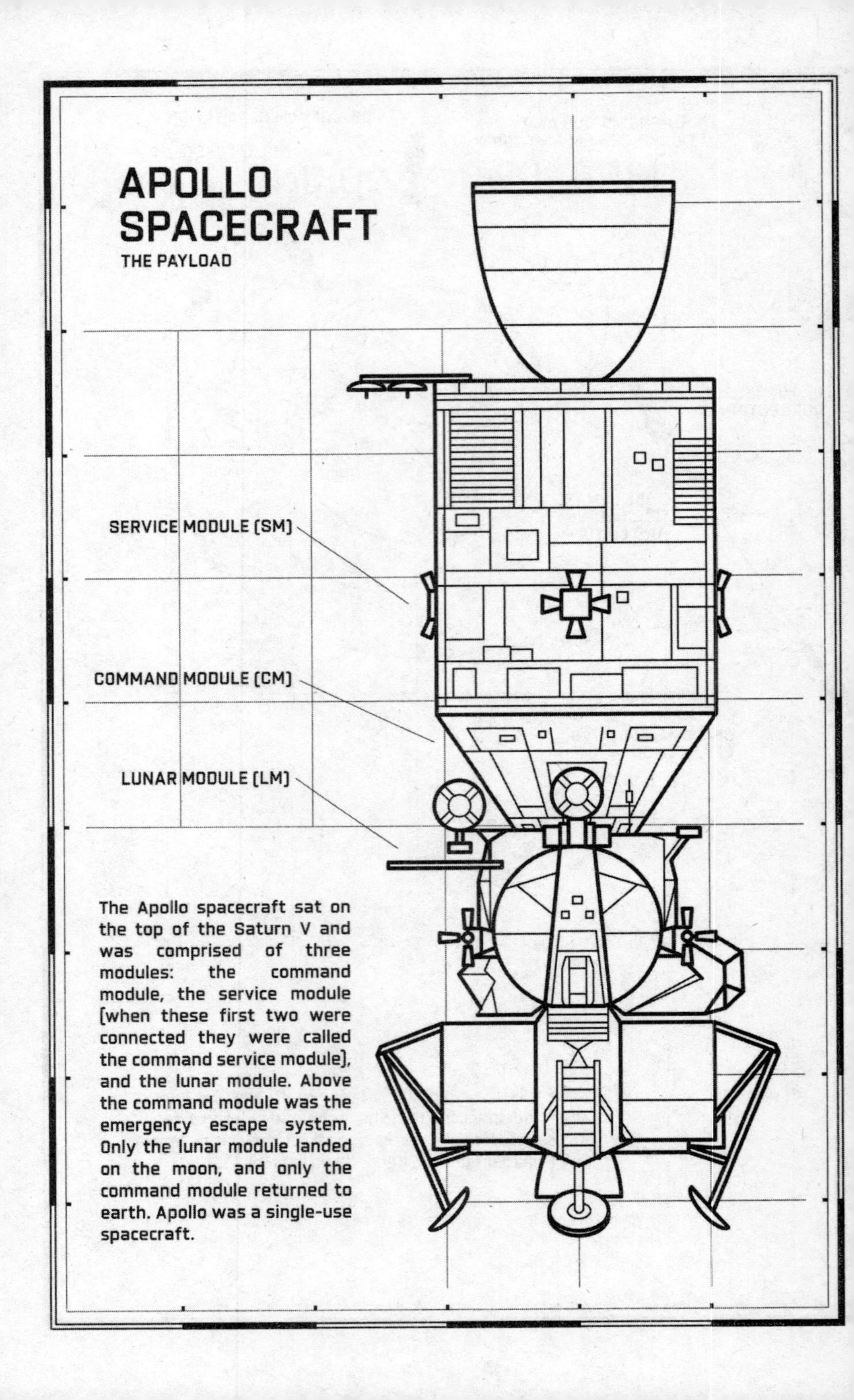
APOLLO
SPACECRAFT
THE PAYLOAD
SERVICE MODULE (SM)
COMMAND MODULE (CM)
LUNAR MODULE (LM)
The Apollo spacecraft sat on the top of the Saturn V and was comprised of three modules: the command module, the service module (when these first two were connected they were called the command service module), and the lunar module. Above the command module was the emergency escape system. Only the lunar module landed on the moon, and only the command module returned to earth. Apollo was a single-use spacecraft.

TRANSPOSITION, DOCKING, AND EXTRACTION

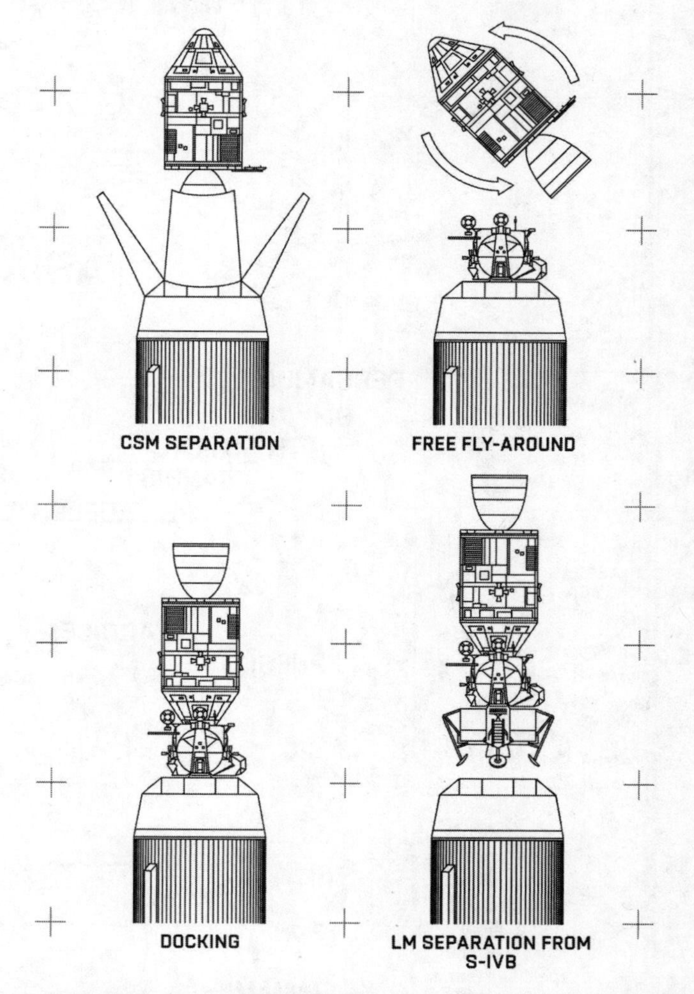

Like a scene from a James Bond movie, the command service module separated from the third stage of the Saturn V rocket (S-IVB) three hours and seventeen minutes after launch. Then it spun around and docked nose to nose with the lunar module, which had been stored in a compartment behind four silver petals that had fallen away. Once they were docked, the LM was released from the third stage and the LM and CSM merged into the Apollo spacecraft.

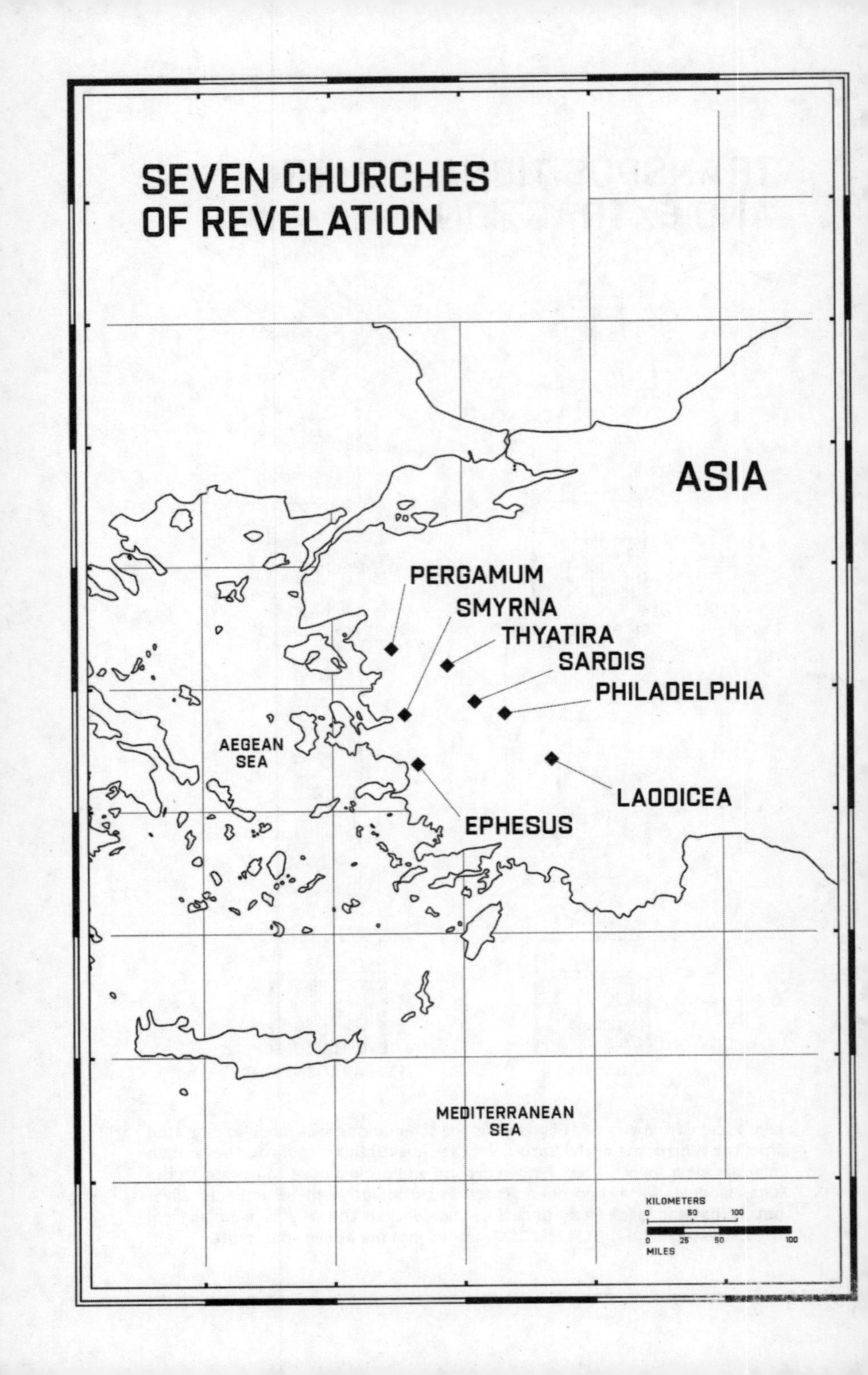
SEVEN CHURCHES
OF REVELATION
ASIA
PERGAMUM
SMYRNA
THYATIRA
SARDIS
PHILADELPHIA
LAODICEA
EPHESUS
AEGEAN
SEA
MEDITERRANEAN
SEA
KILOMETERS
0
50
100
0
25
50
100
MILES

JESUS' LETTERS TO THE SEVEN CHURCHES

Letter #1 to Ephesus (Rev. 2:1–7): "You have left your first love" (v. 4).

Letter #2 to Smyrna (Rev. 2:8–11): "Be faithful until death, and I will give you the crown of life" (v. 10).

Letter #3 to Pergamum (Rev. 2:12–17): "To him who overcomes I will give some of the hidden manna" (v. 17).

Letter #4 to Thyatira (Rev. 2:18–29): "Hold fast what you have till I come" (v. 25).

Letter #5 to Sardis (Rev. 3:1–6): "Remember therefore how you have received and heard; hold fast and repent" (v. 3).

Letter #6 to Philadelphia (Rev. 3:7–13): "I have set before you an open door, and no one can shut it" (v. 8).

Letter #7 to Laodicea (Rev. 3:14–22): "You are neither cold nor hot. I could wish you were cold or hot" (v. 15).

JESUS' SEVEN STATEMENTS FROM THE CROSS

Last Words #1: "Father, forgive them, for they do not know what they do" (Luke 23:34).

Last Words #2: "Assuredly, I say to you, today you will be with Me in Paradise" (Luke 23:43).

Last Words #3: "Woman, behold your son! . . . Behold your mother!" (John 19:26–27).

Last Words #4: "My God, My God, why have You forsaken Me?" (Matt. 27:46).

Last Words #5: "I thirst!" (John 19:28).

Last Words #6: "It is finished!" (John 19:30).

Last Words #7: "Father, 'into Your hands I commit My spirit'" (Luke 23:46).

JESUS' SEVEN "I AM" STATEMENTS

"I Am" #1: "I am the bread of life. He who comes to Me shall never hunger, and he who believes in Me shall never thirst" (John 6:35).

"I Am" #2: "I am the light of the world. He who follows Me shall not walk in darkness, but have the light of life" (John 8:12).

"I Am" #3 and #4: "I am the door. . . . I am the good shepherd" (John 10:7, 11).

"I Am" #5: "I am the resurrection and the life" (John 11:25).

"I Am" #6: "I am the way, the truth, and the life. No one comes to the Father except through Me" (John 14:6).

"I Am" #7: "I am the vine" (John 15:5).

BEGIN HERE

THE MAN READ FROM A THREE-BY-FIVE INDEX CARD HE HAD taken from his pocket, on which the words of Jesus were written:

> I am the vine, you are the branches. Whoever remains in me, and I in him, will bear much fruit; for you can do nothing without me.

Hands shaking, he took out a flask and prepared a drink to steady himself. To mark the moment. He poured the liquid from the flask into a chalice he had brought with him for this very occasion. He had come a long way.

The wine glanced off the bottom and mysteriously curled up the edges of the glass.

He paused, and then lifted it to his lips.

The tart sweetness touched his tongue and millions of receptors fired sensations all over his brain like lights in a cockpit.

Forgiveness.

Quiet.

Jesus' hands held a cup. He lifted it toward his friends.

You will betray me.

Deny me.

Abandon me.

This is my blood of the covenant, which is poured out for many for the forgiveness of sins (Matt. 26:28 NIV).

Taste and see that the Lord is good (Ps. 34:8).

Jesus' hands held a cross.

Father, forgive them, for they do not know what they do (Luke 23:34).

His hands did not shake. For this purpose he had come.

Finished with the wine, the man opened another pouch, mindful of the paper-thin walls—thinner than a Coke can, and easily punctured if even the tip of a pen poked them—that were keeping him alive. Every movement was measured, studied, and precise. He was being watched. Listened to. Recorded. Analyzed. There were probes all over his body.

He looked up and noted the time. It wouldn't be long now.

Millions were waiting. It had taken years to get to this moment. The one who had the idea had died. The man's family was far away.

With the bread in his hand, he looked across the small, crowded compartment and saw the face of an amiable stranger. He lifted the bread, said a prayer, and ate alone.

This is my body.

Broken for you (1 Cor. 11:24).

What I am doing you do not understand now but you will know after this (John 13:7).

While he chewed the food, he opened his eyes. What he saw filled him with wonder.

Jesus' hands hold the world.

"Through the Son everything was created, both in the heavenly realm and on the earth, all that is seen and all that is unseen. Every seat of power, realm of government, principality, and authority—it was all created through him and for his purpose! He existed before anything was made, and now everything finds completion in him" (Col. 1:16–17 TPT).

The meal was pregnant with meaning.

Full of new life.

"Let not your heart be troubled" (John 14:1).

Jesus' feet, though dirty from when he sat unwashed at the table, still smelled of myrrh from when they were anointed by Lazarus's sister, Mary. Those feet would soon be nailed, along with his hands, to a tree.

He had come for all mankind.

The supper ended, and he sang a hymn with his disciples.

He would not eat again until he had risen.

As his body began to assimilate the bread and wine, making them a part of his cellular structure, he prepared for the mission he had come for.

As Eugene Peterson paraphrased, "He was supreme in the beginning, leading the resurrection parade, and he is supreme in the end. From beginning to end he's there, towering far above everything, everyone. So spacious is he, so expansive, that everything of God finds its proper place in him without crowding. Not only that, but all the broken and dislocated pieces of the universe—people and things, animals and atoms—get properly fixed and fit together in vibrant harmonies, all because of his death, his blood that poured down from the cross" (Col. 1:18–20 MSG).

Moments after he swallowed the last of the meal, the man's radio crackled to life. Snapping back to attention, he focused on the hundreds of switches and fuses covering every visible surface.

He, too, had a mission. For this purpose he had come.

In just under two hours, he would leave the crowded space he was in and step into the ages. History had already been made.

The trip they had taken.

The distance they had traveled.

The price that had been paid.

The race that had been won.

The Eagle that had landed.

And the meal he had just eaten.

His name was Buzz Aldrin, and he had just celebrated the Last Supper on the moon.

It wasn't an end.

It was a beginning.

The Communion elements, supplied by his church in Houston, had been approved by the National Aeronautics and Space Administration (NASA) to be taken on the lunar landing. They amounted to the first meal ever eaten on this heavenly body. Before Neil Armstrong walked on the moon, Buzz ate and drank on it. No one has ever traveled so far to eat so little that said so much.

"The last will be first" (Matt. 20:16).

Where Jesus' life ended, yours began.

His final breath put new air in your lungs.

He did not come to make bad good, but to bring the dead to life.

That spark of the gospel, that victorious resurrection-life that comes into the human heart at salvation, is just the start. Where it ends only you can decide.

Today is my birthday.

I am thirty-eight years old, and tonight the final supermoon of the year will appear in the sky. My mom texted me yesterday and told me that it is going to be called a flower moon. She told me that it is a birthday present for me.

When I look at the moon tonight, I will marvel.

I will marvel because it is so far away, and yet it shines so bright.

I will marvel because its brightness is not its own but belongs to another.

I will marvel because even if hidden by cloud cover, it's still there. Veiled but present. Obscured but ever vigilant.

And I will marvel because in the summer of 1969 a man in a tiny spaceship, who flew 240,000 miles with bread and wine in his possession and then thought of the body and blood of Jesus, ate the Last Supper for the first time in the history of the universe on that celestial body. All before filming his friend's descent onto the lunar surface, and then stepping out onto it himself.

I feel the pull of the moon.

It speaks to me. Calls to me. Dares me to dream. Invites me to live fully.

Doesn't it control the tide?

When I look at the sky at night and see the moonlit wonder, the explosions of stars, I ask God, *How is it that you even notice me?*

Are its movements not somewhat hypnotic as it shrinks to a sliver and then grows full-size?

When I see it, I find myself calming down. I take a breath.

You sent us a companion to always be with us, reflecting the light of our sun in the darkness of night.

When I am afraid, I will trust in you and take great comfort from the night-light you left on for me.

When we lift our eyes and see the moon, we are connected to every human who has ever lived. Our lives look different from those of Abraham, Winston Churchill, Aretha Franklin, Chris Farley, George Washington Carver, Amelia Earhart, or Barack Obama. Yet when we look up to the moon in the night sky, we see exactly what they've seen.

Not just similar. The same.

We are not on Endor. There has only ever been one moon above our earth.

When I am stressed or scared, I like to look at the moon and remember those who have gone before me, and the fact that on many nights they probably looked up at the moon while they worried.

David's nauseous guilt over what he did to Uriah (2 Sam. 12:13).

Peter's bitter sorrow over his denial of Christ (Matt. 26:75).

The fear Esther felt as she prepared to step into the chamber of the king (Est. 4:15–16).

Daniel's agony as he defied the order and chose to pray anyway (Dan. 6:10).

The problems they faced have come and gone, and the moon shines on. This is now your time to shine.

Fleeting.

Focused.

Fantastic.

Long after you and I leave this world, the moon will rise in the sky and shine. David called the moon God's "faithful witness in the sky" (Ps. 89:37). Its predictable orbit and nightly glowing point to the one who spoke it, hung it, and calls it out by name. He has a plan for you to shine in the midst of your struggles. Not someday when you sort out all your issues, or when you have a better job, or are out of school. Right now, in the middle of the suffocating smallness of your situation—he wants you to shine.

The moon is not an end but a beginning. Not long from the time of this writing, NASA plans to land on it once more—only this time it will not be as a destination, but in preparation for the next steps of space travel. SpaceX was awarded the contract for the lander with their Starship, and the mission has been named Artemis—the sister of Apollo.

The moon is going to be a base camp for what lies beyond. When we land humans on Mars, it will be because the moon was a springboard for deep-space travel.

This is what I hope this book can be for you: Artemis. A springboard. A base camp. Scaffolding.

Not an end but a beginning. A beginning to a new way to be human. A fresh way to be you—as you were intended to be and are capable of becoming. Childlike but not childish. Without insecurity and toxic thoughts driving how you behave. With vulnerability and empathy. Tapped into kindness and selflessness. Noble, light, and free. Transparent, triumphant, and tender.

In the final book of The Chronicles of Narnia, Aslan said to the children, "You do not yet look so happy as I mean you to be."

He had died for them, risen to give them new life, and begun to unleash the reconciliation of all things. He had done all that was necessary for them to be fully joyful, yet they didn't fully understand or appreciate what was theirs.

Those words haunt me as I think about what was done by the one whom Aslan points to—the true Lion of the tribe of Judah who shed his blood as a lamb.

I fear we think of the cross and resurrection too little and too lightly.

It comes up in the days leading up to Easter. At our best we take a Lent journey or Passion Week trip in our minds in the same way we count down to Christmas.

Five golden rings.

Easter is the end of Holy Week. The payoff of Passion Week, the final seven days of Christ's earthly life.

Then we pack it all up and put it into the boxes in the garage, between Halloween scarecrows and Christmas lights, until next year.

We do this not only with decorations, but also in a deeper way within our hearts. When Jesus died on the cross and rose from the dead, it wasn't the last chapter but the start of a whole new story. The beginning. Artemis. It wasn't the termination of something but the genesis of everything. Through it he triggered the start of the reconciliation of all things.

Theologically this is articulated as *the already but not yet*.

The first domino fell when Christ rose from the dead. The last domino will fall when all things are made new.

"The mountains will drip with sweet wine" (Joel 3:18 NASB).

"Trees . . . will clap their hands" (Isa. 55:12 NLT).

"The wolf and the lamb will graze together" (Isa. 65:25 NASB).

In the meantime, he is working to restore you to who you originally were before sin clouded and pain distorted. Rescuing you from you. Delivering you from the damage done to you when you were . . .

Abused.

Forgotten.

Abandoned.

Molested.

He longs to heal your heart and to enlist you as a change agent to extend that healing to others. He wants you to experience all the happiness he won for you at Skull Hill.

Isn't it so strange how you can't get away from the moon, try as you might, and yet you can't touch it? It remains perpetually in front of you as though you could stretch your arm out and grab it with your fingers, but it is just out of reach. Happiness can feel like that. Like it is something other people experience but it only flirts with you and haunts you every time you glance out the window. You've tried everything, haven't you? You've read books and listened to sermons and gone through small groups and conferences. Something is still missing. Perhaps in all your investigation and accumulated information and revelation you sense that void. This book will help you with the piece that has always been missing, with core transformation—and that comes through exploration.

I want to show you that the cross has the ability to change you. It's a launchpad that will enable you to plant a flag on what otherwise remains unobtainable and unrelatable. Through this voyage you can expect true change in significant areas:

Your past. Not a single one of us can change history. Jesus can. He can forgive and redeem and restore what the locust has eaten (Joel 2:25) and sin has stolen. And the harm others have piled on you? It can become fertilizer for the good he wants to produce through you.

Your future. He offers you the promise of a destiny. This future spills over into life after death and life after that too. This will culminate in a new heaven and a new earth (Rev. 21) in which we will serve and explore into the ages.

And most significantly, your present. I'm talking about participation in his work on the earth today. Plans he has been dreaming for you that begin now. Your mission awaits.

This is the fantastic quest to conquer inner space. Your wake-up call. I am here to tell you that enough time has been spent asleep. Your salvation is closer than when you first believed.

It is an honor to be the crackly voice in your radio as you journey back in time and forward in hope, so you can land anchored, healed, and whole in the present, poised to report for duty as a servant of the King. We will search carefully in the words, wounds, and wonders of the one Buzz Aldrin's wine and bread pointed to—the seven times Jesus used the phrase "I am" and then told us what that means, the seven times he spoke while he hung on the cross, and the seven letters recorded in Revelation that he wrote to the seven churches in Asia Minor. These three groups might seem random, but I assure you they are an incredible window into who Jesus is, what the benefits of following him are, and what his plan for your life is.

This is an opportunity to lose and find yourself in the remarkable sevens that guide us through the death, life, and heart of Jesus and what he is calling you to right here and right now.

Seven has always been my favorite number. I relished seeing it on my birth certificate and my T-ball jersey. I didn't discover until later how truly magnificent a thing it really is. We are surrounded by this number.

There were seven original astronauts in the space program—the Mercury Seven. There are seven stages of grief, seven wonders of the ancient world. Seven days in a week. Seven colors of the rainbow. Seven notes in the Western musical scale. Seven continents. When rolling a pair of six-sided dice, the most common roll is a seven.

THIS IS AN OPPORTUNITY TO LOSE AND FIND YOURSELF IN THE REMARKABLE SEVENS THAT GUIDE US THROUGH THE DEATH, LIFE, AND HEART OF JESUS AND WHAT HE IS CALLING YOU TO RIGHT HERE AND RIGHT NOW.

Most mammals' necks have seven cervical vertebrae. The common ladybug has seven spots.

Coincidence or not, there are seven books in the Harry Potter series. It also happens that the upper limit of our working memory is seven, as recalling a sequence of numbers past seven is difficult. This is why American phone numbers (minus the area code) are seven digits. (The Bell Telephone Company wanted them to be long so they wouldn't run out of possibilities, but if phone numbers were eight digits instead of seven, wrong-number calls would skyrocket.)

"There seems to be some limitation built into us either by learning or by the design of our nervous systems, a limit that keeps our channel capacities in this general range," psychologist George Miller wrote in his famous article "The Magical Number Seven, Plus or Minus Two."

It also seems to be God's favorite number—a number that signifies completion. It's all over the Bible, from the seven days of creation in Genesis to Revelation's seven bowls, seven seals, and seven trumpets (Rev. 6; 8; 16). Peter discovered that we are to forgive seventy times seven (Matt. 18:22). In fact, seven pops up right around seven hundred times in the Scriptures.

It is also intentionally woven throughout Jesus' ministry. And that is what we will explore.

Along the way we will use NASA's 1969 lunar mission as a metaphor for the metamorphosis that has been unleashed by the man from Galilee, because the moon and the cross cannot be separated:

The moon landing is, hands down, the greatest thing mankind has ever done.

The cross is the greatest thing that has ever been done.

There has never been a more important feat of exploration than traveling to and landing on the moon.

There will never be a more important feat of salvation than God hanging on a cross.

Landing on the moon was the result of a visionary president stunning his own fledgling space agency—and the world—with a promise and a prediction that committed them to a project they did not have the ability or technology to accomplish at the time.

Jesus' death on the cross fulfilled ancient prophecy, but it similarly stunned those who followed him.

Jesus' being lifted up on the launchpad of the cross was the biggest rescue mission in history; it had been initiated even before it was needed. It was in motion in God prior to the creation of the sun, moon, or stars.

Both John F. Kennedy and Jesus Christ would galvanize their hearers, and both would die before their predictions were fully realized. But only one of them rose from the dead on the third day.

Passover always takes place in connection to a full moon. (The entire Jewish calendar is lunar instead of solar, meaning it is tied to the cycle of the moon, not the sun like the Gregorian calendar is.) That means that Jesus walked through the Kidron Valley after the Last Supper and prayed in the garden of Gethsemane under the light of a full moon.

On top of the *same* moon, Buzz would eat bread and drink wine almost two thousand years later and nearly two hundred and fifty thousand miles away.

Poetry.

The moon and the cross cannot be separated.

The moon calls to lovers. Dreamers. Painters. Fiends.

The cross whispers to the broken, the needy, the dying.

Through them God calls to sinners and to saints.

May all the godly flourish during his reign. May there be abundant prosperity until the moon is no more.

You might be thinking, *Levi, this is more than I bargained for*, or *I've never really considered myself a space person or a NASA nerd*. That's okay. I'm obsessed with it enough for the both of us. And I promise there will be times you will wonder, *Why exactly do I need to know this?* But stick with me! When I feel your eyes glazing over, I will try to help you see *why* you should care. By the end, I believe you'll think it is a tragedy that you ever thought space wasn't your thing. Remember: exploration is exploration, and you will be able to learn much about yourself through it. If you're anything like me, your inner space feels just as deep and dark and inaccessible as outer space was in 1969, but if you can discover how we accessed outer space, you might just be able to use those lessons to conquer your inner space.

I have included what I hope to be helpful glossaries, a cast of characters, and some diagrams at the front. Don't be afraid to refer back to them when a name pops up and you aren't sure whether something I'm talking about is from that ship that caught on fire or the Christmas mission.

And if you are already a card-carrying space nerd like me and have the T-shirt to prove it but aren't so sure about all this Jesus stuff, please stay with me. I am especially excited to tell you about him. Jesus not only changed history, he also changed my life. He has given me hope when life has hurt the most. I believe that when you see him for who he truly is, you will gasp with joy and amazement.

After the Apollo 11 mission, Buzz Aldrin expressed some remorse over his decision to take Communion for fear of whether the act alienated an atheist or a Muslim.

But such regret is unnecessary. When Jesus died on the cross, he didn't do it for Christians. There weren't any. He did so for all mankind—Muslims and atheists included.

As an ardent, avid fan of history in general, I am especially enamored by the story of the space race, and I have been profoundly motivated to grasp the staggering reality of what was accomplished in this impossible, against-all-odds, extraordinary chapter. In the same way, you'll find your pulse quickening and curiosity awakening as you move closer to the wild blue yonder of all God has for you.

We tend to take the moon for granted. It was there yesterday, and it will be there tomorrow, so we can become blind to it. Just like we do to God. It is so easy to become high-functioning atheists, professing spiritual beliefs we haven't set foot on in years.

The spiritual cold war can end. Not someday but right now. Eternal life is not a quantity. It is a quality. It's not just a destination; it's also a motivation. I want you to believe that your joy can be full. Vibrant. The torture can end, be it the torture of shame, or self-imposed isolation, or debilitating fear. Everything you need to truly live and love you already have; you just need to possess your possessions and lay hold of your birthright. You can't have more than Christ.

What blasted off from the cross can help you become grounded.

What Jesus did wearing skin and bones can help you become comfortable in yours.

So come with me. We will travel in our imagination on a lunar voyage and believe God will give us grace to conquer that ever-elusive, easy-to-miss, often ignored but never silenced secret inner space.

Ten.

Nine.

Eight point nine (ignition sequence start).

Eight.

Seven.

Six.

Five.

Four.

Three.

Two (all engines running).

One.

Zero (liftoff).

And away we go.

A PROMISE

I WRITE BECAUSE JESUS HAS SHOWN ME I AM SUPPOSED TO.

I've just been through the longest time gap between books in my writing career. (At the time of this writing I don't currently have a book contract, and I haven't told my publisher or even my agent that I am writing—and I will not until this book is done. I didn't want to write because I needed to but because Jesus told me it was time.)

I will tell you the truth and not pretend I know more than I do. If I discover something that challenges my assumptions or what I have been told, I will follow the truth and present what I find. I yearn to find new inspiration in Christ and believe both you and I will be changed by this adventure.

There are things in my soul that are not as happy as they are meant to be, and I want to watch them change as much as I want to help you find what you are looking for.

LEVI LUSKO

May 7, 2020

Whitefish, Montana

CHAPTER 1

EVERYONE IS A MOON

I LIKE TO COUNT THE NUMBER OF SYLLABLES IN EVERY sentence. (*That's sixteen.*) That's not exactly true. (*Six.*) It's more compulsive and comforting than it is convenient or conscious. (*Eighteen.*) I find the *need* to count the syllables in a sentence. (*Fourteen.*) It's obsessive, automatic, and even a touch annoying, really. I have many memories of lying in bed as a child and replaying the day, my fingers hypnotically ticking as words flowed through my head, and needing to change the number of syllables in a given sentence to be even and not odd. Never odd.

I would also catch myself in waking hours silently repeating sentences after speaking. No sound would come out, but my lips would be repeating myself like a mute parrot. I think it was to test the rhythmic quality and edit what had just been said back to myself. Occasionally I still find myself doing that, though it has become rarer. My wife catches me and says it's cute. More often it feels like a curse. And every now and again I still count syllables. (*Thirteen.*) Drat. Let's try that again. *And every great once in a while, I still count syllables.* (*Fourteen.*) Phew.

If only my neurosis ended there. In the nearly four decades I have lived with myself, I have continually been amazed at how weird I am. I mutter in the shower nonstop. Sermons. Speeches. Diatribes. Chapters. Rants. Brilliant ideas and less brilliant ones. They all come out in spurts, starts, stops, fits, and sometimes songs. Early on in our marriage my wife admitted to hearing me talking from the other room. Wondering why and how I was on the phone in the shower, she heard what she was positive were many sides to an ardent conversation happening all within the cramped quarters of myself.

The wonderful thing about being weird is figuring out how you can channel your idiosyncrasies in a positive direction. Because when you can focus your uniqueness like light through a lens, you can become a laser.

The truth is, we all are kind of weird.

THE MOON'S DIRTY LITTLE SECRET

Did you know that you have only ever seen one side of the moon? Though it orbits around the earth, the same side faces toward us the whole time. It technically does rotate on its twenty-seven-day journey around our planet, but because it is tidally locked to earth, it manages to keep its rear away from us at all times. Scientists call this synchronous rotation. Like a servant before a king, the moon never turns its back on us.

What is called "the far side of the moon" or "dark side of the moon" is always hidden. Humans laid eyes on it for the first time during Apollo 8. This side of the moon is pocked by craters far more extreme than those on the side we see in the sky every night. One reason for this is obvious: as long as it has hung in the sky, the side pointing away from earth gets the brunt of any asteroids coming our way, like a bodyguard taking a bullet for the one under their protection. And the earth returns the favor, acting as a shield for much of the space debris coming from our direction that would slam into the moon.

Here's how astronaut Bill Anders described the moon's forbidden back side when he first laid eyes on it during Apollo 8:

> The back side [of the moon] looks like a sand pile my kids have played in for some time. It's all beat up, no definition, just a lot of bumps and holes.

Here's fellow Apollo 8 astronaut Jim Lovell's description:

> The moon is essentially gray, no color; looks like plaster of Paris or sort of a grayish beach sand. We can see quite a bit of detail. . . . The craters are all rounded off. There're quite a few of them, some of them are newer. Many of them look like—especially the round ones—look like hit [*sic*] by meteorites or projectiles of some sort.

And this from Frank Borman, also of Apollo 8:

> I don't think anything I'd studied prepared me for the really troubled nature of the lunar surface—it was messed up beyond belief . . . It was terribly distressed with holes, craters, volcanic residue, so it was a very interesting first view of a different world.

The moon is one-fourth the size of the earth but is big for a moon. Its mysterious back side has mountain ranges taller than the Himalayas. And there are craters. Lots of craters. There are craters within craters within craters. Some are only an inch, and some go down a mile. There are craters so big the Grand Canyon could fit inside. One known crater "is eighty-five miles wide and almost thirty thousand feet deep."

I find it extremely symbolic that the moon, like so many of us, keeps the most damaged (and interesting) side out of sight.

Mark Twain once said, "Everyone is a moon, and has a dark side which he never shows to anybody."

The irony, of course, is that even though the dark side is where damage has happened, it is also where the highest heights are located. Highlands rise higher above the surface there than on the near side. And craters plunge low. Craters, after all, are just mountains in reverse.

The pockmarks and wounds you carry around inside and show to no one are not only where the most damage has happened; they are where your greatest potential strengths lie. As they say, the cracks are what let the light in.

Jesus gave us a brand-new way of looking at the textured dark sides we are tempted to keep hidden:

> Blessed are the poor in spirit,
> For theirs is the kingdom of heaven.
> Blessed are those who mourn,
> For they shall be comforted.
> Blessed are the meek,
> For they shall inherit the earth.
> Blessed are those who hunger and thirst for
> righteousness,
> For they shall be filled. (Matt. 5:3–6)

Extreme situations can lead to deeper happiness and more profound joy. What makes you weird is also what makes you wonderful. Your quirks and curiosities are packed full of potential. Your pain can make you powerful. Extreme gratitude will be the end result of learning to embrace—and trusting God to unleash—the hidden potential of the dark side of your moon.

I have found a way to vent my obsession with words, ideas, and thoughts through books and talks. A carefully controlled weakness can become a double strength.

I don't want to ask you if you are strange. I already know the answer. I don't think I have ever met anyone who isn't. I want to inspire you to see your asteroid-streaked landscape as wonderful. God didn't create just one kind of person any more than he made only one kind of animal. Porcupines are weird. So are sea turtles. And geckos.

THE POCKMARKS AND WOUNDS YOU CARRY AROUND INSIDE AND SHOW TO NO ONE ARE NOT ONLY WHERE THE MOST DAMAGE HAS HAPPENED; THEY ARE WHERE YOUR GREATEST POTENTIAL STRENGTHS LIE.

The obstacle to liking who you are and being comfortable with yourself is *comparison*. You make a really lousy photocopy of your sister. I am a horrible clone of Steven Furtick or Louie Giglio. Anteaters can't fly like eagles.

(Okay, you are never going to believe this: I am writing while sitting in my backyard by our firepit. As I wrote that sentence just now, I looked up and saw two bald eagles circling and seemingly dancing, or maybe play-fighting or mating in the sky. Those eagles can't get ants out from deep inside the earth, but they are doing a terrific job within the narrow limitations of who they were made to be.)

For your sanity, I beg you to stop doubting your value or worth based on who you aren't and what you can't do. Comparison is the thief of joy—even if my man crush Teddy Roosevelt wasn't really the one who said so. There isn't only one kind of beautiful, just like there isn't only one kind of flower. Your shape and size and personality type and spiritual gifts and life experiences all have been carefully calibrated and crafted to create your calling. Even the painful situations you have faced and the mistakes you have made are meant to be redeemed and worked into the tapestry of your story. God will do so in an elegant and redemptive way that will both shock you and warm you from the soles of your feet to the top of your head. Don't misunderstand me; I'm not talking about dismissing sins, selfishness, or faults or tolerating personal laziness or unwillingness to grow on the basis of "that's just who I am!" I'm saying God didn't make a mistake when he fearfully and wonderfully made you in his image.

Aren't you exhausted from sucking in and spinning around in every different situation and social media post to keep the dark side of your moon safely hidden and out of sight? I'm not talking about putting your best foot forward; I'm talking about that hidden, horrible secret you have carried for so long: that there is something wrong with you. If people saw the real you, you think, they wouldn't want anything to do with you. So rather than being rejected, you just pretend.

You don't have to. The craters and chasms and canyons of your character give texture and color and complexity that sparkle in your heavenly Father's eyes.

One of the goals of this book is to help you understand your true identity. I firmly believe that identity determines activity. You will never wake up knowing what you are supposed to do until you have a firm grasp on who you are.

This is nothing new. Many people are trying to find themselves. You can read one of a million self-help books or watch inspirational YouTube videos that will tell you how you need to look within for the answers so you can become your own kind of beautiful.

Tried and failed.

The problem with this is that you make a tremendously bad God for yourself. I have looked within, and I didn't find answers. I need answers for what I find within myself. And so do you.

The answers you seek aren't going to come from spending a summer backpacking through the hostels of Europe, or scuba diving the Great Barrier Reef, or taking a road trip from LA to NY. All those things sound tremendous, and if you get the opportunity and have the inclination, I say go for it. But in and of themselves, these experiences can't do anything for you that you can't also find making french fries in the back of a Wendy's. Moses didn't first encounter God on some spiritual retreat or epic adventure. It was in the middle of nowhere doing the same old, same old he had been doing for decades. God doesn't need to take you halfway across the world to change your world. He can do it in your apartment.

GOD DOESN'T NEED TO TAKE YOU HALFWAY ACROSS THE WORLD TO CHANGE YOUR WORLD. HE CAN DO IT IN YOUR APARTMENT.

We are going to go about it differently—by looking not to ourselves, or to novel experiences, but to Jesus. We will learn who we are by first discovering who he is. His identity will help you figure out yours. We all need something to orbit around, and relationship reveals identity. You can't understand the moon without understanding the earth. And no study of the earth would make sense without looking at the sun. The further you zoom out, the more you understand.

Jesus introduced this counterintuitive approach as the paradigm-shifting ethic that rules his upside-down kingdom. The way up is down, the Last Supper is the first of its kind, and if you want to find yourself, you must lose yourself.

The great thing about this is it really works. If we discover that he is love, we know we are loved. If he is the Great Physician, we have a solution for our brokenness. When we hear him say he is the Creator, we then know we are not a mistake.

And Jesus makes it really easy to go down this road. He's not playing *Where's Waldo?* with his identity. Your eyes will not get tired looking for his red striped shirt in the crowded scene on the page. All you have to do is read the gospel and the words in red come leaping off the page.

As you look full in his wonderful face, I believe you will stop feeling the need to keep the comet-strewn darkness of your shadow side hidden from sight. You'll step into the light. Then, and only then, will you discover that God is not afraid of what is behind you. He has a plan to address what is sinful, mend what is broken, and fill what is empty. And he has a plan to use all of you—not just the curated parts you currently feel comfortable letting the world see, because (news flash) he is all around you and already knows about the dark side of your moon.

CHAPTER 1.60769

WHAT THE HECK?

MAYBE EVEN AFTER READING ALL THAT YOU STILL AREN'T convinced you can make it through a four-hundred-eighty-page book on the Apollo program that walks you through the cross and what it means for you today. I must confess this is a different kind of book. That's on purpose. Where has the same old, same old gotten any of us? Maybe something different is exactly what you need. If you want different results, maybe you have to access the message in a different way.

Yes, this book is a little bit like a Christian-living book had a baby with a history book—the Frankenstein love child of Max Lucado and David McCullough. But, again, I'm convinced that as you gain clarity on outer space, you will somehow be able to untangle the mess of your inner space.

I'm sure there are a million objections running through your head. *This won't apply to me. I don't have time for this. It's too long.*

I get it.

The sheer size of this book might freak you out. So let me give you a hacker's formula. Read chapters 3, 5, 7, 10, 13, and the last one, chapter 24, "Blood, Sweat, and Tears." The publisher wouldn't want me to tell you these will give you a pretty good gist of what I have to say, but I'm so confident you'll want the rest that I don't mind giving you a shortcut through the book. All the chapters work together and build, but they also can stand alone and help you no matter what order you read them in.

Wonder is waiting. The moon is calling. Why live with the shades drawn when God says, "Go forth. Be fruitful"? It would be a shame to stay where you are when so much more is waiting.

What do you have to lose? In other words, what the heck?

CHAPTER 2

THE CROWDED HOUR

YOU WANT YOUR LIFE TO COUNT, DON'T YOU? TO BE GLORIOUS. You want to know why you are alive and to go for it. This desire has been put inside you for a reason.

In 1927 aviator Charles Lindbergh flew his single-engine, single-seat, custom plane, the *Spirit of St. Louis*, from New York to Paris, shocking the world by being the first person to ever successfully cross the Atlantic solo and nonstop by air. Few believed it could be done. Around the clock, people followed the progress of his flight, which lasted thirty-three hours, twenty-nine minutes, and thirty seconds, with fervor. Upon landing in Paris, he instantly became the most famous person in the world.

Forty-two years later, the world was stunned when Neil Armstrong, Buzz Aldrin, and Michael Collins flew a quarter of a million miles to the moon in their Apollo spacecraft. Six hundred and fifty million people tuned in live for the landing, making it one of the most viewed events in history. President Richard Nixon sent the three astronauts on an around-the-world publicity tour after the mission. They visited twenty-four countries in thirty-eight days and were paraded through thronging crowds all across the globe.

A staggering connection emerges when we compare the two vehicles and their fuel consumption. The first stage of the Saturn V launch vehicle that took Neil, Michael, and Buzz from the earth burned exactly as much fuel in one second as Lindbergh's plane, *Spirit of St. Louis*, used on its entire voyage across the Atlantic Ocean.

Thomas Mordaunt wrote a poem that I find ultrainspiring. Haunting even.

> One crowded hour of glorious life
> Is worth an age without a name.

Poetry is personal. Here's what his words mean to me: It would be better to live fully and furiously, however briefly, than monotonously and tediously apart from purpose, disconnected from destiny and design. An age of years so bland and nondescript that no one would even bother to give it a name.

King David said something similar in Psalm 84:10: "Better is one day in your courts than a thousand elsewhere" (NIV).

In Mark 8:36 Jesus put it this way: "What will it profit a man if he gains the whole world, and loses his own soul?"

Something stirs in all of us—especially in youth—to do something great, and that is often why there is such despair, resentment, and anger in middle age. As Henry David Thoreau put it, "The youth gets together his materials to build a bridge to the moon, or, perchance a palace or temple on the earth, and at length the middle-aged man concludes to build a woodshed with them."

It is a tragedy to lose your fire. It is an even bigger tragedy to win at the wrong thing. To spend your whole life climbing a ladder only to get to the end and find that it was propped against the wrong wall. What good is a cathedral or a palace if what is actually needed is a woodshed?

The idols of this world that so easily intoxicate our hearts and trick us into worshiping them are so deceptive. Their goal is for you to turn your life into "an age without a name."

Jesus wants to free you from living for things that can't satisfy your heart's deepest longing or support you when storms inevitably come.

One crowded hour of glorious life . . .

When I read that poem, I think about the disciples of Jesus, who were beaten and told to not preach in the name of Jesus but still went out, rejoicing that they were counted worthy to suffer for his name, and preached all the more. They were thrilled by their crowded hour.

I think about the martyrs throughout church history who have gladly chosen death rather than renounce faith in Christ; they went to the stake, or the coliseum, or the sword before they would betray him. "They did not love their lives to the death" (Rev. 12:11).

I think of the demoniac of Gadara, who badly wanted to join Jesus' ministry team and go on the road with him, but Jesus wouldn't let him. Jesus told him that he needed to stay home and testify about Jesus right there in his hometown (Mark 5:1–20). The man thought it would take faith to leave, but he learned the beauty that comes when you have faith to stay. Getting and holding a job, shining a light among his coworkers, and sharing the good news of Jesus would allow him to uniquely impact people and turn his life into a glorious story of grace and redemption that spilled out from the town of Gadara to the ten cities of the Decapolis.

I think of my friend Shane, who works at NASA. As of this moment he is the only living American astronaut who has flown to space in three different ships. He flew in the Space Shuttle, a Russian Soyuz, and a SpaceX Crew Dragon. His love for Jesus has led to a quiet, excellent strength and passion that is lived out both on and off this planet.

I think of Mary Magdalene who, along with other women, supported Jesus with her means.

And I think of you and the unique calling on your life. There is something God has planned for you to participate in—whether it's leaving what you are doing now or continuing to do what you are doing right now, fully alive, and no longer in your own way.

There is a unique plan, written before time, that God has been dreaming of for you, and it was worth his Son going to the cross for it to be possible.

He died for you to live.

This should be the number-one priority for you to discover and participate in.

Once you find your unique sound, play it loud.

If you're amazing with a slingshot, don't let Saul (or anyone else) choose your armor for you when you confront your Goliath. Don't hide the dark side of your moon.

Be weird in Jesus' name.

Me? I have discovered that I am a nerd. I used to try and hide it, but I am increasingly okay with it. What is God calling you to do to make his name famous?

NOT THROWING AWAY HIS SHOT

Jesus gives us a tremendous example to follow as we learn to see our life as one crowded, glorious hour instead of an age without a name. All throughout his life he referred to his "hour." And he meant it both metaphorically and literally. In a very real sense, Jesus came to die on the cross, so that was his hour. And that hour was crowded.

Not only did he speak seven times as he hung on the cross, but the Passion Week in general is the focal point of his ministry and was top of mind for his first biographers.

We know very little about Jesus' childhood—just some info about his birth that was necessary to establish the messianic credibility he required. And we know little about his teenage years except that he got lost and ended up preaching in the temple during a family vacay. (Classic Jesus: the one opportunity for angsty teen drama and the closest he got to rebellion included him giving a sermon. LOL.)

After that we know nothing of the next two decades until he showed up to get baptized. What did he do in between? All we know is that he lived in Nazareth and would have apprenticed under his stepdad, Joseph, as a carpenter and mason.

It gives me comfort that he would have worked with his hands. He was a tradesman—probably an apprentice at first. This meant manual labor. Detailed, probably hot, and sometimes monotonous work. And since we know he does all things well, I envision him doing a masterful job with all the things he built: chicken coops, chairs, tables, whatever.

What a beautiful thought.

From then on, we have some limited information about Jesus' three years of ministry, some of his sermons, and a selection of his miracles—though John expressly tells us that not all were told, and that if every miracle Jesus ever did was recorded, it would fill all the books in the world (John 21:25).

Epic.

In fact, a great majority of the Gospels centers on the last week of Jesus' life, with an even bigger focus on his final twenty-four hours. You can grasp the extreme importance of the cross compared to anything else in his life just by looking at the amount of gospel material focused on it. As you might know, there are four Gospels. But what you might not know is that they are not sequels. There *is* a sequel to the Gospels' story, but it is the book of Acts. The Gospels are different camera angles covering the same events from different perspectives. They highlight different aspects to reach different audiences.

Matthew wrote with primarily a Jewish thinker in mind.

Mark wrote taking into consideration the classic Roman person.

Luke wrote to a Greek mindset.

And John wrote so anyone anywhere could get the point.

They each have a different emphasis on Jesus that lines up with their intended audience—King of the Jews, Suffering Servant, Son of God, and Son of Man. All four Gospels, regardless of their aim, accomplish their intended goal by spending a disproportionate amount of time on a relatively small slice of Jesus' life.

Passion Week is described in:

- 7 out of 28 chapters of Matthew
- 5 out of 16 chapters of Mark
- 4.5 out of 24 chapters of Luke
- 7.5 out of 21 chapters of John

"From the very beginning the redemptive drama moves toward the cross," observed scholar Don N. Howell Jr. "The final week of Jesus' life, Passion Week, takes up nearly 40% of the entire narrative."

What is the point?

Jesus knew what he was about and was laser focused on it. He set his face like a flint (Isa. 50:7). A man on a mission. He never lost sight of the purpose for which he came. He didn't let anyone rush him into it early or do anything that would take him away from it.

This was what gave him confidence to reject the shiny bells and whistles the devil dangled in front of his face in the wilderness temptations.

This is what allowed him to turn down the crowd that would have made him king at the outset of his ministry.

Looking forward to his crowded hour was how he handled Peter, his best friend, telling him he shouldn't go to the cross. And in the end zone, in the garden of Gethsemane, with all the weight of your sin and mine and the whole world's, he was willing to pick up the cup and bring it to his lips, trembling as he was (Matt. 26:39).

It was because he had bit down and set his jaw like a pit bull, and predecided to go to the cross, that he prayed in John 17 and said, "I have finished the work which You have given Me to do" (John 17:4). All the lepers weren't healed; there were still poor people to feed and sermons left unpreached. But he would sign off, and die, and then ascend, having not finished all the work to be done but all the work that was *his* to do.

He refused to let his life be an unnamed age. Because of him, you can too. When you stop holding yourself accountable for what someone else has been called to, you finally will be able to lay your head down on the pillow at night without that nagging feeling of being perpetually behind.

The beauty of our journey, as we follow Jesus through his crowded hour, will be how it gives you power and clarity of purpose to make your way through your own, one load of laundry, one car pool, one Zoom, one Sunday school lesson, and one more bleary-eyed opening shift at a time. The cure for the fear of missing out is eye contact with your Savior. If the most important person in the universe is in the room where you are, why would you worry about being anywhere else?

THE CURE FOR THE FEAR OF MISSING OUT IS EYE CONTACT WITH YOUR SAVIOR.

As you show up at school or work or in your home tomorrow and in the coming days, you will have purpose so firmly etched in your heart that you will refuse to allow your life to digress into an age without a name.

It's been said the two most important days in your life are the day you were born and the day you find out why. If I could be so bold as to add a third, it would be to suggest the day you try the shaved ice at Ululani's in Maui. (Just promise me you get the snowcap on top and the macadamia nut ice cream on the bottom. Trust me.)

There's a reason the book *Start with Why* by Simon Sinek sells year after year. It's the same reason *The Purpose Driven Life* by Rick Warren is one of the most-read Christian books of all time. People are desperate to figure out their *why*.

One of the most life-depleting experiences in this world is to encounter someone who is doing a *what* disconnected from a *why*. Buying a coffee from someone who has a *why* ignited inside them can be one of the most invigorating experiences in the world. Their eyes shine as they describe the different beans, and they take a moment to appreciate your first sip and the anticipation of the caffeine. They know their regulars and can predict their orders, and occasionally they have their drinks ready as they enter the door. What a beautiful thing it is to see someone doing their *what* powered by *why*.

On the other hand, we have all had someone go through the motions of the drudgery of a job that you can tell they resent. To be treated by a nurse at a doctor's office, interact with someone at the DMV, or buy a gardening rake, a car, or a foam roller from someone who you can clearly tell would rather be somewhere else—*anywhere* else—is so uncomfortable. Awful, even.

This extreme, exaggerated, and intentionally uneven focus on Jesus' final minutes and death, to the exclusion of all the lesser-known threads, really communicates what Jesus' life was all about.

His miracles were awesome. But he didn't come to do miracles, primarily. That wasn't his mission. His sermons were better than any that have ever been given. But he didn't come to give inspiring messages. His life example and personal ethic were off-the-charts epic. He elevated women, defied racist ways of thinking that were the norm in his day, and was kind to children. But he didn't come so we could ask ourselves, "What would Jesus do?" and follow in his footsteps.

He came to die.

That was his crowded hour that he never lost sight of. He lived, preached, and prayed like a man running out of time.

The gospel is the good news that Jesus died for our sins and rose from the dead so we could be forgiven and born again, thus reconnected to God. It was the only way for him to save us. Once taken root, this has implications for every area of our lives, but that's what it leads to, not what it is or where it begins. In other words—anything added to the gospel is not the gospel.

Why was the cross so important to Jesus that he refused to be distracted or deterred? And what exactly took place in that transaction where the veil tore, and the sky turned black, and dead people gasped for air and ran around for a few days (Matt. 27:51–53)?

This is what I intend to find out. And you're coming with me. Slowly.

We are often in a hurry to run through Holy Week so we can get to Easter. As it is the crown jewel of our faith we can't spend *too much* time focusing on the resurrection, but we can certainly focus *too little* on the cross. The cross was Jesus' crowded hour, and it is our only link to glorious life.

Through the golden words he spoke at Golgotha, we will explore the seven things he accomplished that can turn your pain, guilt, thirst, regret, relationships, fear, and despair inside out and upside down.

It's not as though they never happened, but Jesus can make it so that your life is even better, richer, and fuller *because* they happened, even though they were terrible. C. S. Lewis said we will find that heaven works backward. In light of the cross, in heaven I will not only find great joy in seeing my daughter Lenya again but even greater joy at being reunited with her than if she had never been taken from us in the first place.

So bring your sadness, conflict, and broken heart, and let's go to the stream that makes glad the city of God (Ps. 46:4). There is a river. And it is wild.

In Netflix's *The Crown*, Neil Armstrong, Buzz Aldrin, and Michael Collins sit at a press conference. The writers of the show reimagined the scene this way:

REPORTER: Now, if successful, you will be the first men to walk on the surface of another heavenly body. What exactly do you hope to discover?

NEIL: I think even more important than the answers that we'll be able to find will be the fact that we get a whole bunch of new questions to ask.

The mission is just beginning.

I hope you are ready not just for new answers but better questions.

CHAPTER 3

SO THIS IS WHAT ELEVATORS LOOK LIKE

WHAT'S WRONG WITH ME?

Have you asked yourself this question a thousand times? I know I have. I often expect unrealistic perfection from myself, so I am inevitably disappointed when I don't stack up.

It's easy to imagine everyone else has an easier time than we do. This is fueled by the fact that we are comparing the dark sides of our moons to the front sides of everyone else's. Remember that we all have like fifty photos on our phones that we don't post for every epic shot that we do post, and even that one required some serious fiddling with before it was grammable.

So how do we get from where we are to where we are meant to be?

Let me answer that question with a question. How do you get a spaceship to the launchpad? Especially when that spaceship is 363 feet tall? That was the problem when humans went to the moon. They had to get the Saturn V rocket from the Vehicle Assembly Building (VAB) to Complex 39A—a distance of three-and-a-half miles.

I have seen lots of packages and parts come over the years. It doesn't help that I have a mild addiction to online shopping. And by mild, I mean real problem. I would be ashamed for you to know how many reviews and YouTube videos and websites I visited before buying a new waffle iron. (Go Breville, by the way.)

The largest delivery I have ever personally witnessed was an elevator. The delivery did not come from UPS; the elevator showed up on an eighteen-wheeler and took up an entire flatbed trailer. It was on my radar that I would be seeing it because the builder doing our Fresh Life church building's transformation let us know it was coming a few days before. We had purchased a couple of hundred-year-old buildings a block apart from each other and joined them together with an infill building in between. As a part of that project, we ordered this elevator at the end of 2019, and these things take time. "These things" being construction and buying elevators. I had never really given much thought to where elevators come from (a company called Otis) or how you order one (from an elevator dealer), but I learned the hard way they are stupidly expensive.

The elevator had to be custom ordered because one of the three floors it would service opened to a different side than the others, and because I guess all elevators are custom ordered. Since the elevator wasn't going to be ready right away, we built the empty space where the elevator would go and drywalled over the entrances to it so we could open the building without it. We also received a temporary permit to add it when the elevator finally arrived. It took longer to come than we anticipated because of COVID-19 causing the entire world to shut down for a hot minute.

When it arrived in May 2020, I was in my office trying to film an emotionally charged video. It is hard to describe what it sounded like, but it involved beeping and engines revving and voices calling out instructions to each other like "swing it wide" and "back that thing up." *What on earth?* I looked out the window, half expecting a Juvenile concert, and was shocked to see a trailer, a crane, and a plethora of hard hat–wearing individuals running around with what seemed like an explosion of materials and parts and pieces covering our staff parking lot.

Then it hit me. *It's the elevator.* The bizarre thing was it looked *nothing* like an elevator. If I had wandered into the scene with no clue and was told to guess what I was looking at, I would have said a thousand things and never even thought about an elevator. I was shocked and highly intrigued. And of course, I did the only reasonable thing a grown boy would do: I sat there and gawked. Load after load was swung from the flatbed to the ground by a crane. And none of it made the least bit of sense to me.

I have been in a lot of elevators in my life, but only on the inside of the actual passenger compartment. I had never seen the components necessary to make a metal box ascend and descend in a building full of people. My only knowledge of the required pieces, and all that makes an elevator tick, comes from action movies where Tom Cruise or Tom Holland is inexplicably forced into the shaft to hang by the wire, usually moments before the entire thing is engulfed in a ball of fire.

I was standing there, staring, when I had an epiphany: *So this is what elevators look like*. I thought I knew before, but only through this experience could I see what is necessary behind the scenes.

This is exactly why you and I are going to be taking our time as we explore and examine what Jesus accomplished during the six hours he hung on the cross. Please don't rush. As we see him hanging there, speaking seven times, do what I did looking out the window at the elevator truck. Take it all in.

We will try to smell the coppery tang of the blood that ran through the streets from all the Passover lambs being killed for sacrifice. I want you to try to taste the bitterness of the sour wine that was brought to Jesus' lips on a sponge raised on a reed. Hear the jeering of the religious leaders ("Come down if you are the Messiah and we will believe you") that inadvertently fulfilled biblical prophecy in their mockery and derision. Some of the moments might seem strange. We will study them anyway. Theologically you might not even understand how and why the Father forsook the Son as he hung there. Join the club. But I want you to grapple with it. Forget what you think you know and try to see it again for the very first time.

Whether you are hiding in the crowd still unsure of where you stand with Jesus or are brave enough to come close, take it in. All of it. Follow the truth wherever it leads you.

So, how do you move a spaceship from the hangar to the launchpad? The answer to the question I asked at the beginning of the chapter is: Slowly. Really slowly. That's how you get a rocket that weighs a half million pounds, and its launch tower that weighs just shy of eleven million pounds, to the launchpad three and a half miles away. The task is daunting. Gargantuan.

Have you ever tried to move a completed LEGO block creation without breaking it? Impossible. How do you do that with a structure the size of a thirty-six-story building? For that task NASA invented something that is almost as impressive as the rocket it was built to carry: the Crawler, which is itself an engineering marvel. The rocket is built on this machine, which drives the rocket and its tower vertically to the launchpad, where it can be fueled and then launched.

Its official name is the NASA Crawler-Transporter. Its latest iteration, the Crawler-Transporter 2, weighs 6.6 million pounds (that is fifteen Statues of Liberty), and it's able to transport up to 18 million pounds. That is the weight of more than twenty fully loaded 777 airplanes.

Crawler carrying Apollo 14 Saturn V on November 9, 1970

NASA; scanned by Kipp Teague

- The flat top of the original Crawler measures 14,934 square feet, which is bigger than a baseball diamond.
- When it was built, it was the largest land vehicle ever made by humans.
- It moves like a tank on enormous treads and can do so forward and backward.

- The real secret sauce is the fact that it can keep the rocket completely level while making its way up the five-degree climb over crushed rock to the launchpad.
- Its enormous, locomotive-sized diesel engines consume 165 gallons of fuel for every mile it travels. That's thirty-two feet per gallon. Not great fuel efficiency. But impressive. It was perfectly assembled for its assignment. (Just like you.)
- The ground this beast travels on took as much work as the machine itself. It was precisely engineered to a depth of thirty feet to ensure it could handle the strain of many million pounds moving across it without giving out or shifting.

It's happening, isn't it? You are wondering, *Levi, why are you telling me all this?* Because you can't get to space if you can't even make it to the launchpad.

This might seem like overkill, but know this: there was no other way. The Crawler transported the most complicated, powerful engine built up until that point by humans—the Saturn V rocket. And it carried it slowly. Very, very slowly.

The irony is that on the lunar missions, the star sailors (that is what *astronaut* means) would reach a top speed of 25,000 miles per hour, but they did so in a vessel that makes its journey to the launchpad at less than a mile per hour. That is one foot per second. And hilariously, the Crawler's driver still wore a seat belt for this comically slow-motion journey.

YOU GOTTA CRAWL BEFORE YOU CAN FLY.

Here's the point: the rocket would shake the ground and shatter windows miles away on its way out of this world, but it got to the launchpad at a snail-like crawl. This is why you and I need to be patient with ourselves as we grow. You gotta crawl before you can fly.

You have to know how things get built and how they get to where they're going before you shoot upward. This is what elevators look like.

SOME ASSEMBLY REQUIRED

I want you to hear this next sentence in your head in an ancient man's voice, gravelly and shaky.

"This Child is destined for the fall and rising of many in Israel. . . . Yes, a sword will pierce through your own soul also" (Luke 2:34–35). These words were spoken by a man named Simeon who, strangely enough, was given the promise that he wouldn't die until Jesus was born. It's a strange take on the normal holiday gift tag: "Do not open until Christmas."

God gave Simeon the gift of not dying until the Messiah arrived. And so, Simeon sat and waited and worshiped and grew old. I imagine him attending the funeral for one friend after another until nobody he knew was still alive. There's definitely some *Lord of the Rings*–style, Aragorn-and-his-elf-bride-Arwen ramifications to this story when you think about the downside of immortality—bidding adieu to everyone you know and love and then waiting and waiting and waiting.

But one day his ship came in. And by ship, I mean a carpenter, that carpenter's young bride, and an eight-day-old baby, who just so happened to have hung the heavens and created the world.

He cradled the child in his arms as Mom and Dad prepared two turtledoves, the requisite offering poor people gave at their baby boys' circumcisions. God whispered into Simeon's heart that this was the moment he had been waiting for. Simeon was holding in his trembling hands the one who all his life had been holding him. God with us. The one who had come to crush Satan underfoot and ascend to heaven, from whence he would return brighter than the sun, riding on the wind, adorned by lightning . . . only he would start out so helpless he couldn't even crawl.

The long-awaited King had come, and he was a baby.

There is so much about the story that is worthy of second thought. For instance, if I were Joseph, I think I would have balked at the prospect of giving an offering to God on the occasion of the baby dedication of God's Son. *It's one thing for me to raise him for you, but how about* you *give an offering to* me*? Haven't you ever heard of child support?* And it's not like Joseph was loaded. Otherwise, they would have given the normal offering of a lamb rather than the doves. But we get no trace of that from Joseph. He honorably gave a gift that had cost him and Mary something, and he did it cheerfully.

The cool thing to me is what happened next with Joseph and Mary. I imagine that money concerns were on Joseph's mind. They had just paid taxes; then there was the expense of the trip and the doves. Diapers aren't cheap. How was he supposed to make this work?

Then the doorbell rang. (Yes, I know they didn't have doorbells back then, but stick with me.) Mary and Joseph opened the door, and there on their stoop were three men. "We three kings of Orient are; bearing gifts we traverse afar."

What's this? And these three kings (we assume there were three because three different gifts were given—it could have actually been thirty or three hundred wise men) of course gave gifts. Love always gives. You can give without loving, but you can't love without giving. Frankincense and myrrh (whatever those things are) and a gift we do understand: gold.

Cash money.

With their visit came some advice: Herod is *loco en la cabeza* and means murder for the child. Don't stay here. Get out of Dodge.

Now, how exactly was a couple so poor they couldn't even afford a lamb going to afford a trip to another country? The answer was in the gold. This gift funded the expedition. It's remarkable that, because the three kings had come from the East (Babylon or Persia), they most likely were already on the road when Joseph and Mary were dipping into their emergency fund to pay for the turtledoves. God must have smiled as he saw their faithfulness, knowing the resources were already on the way.

I think this is what Paul meant when he told the Corinthians that as they gave out of their lack, God would be able to make all grace abound toward them (2 Cor. 9:8–15). The vital point to not miss: the two turtledoves came before the five golden rings. Mary and Joseph sacrificed before they were supplied.

So it always is. You must step out in faith before you will watch God miraculously work in your life. Nothing ventured, nothing gained. If you had everything it took to start the business, or if you knew exactly who you were going to marry and when you would meet them and what color their hair was going to be, or if you knew that when you planted that church it would be two hundred people at launch and six hundred by Valentine's Day, you wouldn't need faith. Therefore, you wouldn't really need God. He seems to like the deck stacked against us so we will trust him and watch him move in power. (Examples: Gideon versus the Midianites, David versus Goliath, you versus all the challenges that will go into raising your kid to know and love God in this crazy world.)

You will never see your divine purpose become a reality if you insist on understanding exactly how things are going to work. The book of Ecclesiastes noted that he who watches the wind will not plant (11:4). Or to quote a more modern expression of that sentiment, let me point you to Richard Branson's business plan that has worked all across his vast Virgin enterprises: "Screw it, let's do it." He has used this mantra many times over his storied career when "experts" and the voice in his own head have told him that something couldn't work because it was too crazy or had never been done before.

Perhaps you have had people around you try to talk you out of something you knew deep down you were supposed to do. I suggest you think twice before you change your course. "Make no little plans; they have no magic to stir men's blood." If you can explain it, and everyone understands it, then maybe God's not in it.

And in case you are about to object and tell me how daunting your idea or dream is . . . may I remind you that *Humans. Landed. On. The. Moon. In. 1969.* With no Wi-Fi. Andy Andrews likes to say, "We put men on the moon before we thought to put wheels on luggage."

Our biggest problem is not ability or resources; it's often our imagination. We need the jolt this exploration of outer space will give us so we can wake up our wonder. Because if the moon does anything, it cordially invites us to a life full of imagination and exploration.

You will need that wonder as, fueled by the cross, you navigate the depths of your inner space.

Back to Simeon. That day in the temple he articulated an ultra-significant prophecy about Jesus. This child would be responsible for the rise and fall of many in Israel. Translation: *Going to heaven has everything to do with him.*

OUR BIGGEST PROBLEM IS NOT ABILITY OR RESOURCES; IT'S OFTEN OUR IMAGINATION.

So this is what elevators look like.

The baby in his arms didn't look like much. The son of peasants. Eight days old. Feeble and helpless. If Jesus had been a horse or a giraffe, he would be walking already. How could this defenseless little child staring up at Simeon be the linchpin of salvation and the difference between life and death for the souls of humanity?

He had his answer within moments.

A rule.

On the eighth day, every male child that opens the womb was to be circumcised (Luke 2:21, 23).

I watched my son's circumcision. I felt I needed to be there for it, to be there for him. It left a lasting impact on me. And on him.

The doctor fed him sugar water and numbed him. I resisted the urge to ask for something to medicate me from the experience. Moments after it was done, my son's eyes darted up with a look of surprise and betrayal for just a moment, and then he began to cry. Then more sugar water. For a split second he seemed conflicted as to whether he was willing to accept this bribe or continue to protest. The sugar water eventually won the day as my son, like a giant hummingbird, sucked it eagerly from the bottle.

I can imagine Simeon connecting the dots as he saw the drops of blood.

A sword will pierce through your own soul also.

In a prophetic flash he leaped forward thirty years to a hill outside the city where he stood.

Fear and loathing.

Sin and shame.

Guilty conscience.

Hateful leering.

They know not what they do (Luke 23:34 ESV).

Circumcision was like a wedding ring, a visible symbol of an invisible promise, a sign of a covenant with God who vowed to Abraham that he would send a Savior—himself—as a lamb. Abraham's obedience to be circumcised was rewarded with the birth of Isaac, but it was once again tested when God asked Abraham to present that same boy back to him as a sacrifice. In reality, it wasn't needed. A ram was provided in the thicket near the altar (Gen. 22) and the promise would be kept on the day of Christ's crucifixion.

I've always had a problem with that story. Something doesn't sit right. We are supposed to applaud the faith of Abraham, who was willing to sacrifice his son? That sounded pretty sick to me, until I learned that the culture of child sacrifice saturated the people he lived among. It wasn't out of the ordinary to do such a thing as an act of worship. A god requiring the blood of a son was normative. What wasn't normal was God telling him *not* to sacrifice Isaac. Instead, God promised that he would send his Son. "Behold the lamb of God who takes away the sin of the world!" John the Baptist thundered (John 1:29 ESV), answering the question Isaac had asked thousands of years previously: "Where is the lamb for a burnt offering?" (Gen. 22:7).

Circumcision pointed to the fact that God would do what was necessary to cut away not the physical flesh but the spiritual sin. An outward sign pointing to an inward reality. Because of the forbidden fruit the first Adam took from the tree, Jesus—the greater Adam, as the New Testament reveals him to be (Rom. 5:15–17)—had come to die on one.

If anyone believes, he has life; if anyone does not believe, he does not have life (1 John 5:12).

Some people take offense at the gospel's exclusivity in that only one road leads to God. They feel that people shouldn't be banished to hell because they picked another team.

Let me suggest a different way of looking at it.

If you were drowning in the ocean and someone saw you and came to your rescue, would you bristle that only one option was presented to you?

Let me humbly suggest that there is only one name that can save. When you were lost in your sins, only one person was willing to come and die to set you free and give you life spiritually. Instead of protesting that there is only one way to God, celebrate that there is a way to God at all.

Jesus is that way. Be rescued.

As we journey and venture through the things Jesus said and did on the way to his crowded hour, we are seeing all the pieces and parts of the elevator he came to assemble.

He descended from heaven to earth—and not just so we could go to heaven when we die. He did it so we could be raised to newness of life here on this earth (first spiritually and ultimately physically, but, between now and then, mentally, emotionally, sexually, relationally, and professionally). He came so we could be witnesses of and party to his bringing heaven back to this earth as he resurrects and restores all things to himself by the power of the cross.

IF YOU WERE DROWNING IN THE OCEAN AND SOMEONE SAW YOU AND CAME TO YOUR RESCUE, WOULD YOU BRISTLE THAT ONLY ONE OPTION WAS PRESENTED TO YOU?

As you look at your life right now and see pieces and parts and things in process, don't despair. There might be Play-Doh smeared in the carpet. Maybe the kids left nonwaterproof toys outside even though you told them to bring everything in before coming in for the night, and now the sprinklers are running. Your marriage might not be where you wish it was. You might feel like you have taken your team or your church as far as you can. *Why aren't we flying quickly?* you might be wondering. Because the only way to the launchpad is a mile per hour on the Crawler.

You are moving up, a foot a second. So be encouraged as you survey the dysfunctional pieces and parts of your life that seem to be in disarray but will one day be configured to raise and lower. Say it with me: *So this is what elevators look like.*

CHAPTER 4

THERE IS NO MOON

HOW DO YOU MEASURE SOMETHING'S VALUE? IN ONE SENSE IT'S all about what someone will pay for the object—for example, what you can move it for on Facebook Marketplace or Craigslist. From a replicability perspective it's all about how rare it is. Is it the only one? But on another level, value can be determined by the price that was paid for it to exist. How much did the materials cost, and how much risk was undertaken in order to come by it?

When you look at it that way, the Apollo mission was worth a lot. In the words of President Kennedy from his now-famous speech before a joint session of Congress on May 25, 1961, "None will be so difficult or expensive to accomplish."

He wasn't joking. In dollars and cents, that moonshot cost $25.4 billion ($152 billion today). More resources were given to the space race than any peacetime objective in history. The International Space Station is the most expensive thing humans have ever built. Many people balk at the idea of this sort of price tag. When I told my mom

I was writing a book about NASA's journey to the moon, she told me that she knows some people who think it excessive to continue exploring space when people on earth are hungry. People thought the same thing in the sixties. But that doesn't take into account the impact the journey to the moon would have on life here on earth.

There were *many* ancillary benefits that came from the mission that we enjoy today. Consider this list Douglas Brinkley compiled of the medical breakthroughs in particular that emerged from NASA's efforts.

> Space medicine contributed to radiation therapy for the treatment of cancer; foldable walkers (constructed from lightweight metal developed by NASA); personal alert systems (devices worn by individuals who might require immediate emergency medicine or safety assistance); CAT and MRI scans (devices used by hospitals to look inside the human body, first developed by NASA to take better pictures of the moon); muscle-stimulant devices (to prevent muscle atrophy in paralyzed patients); advanced types of kidney dialysis machines . . . the implantable heart defibrillator, a tool to constantly monitor heartbeats, was developed by NASA and could deliver a shock to restore heartbeat regularity. . . . Even simple *everyday* health-related objects (such as the special foam used for cushioning astronauts during liftoff, which was starting to be used in pillows and mattresses at hospitals to help prevent ulcers, relieve pressure, and ward off insomnia) . . .

If your mother were on dialysis and her life were spared because of the advanced machinery, wouldn't you be glad we shot for the stars? This is just part of the reason I think it is worth it to continue, regardless of the cost. And on a personal note, I am really, really thankful for my memory foam pillow. Thank you, President Kennedy.

But the mission wasn't expensive only in dollars. It was expensive in divorces too. Many astronauts, engineers, and flight controllers lost their marriages through the obsession they poured into the project. Meeting Kennedy's deadline meant missing family dinners, birthdays, and Christmases. Assigned astronauts and backup crews would typically only have one night a week free to spend with their families.

It had a price in lives as well. Seventeen astronauts have lost their lives in the agency's history. To the horror of all those at NASA, the first of the Apollo missions saw three astronauts perish on the launchpad during a test. They were tragically killed when a fire broke out in their cabin while they were strapped into their seats on top of the Saturn V rocket.

Many of the original Mercury Seven, as well as the next batches of astronauts brought in during Gemini and Apollo, were test pilots who were accustomed to the risks associated with flying experimental aircraft. They had experienced their fair share of loss along the way. Three astronauts died in plane crashes during the space race traveling back and forth across the country. But to lose comrades who had never left the ground was senseless and inexplicable.

So yes. The trip to the moon was one of the costliest undertakings in history. And one of the most important too. And not merely from a propaganda and space race perspective, though that mattered to our national pride at the time.

At the height of the Cold War, we were in a space race with the Russians, who had the upper hand. The world changed overnight when they became the first to put a satellite into orbit. Sputnik was an unmanned satellite the size of a beach ball. It took flight in October 1957, and the race was on. The Russians were also the first to put a dog into space, and the first to put a man, Yuri Gagarin, into orbit. It seemed likely they would stay ahead of our every move. The US's great fear was that the Russians would weaponize the heavens. Rain down missiles from the sky. Spy on everything we did. America's first attempt at a rocket launch didn't even make it four feet into the air before it blew up on the launchpad. It was dubbed a Flopnik. If there really was a race, Russia was winning.

President Kennedy's plan was to leapfrog them by attempting something an order of magnitude greater than what they had done: he wanted to go straight to the moon. He believed a successful mission would win the competition and, in the process, get him reelected. The big problem was that when he presented the idea before Congress—and then again cast the vision for the moonshot in a more eloquent version of the address at Rice University—everyone at the fledgling space agency knew that almost none of the technology they would need to make it happen had been invented yet. Nor did the infrastructure, staffing, or industry exist to support it. America's space program would have to learn how to run a marathon when they had hardly taken their first step. As the broad strokes of planning for Apollo began, they identified more than ten thousand separate tasks that had to be accomplished if they were to put a man on the moon. At the time the president made his decision and publicly committed to it, how to do most of these things was not yet known. But that didn't stop him from going all in.

Kennedy liked to reference Frank O'Connor, who wrote in his memoir *An Only Child* about how "as a boy, he and his friends would make their way across the countryside, and when they came to an orchard wall that seemed too high and too doubtful to try and too difficult to permit their voyage to continue, they took off their hats and tossed them over the wall—and then they had no choice but to follow them." Kennedy said, "This Nation has tossed its cap over the wall of space, and we have no choice but to follow it. Whatever the difficulties, they will be overcome."

Ready or not.

It was crazy. And it worked. It galvanized the American people and injected a spirit of adventure and possibility into mankind in a Lewis-and-Clark kind of way.

The new ocean of space.

The final frontier.

It sends shivers down the spine. Doing hard things because they are hard. Climbing high mountains because they are there. An entire nation mobilized in pursuit of a dream. People working together toward a vision bigger than any one of them. A sense of urgency because of a very clear and rapidly approaching deadline.

HOW CAN YOU KNOW?

Why does that resonate so deeply?

It's intrinsically a part of how God wired us and wants us to approach life. Be fruitful and multiply (Gen. 1:28). Explore. Subdue. Investigate. Name. Improve. This inclination is hardwired into our DNA because we have been made in the imago Dei (image of God). This mandate from the Old Testament is echoed to us in the Great Commission from the New. To resist is to build a Tower of Babel all over again—standing around in a clump when we were told to scatter and bless the world. We "must work the works of Him who sent Me while it is day; the night is coming when no one can work." Life is a timed test, and "pencils down" is coming.

That's why.

He invites us to consider the heavens. To look up. To dream. And yet we live in a strange moment in history where a growing number of people live in a fantasy land where they don't believe that people ever landed on the moon, or that there is a moon, or such a thing as space. Others are "flat earthers" who are fully convinced that the earth is flat.

While my daughter Alivia and I were in Palm Springs for an event where I was speaking, we had the opportunity to see the recently released *Apollo 11* film that included never-before-seen seventy-millimeter footage of the moon landing. This large-format movie is so much crisper than the past grainy footage. The images were spellbinding and majestic. The video was so sharp and clear it felt like it could have been filmed yesterday, but the outfits and hair time-stamped it in the late sixties. You really got a sense for the energy and expectation, especially seeing the million people who camped out on Florida beaches around Cape Canaveral (then Cape Kennedy). There were three thousand reporters from fifty-six countries at the launch, all of them straining their eyes to take in the blastoff of the mighty Saturn V. I felt my heart race and surge like a bull in a chute as the gargantuan beast rumbled and heaved through the final stage of the countdown. A rocket launch is an emotional thing. No matter how many times you watch the videos, you still disbelieve something so big can move. It's like watching the Empire State Building lift off and fly away with a mighty roar. You simply can't watch a rocket launch without a distinct sensation of humility and euphoria rising up inside you. When the astronauts returned eight days later, and the drogue and the parachutes opened, and they survived the fiery trip through the atmosphere in their now-charred vessel dangling from its red-and-white parachutes—when they splashed down in the Atlantic having made it to the moon and back—I wanted to stand and cheer. That's proof that it's possible to know how a story ends but still feel the tension and the emotion as it unfolds when it's told well.

IT'S POSSIBLE TO KNOW HOW A STORY ENDS BUT STILL FEEL THE TENSION AND THE EMOTION AS IT UNFOLDS WHEN IT'S TOLD WELL.

The film was awe-inspiring. Ten out of ten—I recommend you check it out.

We stumbled out of the movie brimming over with superlatives. *Unreal. No way. Can you believe that happened?*

When Alivia and I got back to our hotel outside of town, surrounded on all sides by desert, we asked for an outside table at the restaurant so we could enjoy the night air. We set down our menus and one of us looked up to notice there was a full moon—bright, beautiful, and lighting up the dark desert sky.

Poetry.

The moon was pulling on more than just the ocean; it seemed to summon us to dream.

The moon has always captivated those with eyes to see. It sits in the night sky as a beacon of hope.

We toasted our water glasses and felt like the experience was meant to be.

When the waiter finally made her way over, our metaphorical cups were still brimming over with the joy of the moon above us and we were smiling like lunatics (a word that comes from the Latin word *luna*, "moon," from the belief that phases of the moon caused intermittent insanity—the more you know). As she followed our gaze to the heavens and commented on the beautiful night, I mused enthusiastically, "Can you believe people landed on it?"

The waitress nodded somewhat half-heartedly and said, "Yeah, I guess, though my roommate's brother doesn't think the moon is real. And I'm not so sure . . . he is pretty convincing."

Nearly spitting out my water, I pushed back. "You're surely not serious! If it's not real, what is it exactly? And who put it there and why? What possible gain would there be?"

She said, with no trace of irony or sarcasm, "Well, he thinks that it is a CIA base for surveillance."

I felt like I was going to have a seizure. So many things—words, emotions, and sadness—came washing over me as I listened to her give credence to something so preposterous. I was quiet for a moment, hoping that she would say, "But that's crazy. And obviously . . . there it is right in front of us. Clearly not a CIA base but, in fact, the moon."

Eventually I asked, "How do you explain that in all literature going back to, well, ever, there are references to the moon?"

For example, Aristarchus correctly observed the earth's place in our solar system and estimated the moon's distance in the third century BC. And that's not all. The moon figures into key moments in history:

- George Washington was aided by moonlight when he boldly fortified Dorchester Heights under the cover of darkness to the utter surprise of the British, and in so doing, he gained his first victory of the Revolutionary War by driving the British out of Boston.
- Because of the fierce raids that took place under the illumination of moonlight by the Comanche Native Americans, to this day a bright full moon in Texas is known as a Comanche moon.
- When the Nazis rained bombs down on London during the Battle of Britain in World War II, people knew to be afraid of full-moon nights and the moon's bright waxing and waning gibbous phases, which they referred to as a bomber's moon and which was a source of dread because the Luftwaffe could easily see to line up targets.
- Then again at D-Day, a full moon lit up the sea as the largest amphibious assault in human history readied itself off the coast of the five beaches of Normandy, France, while Adolf Hitler slept.

All of this predates the CIA's establishment. But there's also the fact that we landed on it in 1969 and revisited it multiple times; we left a flag there, drove a car on its surface, and brought hundreds of pounds of rocks home.

Renowned astrophysicist Neil deGrasse Tyson answered with a snort when asked whether the moon landing could be trusted. He reasoned that, first of all, we all saw it take off. Second, there was an immense amount of paperwork for every part that was designed and manufactured—it would have required an immense amount of work to fake all of them. It would be easier to believe it really happened than to assume someone would forge all the documents that exist for the manufacturing of every part involved.

He's right.

Consider this: "By the time of Apollo 11, the number of printed pages, including interface control documents, that were required to check out a space vehicle actually surpassed 30,000." So many pages were printed "that a boxcar would've been required to hold the documents necessary to launch a Saturn V."

Our waiter thought on my verbal eruption for a moment and then said, "That's a good point, but really there's no way for us to know it's there." Then she deftly changed the subject. "Well, anyway, can I get you started on an appetizer?"

Not sure I was ready for this conversation to end with so much left hanging in the air, I sat in silence for a moment, then said, "Well, yes, yes, I do want an appetizer."

As she walked away, my daughter and I looked at each other and then the moon, and then each other and then the moon, and then we shook with laughter until we both had tears running down our cheeks. Thinking of the waiter's roommate's brother, all I could manage by way of fatherly advice was, "Alivia, it is very, very important that you don't take drugs."

The moon landing happened.

And it changed the world.

FROM GRAVES INTO GARDENS

The value and confidence boost gained because people left our planet to set foot on another celestial body cannot be crudely calculated simply by dollars and cents, or even hours spent. It gave us confidence in frightening, anxious times. It galvanized us even as the Camelot era was cracked and innocence was lost with the assassination of President Kennedy. The grief caused by the death of the dreamer was eased by each and every launch. The bitter irony was that he didn't get to live to see his dream come to pass, much like Moses, who died at the edge of the promised land.

But Kennedy's death was not the only traumatic thing to happen in that era. It is easy to think that events in the period you are living in are "unprecedented" and the worst or most out-of-control ever. Everyone feels like that when they are living through turbulence. The truth is, as Solomon said, there is nothing new under the sun (Eccl. 1:9).

The year of 1968 was a very violent one. There was political turmoil so great it seemed like the fabric of the nation would tear in two. It came to a head in Chicago during the Democratic National Convention in August. Haynes Johnson, a reporter who was there covering the event, later said in a 2013 *Smithsonian* magazine story that the convention was "a lacerating event, a distillation of a year of heartbreak, assassinations, riots, and a breakdown in law and order that made it seem as if the country were coming apart."

Many things led to this tipping point:

- The entire situation in Vietnam was exploding, and the reaction to it at home became unstable when reporters, embedded with the troops, began reporting back the ugly truth about the conflict between the North and South Vietnamese and the Viet Cong. The American people had been misinformed about what was really going on. But in the aftermath of what came to be called the Tet Offensive, people realized that in addition to our murky reason for being there, victory was by no means the slam dunk that President Lyndon B. Johnson had sworn it would be. Many people back then decided they didn't have the stomach for what ongoing involvement in the war would cost in both dollars and lives.

- Both Senator Robert "Bobby" Kennedy and the famed civil rights activist Martin Luther King Jr. were assassinated at the height of their popularity and influence.
- President Johnson revealed that he wouldn't run for reelection, in large part due to America's reaction to his handling of the Vietnam War. He realized he had no chance of winning when he lost the faith of famed TV reporter Walter Cronkite.
- Major cities were rocked by riots (more than one hundred cities in all across the country), college campuses were gripped by peace protests, and a clear front-runner hadn't yet materialized to lead the Democratic party.
- Tensions came to a fever pitch in the so-called Windy City—a name that, believe it or not, did not come from the weather but from the constant bickering with Philadelphia after Chicago surpassed Philly as the second-largest city in the country behind New York.
- Agitated protesters converged on Chi-Town by the thousands, where droves of police officers, members of the military, the National Guard, and even the Secret Service were waiting to be squared off against them.
- On August 28, 1968, a bloody confrontation took place that has been dubbed the Battle of Michigan Avenue. Protesters headed to the convention site, and the law enforcement—armed with clubs, tear gas, and rifles—forcefully stopped them in

their tracks. Riots were nothing new, but these were televised. Cameras were rolling as unarmed college students were bludgeoned for trying to make their voices heard.

- In addition to the protestors, innocent bystanders, first responders, and members of the press who were there to report what was going on were brutally beaten by the police. Mike Wallace, who would go on to fame with *60 Minutes*, was punched in the mouth. People watching on television at home all across the country were horrified.

It was fairly easy for some adults to shrug off the hippies as dirty kids needing a bath. But, not unlike the footage of the murder of George Floyd in 2020, the recordings of malevolent violence showed that there really were things that needed to be protested.

There still are.

It is worth noting that the civil rights movement—which called us to move closer to the self-evident truth encapsulated in the Declaration of Independence that all men are created equal—and the space race played out side by side with each other in roughly the same period of time (1950s and 1960s) and both required visionary leadership, tenacity, commitment, and a whole lot of people working together to fulfill the dream.

One of the only things that brought comfort and hope in this tough time was the Apollo 8 mission. It offered the potential that there could be good in the midst of darkness and that you can, and should, shoot for the sky. That imagination and curiosity can give people something to dream about. It showcased what Abraham Lincoln called the "better angels of our nature."

The Apollo 8 launch was scheduled to take place at the end of this dark year, at Christmastime—which was a risk. If something went catastrophically wrong—for example, if the astronauts crashed into the moon's surface and remained enshrined like Egyptian mummies in their spaceship—it could forever taint for Americans the most wonderful time of the year. And people would never truly get over it because, well, you can't really get away from the moon. As NASA's James E. Webb said at the time, "If these three men are stranded out there and die in lunar orbit, no one—lovers, poets, no one—will ever look at the Moon the same way again." But that was true of Christmas too. Since Apollo 8 would take place on December 25, if the astronauts died, then Christmas might never be the same again. It would be a reminder of a mission that went horribly wrong.

These fears were not crazy. After the loss of our daughter and sister Lenya during that time of year in 2012, our family knows all too well how challenging it can be to have the Christmas season tangled up with death. The next year when I first saw holiday cups show up at Starbucks, I remember thinking, "I'm afraid of Christmas." It definitely takes some of the shininess off the tinsel. But in a more profound and unexpected way, it makes the holy day even more powerful; death is the great enemy the long-awaited Messiah came to unseat and defeat. Death and taxes at Christmas are all a part of what elevators look like. We made the painful but ultimately important decision to go even more all-out in our decorations and lights and were fueled by refusing to live in fear.

As it turned out, the launch was a massive success. I'm going to give you more details about the mission in chapter 16, "Eight Comes Before Eleven," but here's a spoiler: not only did no one die, the astronauts were able to report back to earth with, "There is a Santa Claus."

In a similar way, 2020 (admittedly, a tough year) had some beautiful moments—including America's successful return to space when Bob Behnken and Doug Hurley launched from Kennedy Space Center (KSC) in a SpaceX rocket bound for the International Space Station. It was very much a redemptive moment to see Americans leave the world in an American rocket from American soil once again, witnessed in person by a sitting president during a time when so many were in lockdown and suffering.

Apollo flight controller John Aaron recalled that he realized how much Apollo 8 meant to people at large when, in the flood of congratulatory telegrams and letters mailed in from around the world, a telegram came in from a woman that caused him to choke up in an interview years later. She said, "Dear NASA, all of these things have happened so far this year to this country. You took us around the moon on Christmas. Thank you for saving 1968."

LORD, LIAR, OR LUNATIC

I love the sentiment of that telegram, but in truth, salvation for the problems in this world came not when mankind *left* earth but when Jesus came *to* earth. Sin is what drove all the problems in 1968 and today. And because sin is our biggest problem, forgiveness is our greatest need.

This is exactly what the cross is all about and what we will discuss in the next chapter: extraordinary, extravagant, excruciating forgiveness.

But did Jesus really die on a cross? How can you know? Just as some think the moon landing was a hoax, there are people—a philosophy professor at college, a best friend's dad, someone on Joe Rogan's podcast, or perhaps your own brother—who doubt that Jesus is who he said he is. What if it is all just made up? Let me answer your question with a question of my own.

How can we know *anything* that we know from history? For instance, the second president of the United States was John Adams. He was also our first vice president, our first foreign minister to England, and the father of the United States Navy; he wrote the source document that provided the framework for our three branches of government and the separation of power that would end up in our Constitution, and he was the one who decided to start the Library of Congress. (It's now the largest library in the world, with millions and millions of volumes, but he started it with a meager budget of $5,000, so don't ever let anyone tell you that you can't do something great with a small start.)

How do we know any of that? We weren't there.

We know because of what people wrote down during that period. Notes were copied and copied and copied, and eventually they were put in history books. For instance, we have verifiable writings from a couple hundred years ago that document George Washington was our first president, and we know he had fake teeth.

Let's go back further. Take Alexander the Great, a very well-known figure from history—perhaps the man most responsible for Western civilization. What do we know about him? Let's compile a list.

- He conquered the known world (approximately 330 BC) before age thirty, never losing a single battle in fifteen years.
- He died at age thirty-two.
- He was tutored by Aristotle.
- He died after drinking a bowl of wine and developing what some people believed to be a fever. Then his body was likely submerged in a vat of honey.
- He named seventy cities after himself and one after his horse, Bucephalus.
- Oh, and he was played by Colin Farrell in an embarrassingly bad movie.

And we also know he was in touch with his emotions. Like Jesus, Alexander wept. If you've ever watched *Die Hard*, you've heard Hans Gruber reference Alexander's misty eyes.

Very good. Okay, so the facts that I just told you—and anything else we know of Alexander the Great—come to us primarily from two sources: two definitive biographies about him written by Plutarch and Arrian. And listen to me: they were written four hundred years after his death. Yet please tell me, in all the great debates that have broken out in your school classrooms over the years, how many were about whether Alexander the Great really lived? *I don't think he was Philip II of Macedon's son. I don't believe he had elephants in his armies; I actually believe it was zebras. I doubt he wept. I don't believe he existed at all!*

There's not really a lot of that. We know facts about Alexander because Plutarch said so and because Arrian said so. And we don't flinch. No one bats an eye at trusting those as reliable sources, even centuries after the events took place.

So when I tell you that the majority of the New Testament was finished by about seventy years after Christ's resurrection and that there are more than five thousand manuscript copies that are 99 percent identical, not diverging on one doctrinal issue of significance, to attest to the bloody death and resurrection of Jesus Christ, I am telling you that this is head-spinning verification. But that's not all. As Lee Strobel pointed out, "We have preserved for us a creed of the earliest church in 1 Corinthians 15:3. This creed summarizes the essence of Christianity—that Jesus died for our sins, He was buried, He rose again, and then it mentions specific names of eyewitnesses and groups to whom He appeared. This creed of the earliest church has been dated back by scholars to within months of the death of Jesus."

As one expert said, it's the kind of evidence that historians "drool over."

Yes, the moon is real.

And yes, Jesus died and rose again. It is one of the best-attested facts in history. He is not (just) a good person. Good people don't make the claims he did. Jesus Christ is Lord.

And he did it for you.

Just as space travel provided ancillary benefits like memory foam and lightweight walkers in our everyday lives—benefits that weren't the point of the mission but came about as a consequence of exploration—there are by-products of the cross. The primary mission objective was salvation from sin, but there are a whole host of fringe benefits for your daily life, ranging from the boardroom to the bedroom. Jesus didn't come to improve your life but to save your soul. Fortunately, life is much better with a saved soul. If you let him make you holy, you will watch as, day by day, he makes you happier.

JESUS DIED AND ROSE AGAIN. IT IS ONE OF THE BEST-ATTESTED FACTS IN HISTORY.

I know much has been made of the difference between happiness and joy, and how the world only offers happiness but following God leads to joy. The problem with splitting hairs like that is that there is crossover between the two concepts. In fact, the literal translation of the biblical word *blessed*, as used in Psalm 1, is "happy." And that holy happiness doesn't even scratch the surface of the world-without-end the cross of Jesus unlocks if you follow him.

My humble suggestion?

Toss your cap over the wall of faith and see what happens.

CHAPTER 4.5

ON CRUCIFIXION AND CENTRIFUGES

IN SPACE IT'S IMPOSSIBLE TO BREATHE WITHOUT A SPACE SUIT. (Doesn't that make you squirm a little bit just thinking about it?) Even getting to that suffocating vacuum of space means enduring pressure on the body. When the different stages of the rocket's boosters explode, and you go from being static to traveling at 17,500 miles an hour, the pressure, measured in g-force, is the number of times you are experiencing the normal sensation of gravity. If you have been to a state fair and ridden the Gravitron, a spinning ride that causes your body to stick to the wall and rise up from the floor, you have experienced 3 Gs (three times normal gravity).

Every Apollo 11 astronaut oversaw a specific part of the mission, and Michael Collins was the one who oversaw the development of pressure suits that would help an astronaut's body cope with various g-forces in space. To pick the right one from the selection the subcontractor provided, he had to test all the different versions. The downside of Michael's role was that he was repeatedly put into simulations in a spinning centrifuge while wearing different prototypes to see how long it took for him to reach blackout. As a result he spent a lot of time losing his breath.

He wrote of this experience:

> At anything over approximately 8 Gs, I start feeling very uncomfortable, with difficulty breathing and a pain centered below my breastbone. By 10 Gs the pain has increased somewhat and breathing is nearly impossible; in fact, an entirely different breathing technique is needed at high Gs. If you breathe normally, you find you can exhale just fine, but when you try to inhale, it's impossible to reinflate your lungs, just as if steel bands were tightly encircling your chest. So you have to develop an entirely new method, keeping the lungs almost fully inflated at all times, and giving rapid little pants "off the top." In addition to breathing problems, at high Gs the vision starts to deteriorate, and darkness closes in from the edges toward the center.

It's difficult to imagine facing this, period. Let alone facing this and keeping your cool during a launch so you can flip switches while resisting the urge to pull the abort handle. Neil Armstrong managed to model that kind of "right stuff," as Tom Wolfe famously framed it, when a thruster malfunctioned on his Gemini 8 capsule. It took him and David Scott to their physiological limits as they tumbled out of control at about one revolution per second, causing them to experience vestibular nystagmus (involuntary movements of the eyes), nausea, and blurred vision. Fighting off blacking out (which is good, because they would have almost certainly been lost forever if they hadn't), Neil managed to raise his arm (ever try that in the Gravitron?) and flipped a switch above his head so he could successfully use the reentry system thrusters to regain control of his berserk, spinning craft.

Sticking to your mission despite an unfathomable weight on your chest is not just for astronauts. It's what Jesus did as he hung on the cross.

LAST WORDS #1

"Father, forgive them, for they do not know what they do."
Luke 23:34

The first of the seven recorded statements Jesus made while on the cross was this: "Father, forgive them, for they do not know what they do." As he prayed these words, he would have found it difficult to breathe.

To die on the cross is to slowly suffocate.

Before Jesus was raised into the air, he had already been beaten, deprived of sleep, and almost certainly denied water or food. Why care for the condemned? Prior to that, even before his arrest, he had been through as trying an emotional ordeal as one might experience.

A large majority of doctor's visits in our country are in some way connected to stress. When you are facing emotional distress, it can feel like you are being squished.

You might not ever feel the g-forces of space travel, but I'm sure you can relate to the sensation of being out of control. Anxiety can make it feel like you can't breathe. Terror can strike by night and make you feel like you will be lost forever. Grief can paralyze, and in those scary, shaky moments it can be unclear what to do or how to even see the way out. You wake up with the smell of fear and sweat so thick in your room that it seems like you will never find the way out of the dark.

I know this all too well. I've had panic attacks so serious that they threatened to capsize me and wipe me out. I actually had one last night. Each time I've felt like my chest could explode. One time I felt like a necklace I was wearing needed to be removed immediately or it would surely strangle me. It often feels like my body is getting ready to run a marathon and as though my arms and legs are hot. My mind races through thousands of terrifying scenarios and only speeds up when I try to slow it down. Try as I might, I can't calm down. That loss of control is the scariest of all. Breathing exercises are the only thing I have found that are really effective at bringing my screaming autonomic nervous system back from this fight-or-flight mode. And for what it's worth, every time I am really panicking and need to resort to this kind of breathing, my mind tells me it won't work and protests against doing it—all the way until it works.

It is absolutely astounding what physical sensations mental and emotional anguish can produce. I have gained a profound respect for what agony of the mind can inflict on the body.

In the garden of Gethsemane Jesus was under such crushing stress that he actually began to sweat drops of blood—a medical condition called hematohidrosis where tiny capillaries near the sweat glands begin to burst.

The word *Gethsemane* means "oil press," and a more fitting location for his hour of temptation cannot be imagined. It was a stone's throw from an enormous stone that would be rolled over fresh, whole olives, producing extra virgin olive oil. If you have ever bitten down on an olive pit unaware, you know the extreme force it would require to crush both the fruit and the pit, making it give its liquid offering.

Undoing the failure of the garden of Eden, Jesus, the greater Adam, faced the crucible he had come for in this garden. The weight was that of ten million boulders being rolled across his soul. Imagine—flecks of blood mingling with sweat, causing his skin to glisten. It speaks of an agony that is unfathomable.

It was in that condition he was arrested.

Mocked.

Chained.

Repeatedly punched in the face while blindfolded.

Spit on.

Handfuls of his beard ripped out.

And you must remember that though he was God, he became fully man, and as a man he could feel it all. He had the power to command a legion of angels and stop the torture at any moment, but that knowledge would almost make the pain that much worse.

And then there was the flagellum, the cat-o'-nine-tails Jesus was whipped with. Pilate decided that if it was blood the Jewish leaders were after, then blood he would give them. He had Jesus subjected to torment and torture, having him whipped with the leather straps that had little bits of bone and metal embedded in them, connected to a wooden handle. The professional lictor would continuously whip Jesus' back while his arms were chained over his head, leaving his back torn, lacerated, and pulverized into a bloody mess.

If Pilate thought seeing Jesus bloodied would make the Jewish leaders feel sympathy and lessen their hatred, he would soon be disappointed. Like sharks circling chummed waters, Jesus' enemies only became enraged with hatred and bloodlust.

Humans can be so cruel. We can be so cruel. I can be so cruel.

The Galilean would be sentenced to death.

The soldiers had their sport. They put a purple robe on his back, and a crown of thorns was driven into his scalp and forehead. They had to pass the time somehow, and what was the harm? He would soon be dead and forgotten.

By the time he was ready to be paraded through town, the blood-soaked robe, which had dried, was torn off, reviving all the wounds that it had stuck to while it was worn. We know he was depleted and in no condition to carry the upper beam of his T-shaped cross, known as the patibulum, as was tradition on the nearly half-mile-long journey through the old city to the hill called Calvary.

There he was crucified.

With what seems to us a shocking, startling lack of details, none of the Gospels elaborate on what was entailed. But no explanation was given because none was needed. So many had been killed in this way: thirty thousand in and around Judea by the time of Jesus.

Many in Jesus' day would have grown up under the shadow of the cross. Such violence can desensitize you. The nail wounds, as bad as they were, weren't what killed. They were horrifically painful but not the cause of his death. As mentioned, those condemned to die on the cross suffocated.

The Assyrians invented it, the Persians perfected it, and the Romans did what Henry Ford did for the car: they mass-produced it.

When a person's hands were spread out, nailed to a cross, and then raised in the air, it basically paralyzed all the normal muscles involved in breathing: the diaphragm and the intercostals. Hanging in that way allowed the introduction of air into the lungs, but they could not be emptied—the opposite of being in a centrifuge. They were stuck in a permanent state of inhalation. To exhale required shifting one's weight from the nails in the hands to the nail that went through the feet or the small pedestal the toes could touch. Of course, to do this would cause agitation to any wounds as the person's back scraped up and down the rough cross—which would be unimaginably painful, especially after having been whipped.

Today when someone is put to death in the United States, measures are taken to make it as fast and as pain-free as possible. To die on the cross was the exact opposite. The entire process was designed to make it as painful and as slow as possible. The body could hang on for days, slowly sliding up and down the cross for each and every breath.

The entire time the accused hung there, they were a living demonstration of the fact that crime doesn't pay.

The Romans always crucified in highly visible locations beside roads, so there was no privacy in which to suffer. You were essentially being gawked at by people for every minute of this barbaric and cruel torture. So varied was the traffic in and out of town at the location of Jesus' crucifixion that the sign listing his offense had to be translated into three languages (John 19:19–20).

If you visit the spot of Jesus' crucifixion today, you will see a bus station at the foot of the bluff thought to be Skull Hill, which we know of as Calvary. Golgotha. To imagine people traveling through—families, businesspeople, tourists—squabbling, eating packed lunches, asking for a bathroom break while you struggle to breathe in the final minutes of your life, your mind delirious with pain, is to imagine what Jesus was facing in his crowded hour.

This all was so terrible, a word had to be invented just to describe its horrors. And so they did: *excruciating*. A word we toss around somewhat flippantly to describe a bad headache, or a time on pins and needles waiting to receive a callback after an interview. *Excruciating* actually means "from the cross."

GOLD FROM GOLGOTHA

This better understanding of the crucifixion should give us a new, humble appreciation of what we are reading when we come to the seven red sentences spoken during the hours Jesus suffered on the cross. Since the only way you can form words and sounds is to breathe, and since Jesus could only breathe out when he raised his body up on his tiptoes, that means for every precious statement he made he had to endure yet another agonizing and difficult labor of love.

These seven sentences are weighty and should not be read lightly. They should be handled as the costly treasures that they are. Warren Wiersbe said, "These seven last words from the cross are windows that enable us to look into eternity and see the heart of the Savior and the heart of the gospel."

Breathe them in. Breathe them out. Feel yourself release fear and pain and panic. It will feel like it won't work right up until it does.

The first words Jesus spoke on the cross, "Father, forgive them, for they do not know what they do," were not addressed to those crucified with him, to his enemies, or to his friends (Luke 23:34). They were to his Father.

The first time he spoke was to pray.

It would not be the last prayer. The fourth and the seventh times would also be prayers. Nearly half of the words spoken during the crucifixion were to God.

This explains everything.

Namely, the peace and calm he modeled for us all on the hardest day of his life, his care and concern for others, his utter lack of vengeance, his unwillingness to shoot lightning bolts from his fingertips and fry those who had put him there as they gathered to mock him, his ability to face this fiery trial with care and concern for everyone but himself.

He tapped into secret power.

David said that when he prayed evening, morning, and noon, God would hear his prayers (Ps. 55:17). Daniel did the same (Dan. 6:10). When you see David standing over Goliath, Daniel surviving a night in the lion's den, or Job worshiping God on the worst day of his life, it is easy to mistake what you are seeing. These epic hero moments are merely the tips of icebergs. Beneath the surface are souls who were strengthened by prayer.

Public victory comes from private discipline.

Like Daniel, Job, and David, Jesus made a habit of prayer. He spent the morning, noon, and night of his crucifixion the same way he often spent mornings, noons, and nights during his ministry: praying.

What happened between his prayers was possible because of what happened during his prayers. He was praying when the soldiers came for him. Hunched over in agony. His face in the dust. He chose this spot on purpose and he often came there with his disciples (John 18:2). I like to think that it reminded him of the desert where he spent forty days in the wilderness alone and began his ministry.

I realize I have always thought about his time in the wild incorrectly. He was driven by the Spirit into the wilderness for forty days and forty nights, where he was with the wild beasts. Then the devil came and tempted him during this period. It seemed to me almost like being in the wilderness was one more grueling, horrible test to pass, like the fire swamps in *The Princess Bride*. Like the location and the isolation were intended to beat him down so he would be at his weakest when the tempter came. If that was the case, why does it say in Luke 4 that the *Spirit* sent him into the wilderness immediately after his baptism where he was tempted by the devil?

What if he didn't go to the wilderness for those days so he *could* be tempted but because he was *going to be* tempted? What if the Spirit sent him to the wilderness not because it made him weak but because it made him strong? Perhaps that is why the Spirit drove him there.

If you can't change the battle, you can change the battlefield.

Meet the enemy on your terms.

IF YOU CAN'T CHANGE THE BATTLE, YOU CAN CHANGE THE BATTLEFIELD. MEET THE ENEMY ON YOUR TERMS.

This is why Jesus would often have brief wilderness moments before difficult ministry days. I believe he was doing in miniature, in those pit stops, what he did "in maximum" before his ministry began.

It's fitting he would end his ministry as it began, in a location reminiscent of a wilderness setting. He was surrounded by nature and trees that will one day sing (1 Chron. 16:33).

In the garden he was lonely but not alone. He brought friends, but they had fallen asleep and would soon run away. His help came from above.

And so, on the cross, he chose to begin the way you and I should begin our days: in prayer.

When I sat down to write this chapter, I picked out very specific music to play in the background. Claude Debussy's "Clair de Lune" is on repeat. I chose it because I read in a very sad, very well-written book called *The Tender Bar* that it is "Debussy's musical portrait of the moon." It seemed appropriate to write about the moon while hearing someone's impression of it on the piano, and it set the tone for my writing.

Why are track athletes so particular about how they come out of the blocks? They know the power of starting out on the right foot. You can always salvage a day that has begun poorly. But why do that if you don't have to? Beginning my day intentionally often sets in motion a day's worth of activities that are more God-glorifying, personally satisfying, and life-giving. I find myself less thirsty for people's approval. I'm more grounded, confident, and comfortable in who I am. It is easier to be the dad I long to be, the husband Jennie deserves for me to be, the pastor and author God built me to be.

It's much better to live from a well-built start than rebooting on the ashes of a day that you let turn into a dumpster fire.

Starting well makes it easier to end well.

There is wisdom in having at least a portion of your devotional time at the outset of the day.

You can't have a quiet soul without quiet time.

Jesus' steely resolve and unflinching focus all throughout his excruciating time on the cross were the results of lighting his soul on fire through prayer. When Jesus arrived at Golgotha, the guards offered him myrrh—which was a strange full-circle moment from the gift the wise men brought to celebrate his birth (Mark 15:23). He rejected it. The myrrh was a mild act of Roman mercy, a little bit of anesthesia with a stupefying effect. It would be the equivalent of giving him a hit of a joint to dull the edge and take his mind off what was about to happen—the spikes being driven through his hands.

He rejected it.

He would feel it all. He wanted his wits about him.

(William Wallace was following Jesus when Wallace spit the anesthesia out of his mouth.)

I find it interesting that the Bible presents being drunk as the opposite of being filled with the Spirit (Eph. 5:18).

(A side note regarding the liberty to drink alcohol—I love the wisdom of G. K. Chesterton: "Drink because you are happy, but never because you are miserable." This is a profoundly helpful piece of advice that can apply to any gift of God that can take his place in our lives. When you come under the influence of a substance, you are unable to be filled with the Holy Spirit in the same way. And Jesus himself drank wine, so we are not talking about all drinking; we are talking about being drunk. In Ephesians 5:18, Paul reminds us not to be drunk but to be filled with the Spirit. So when you are tempted to have a few drinks or get stoned to "take the edge off," instead pray and allow God to do a much deeper work inside you.)

In this moment Jesus wanted to feel. He didn't want his vision to be dimmed. He wanted his eyes wide open. Jesus could have taken the myrrh and he would have seen less. He instead allowed the Spirit to let him see more.

At that moment the cross was raised, all Jesus' weight landed on those wounds. Pain ripped through his nervous system, sending fiery signals screaming to his brain.

He began to move. Rising for the first time, he spoke: "Father, forgive them, for they do not know what they do" (Luke 23:34).

Even more amazing than the fact that he prayed, I am stunned that he didn't pray for himself. He used his precious breath to pray for someone else.

The soldiers hardly noticed. They had already begun arguing over who would get to keep his clothes.

Five garments, four soldiers. To each of them went a piece. All that was left was his outer garment. Highly functional, it doubled as a sleeping blanket and jacket. And Jesus' was high quality, sewn from top to bottom without a seam (John 19:23–24).

Surely it was a gift. Perhaps from Mary Magdalene, out of whom he'd cast seven demons (Luke 8:2). It has been thought that Magdala was a wealthy city known for its textiles. A garment district. And she wasn't just any Magdalene, she was *the* Magdalene. Prominent as the leading, most well-known of all those in Magdala. So it is conjecture, but not outside the realm of possibility, that she had supplied this jacket.

What's interesting about the guards' arguments over Jesus' clothes is that as he prayed his prayer, clothes were what was on his mind.

Sins were like crimson, covered white as snow.

Righteous deeds as filthy rags . . . he has covered us in righteousness (Isa. 64:4–9).

The clothing cover of salvation to hide our nakedness.

This is what God alluded to in the garden of Eden when he gave animal skins to Adam and Eve. Their best plan was fig leaves. Maybe this is why Jesus cursed a fig tree in Matthew 21.

Religion continues to make pathetic attempts to cover up the nakedness of sin. But it is impossible without the death of an innocent third party. Either the one who did the crime perishes, or a substitute does.

For all mankind, Jesus is the Lamb who takes away the sin of the world (John 1:29). At the cross he was actually paying for what he was praying for. So to think of those soldiers in the coming weeks wearing Jesus' clothing around town is to picture what God wants your life to look like.

Father forgive *them*.

[STAINED] GLASS HOUSES

Jesus was praying for forgiveness, but who was it for?

- Judas, who had betrayed him?
- Religious leaders, who had lied about him?
- The Jews, who yelled, "Crucify him"?
- Pilate, who had sentenced him?
- The soldiers, who had just physically driven nails through him?

Yes, yes, yes; but there's more. Jesus' prayer that day was a prayer for you and me too. Isaiah 53:6 says, "All we like sheep have gone astray; we have turned, every one, to his own way; and the Lord has laid on Him the iniquity of *us all*."

And the Hebrew word translated as "forgiving" in Exodus 34:7, when God describes himself as forgiving, is, wait for it—*nasa*. It means "to lift, carry, take." That's beautiful on so many levels. Jesus was lifted up on the cross in order to carry our sins and take them away so we could be launched into a whole new life. One in which you are forgiven but also in which you could become forgiving.

This is why we are told in Matthew 6:12 to pray, "Forgive us our debts, *as* we forgive our debtors."

To not release debts hurts you.

And what you hold on to doesn't have to be something huge to be lethal. It's the little splinters of unforgiveness that are the most annoying, and if they become infected, they can cause big problems.

For you and me, we're triggered by things that are less like someone cutting off our heads and more like someone cutting us off on the highway. Not so much being stoned by a mob but political arguments over holiday dinners. Left unchecked, like a cancerous infestation, these roots of bitterness multiply, spread, and eventually kill relationships.

Fortunately, the forgiveness that works in big issues also works in small ones. And when you keep your eyes on the cross, mountains turn back into molehills.

Is there anyone in your life you need to forgive? Any grudges you need to let go of?

Well, yeah, you might be thinking, *but they don't deserve it.*

Kindly, gently, I believe the Holy Spirit wants us to sit with this uncomfortable truth: neither did you. Neither do I. Consider this: it didn't stop Jesus from wanting to forgive you or me.

To be an unforgiving Christian is an oxymoron. Like *jumbo shrimp*, *Icy Hot*, *mutual differences*, *only choice*, or *tiny elephant*. We are, after all, the followers of Jesus—a man who used his dying breath to ask forgiveness for those who killed him. *A Christian is someone who is both* forgiven *and* forgiving. Everywhere we go we ought to show love. I'm convicted just typing that.

Matthew 5:46 says, "If you love those who love you, what reward have you?"

Translation: *Anyone can do that.* And yet it is easy to start being stingy with the grace we give, even though, if you and I are honest, we have a never-ending appetite for the grace we want to receive.

We love that God's forgiveness for us is an *ocean*. Seriously—we get high on all the superlatives. "East to west," "depths of the ocean" . . . yay (Ps. 103:12 CEV; Mic. 7:19 NLT)! But when it comes to the betrayal we have sustained, the backstabbing we have endured, the treachery our relationships are riddled with, the person we served and gave opportunity to who bolted the moment things got hard or something better came along . . . then all of a sudden we want to be in control of exactly how much forgiveness is being doled out. Like toilet paper in a pandemic: "Sorry, one per customer. Only on Tuesdays. Fresh out. Better luck next time."

We have a scarcity mindset.

This clearly was on Peter's mind when he asked Jesus, "Lord, how often shall my brother sin against me, and I forgive him? Up to seven times?" (Matt. 18:21).

I relate to Peter's desire for things to be firm and equitable and, most of all, controllable. He wanted to know when a forgiveness quota had been reached so he could turn the lights off, lock up, and go home. He probably thought Jesus would be impressed by the big number he had come up with. Seven. *Wow. Jesus will think I am a magnanimous guy.*

No doubt he was stunned by what came next. Jesus said to him, "I do not say to you, up to seven times, but up to seventy times seven" (v. 22).

Jesus blew his number out of the water. Peter was probably both scratching his head and trying to do the math. I can see his toes wiggling in his sandals as he ran out of fingers trying to hit 490.

But that is not what Jesus meant at all. He was saying you should never be able to reach the end of the forgiveness you give, because there is no end to the forgiveness you have received.

Then he told a parable to give a picture of how treacherous and horrible it is to build a dam on the river of forgiveness that flows from the wound in Jesus' side. It is the most uncomfortable story (uncomfortable to me because it looks so much like me—I do it every day) of a man who was forgiven a bill of a few million but then straightaway threw someone into jail who owed him only a few dollars (vv. 23–35).

When you receive the revelation of how much forgiveness costs, and how preposterous it is to withhold it, you are changed—opened up to a miraculous life with a triumphant spirit. Stephen understood that. That's why he died with a prayer on his tongue for the men who were stoning him to death. "He knelt down and cried out with a loud voice, 'Lord, do not charge them with this sin.' And when he had said this, he fell asleep" (Acts 7:60). He died as Jesus did—not with his own condition on his mind but concerned for those who were attacking him.

What a revelation of the forgiveness of God! He will transform self-pity into compassion for others. We realize that instead of being an easily agitated victim we get to be a victor, an agent of grace.

Many other people throughout history have followed in Jesus' and Stephen's footsteps and died martyrs' deaths, full of empathy for those who were ending their lives. But I don't want it to seem like this level of forgiveness is something you can just choose to muster up. That's an exercise in futility and will be as short-lived as the JNCO jeans' heyday in the nineties.

We have a much more reliable power supply than just our good intentions and willpower. Ephesians 4:32 says, "Be kind to one another, tenderhearted, forgiving one another, *even as* God in Christ forgave you." That is where the power comes from. You don't have to conjure forgiveness mojo out of the atmosphere when someone has sinned against you. Forgiveness is not "water under the bridge" or just "letting bygones be bygones." Rather, it's remembering that when Christ hung on the cross he paid for all sins—including those that were committed against you.

NEWS FLASH: FOR THIS FORGIVENESS THING TO HAPPEN WE MUST STOP LIVING IN DENIAL AND PRETENDING WE DON'T NEED HIM.

News flash: for this forgiveness thing to happen we must stop living in denial and pretending we don't need him.

In 1830, President Andrew Jackson

pardoned a man who was sentenced to death. Incredibly, the man refused the pardon. The case went all the way to the Supreme Court, where it was determined that a pardon is not valid without acceptance.

Jesus' prayer is not enough for you to be pardoned. You must accept it.

There is a moon.

And there is forgiveness.

Both for you and for those who have hurt you.

CONQUER YOUR INNER SPACE

Before you end this chapter, I encourage you to sit in silence for a moment and feel your breath while you consider Jesus' pain and his first prayer from the cross. Allow God to help you see scenarios in which you have been wronged and need help to get to a place where you are willing to forgive. Don't get wrapped up in all the details. You can't actually experience reconciliation unless the other person accepts it, but you can come to a place where you are willing and not stuck in bitterness.

You can pray this prayer:

Father, thank you for being willing to forgive me. I pray that, by your power in me, you would allow love and grace not to get stuck in me but to flow through me. Help me be not just forgiven but also forgiving, amen.

CHAPTER 5

LET THE PARTY CONTINUE

LIFE HAPPENS AROUND THE TABLE.

The table in our house is built on a cross. Literally. We commissioned it from a fabulous furniture maker by the name of Levi who runs a company called Birch and Bennett Co our friend Eric introduced us to. We wanted reclaimed wood on the top, but we also wanted the structure underneath to be metal and have a modern feel. He sourced the wood from a hundred-year-old church building, the surface buffed smooth by ten decades of footsteps. And he built a cross into the design—an enormous, seven-foot, stainless-steel cross forms the base for the table. Jennie and I love this juxtaposition of something old mixed with something new; it has also been the guiding principle of our church. Old truth, new presentation.

When Levi delivered our cross table on a cold winter night, it far exceeded our expectations. We were speechless when we saw what he built.

This table is now the heart of our home, and the cross is the heart of the table. If you came over for a meal, it is where we would sit. I would warn you to watch out if you chose to sit on the booth side; many knees and shins have come into contact with that very sturdy cross in a bad way. We have had as many as nineteen people squeezed around it for a meal. Games are played, prayers are prayed. Hands are held. Tears are shed. Life is lived. And of course, food is eaten.

Food is intrinsically social. It is connected to relationships like air is to breathing. Many foods capture my attention—quesadillas, chips and guacamole, pasta, pralines-and-cream ice cream, anything that combines chocolate with peanut butter. Beside the fact that these foods betray me, giving happiness now and regret later, they are all so much better when eaten with other people. There is nothing quite so fun as hearing the *oohs* and *aahs* of a table full of happy friends and family when a sizzling plate of fajitas or a gooey dessert shows up with a round of coffees. (Next time you're in a restaurant, close your eyes and listen. It really is a beautifully happy sound.) There's just something about coming together and eating with someone that is both intimate and endearing. It can also be a humbling exercise because it's tough to look cool eating buffalo wings and challenging to not appear clumsy eating lettuce wraps.

It is with this rich, sensual, symbolic imagery in mind that we are ready to begin our approach to the concept of Jesus' first "I Am" declaration recorded in the Gospel of John.

"I AM" #1

"I am the bread of life."

John 6:35

In an obvious and practical sense, Jesus is saying that he is everything we need: "I am the bread of life. He who comes to Me shall never hunger, and he who believes in Me shall never thirst." Bread, in his day, was more a necessity than a luxury. It was a staple of the diet, like rice is in many Asian countries. Not a side item like a roll at the edge of your plate; it was part of the main course. In declaring himself the bread of life, Jesus was saying that he was the difference between life and death, for to not eat is to perish.

There were then, as there are now, many bread alternatives. Lots of things can fill your stomach for a moment, but you will get hungry again. And I'm not just talking about chips and queso. Sex, drugs, accomplishments, money . . . all these things are expected to give more meaning than they are able to. The problem isn't that they don't work but that they don't last. No one ever has been high, and the next morning declared that the high was so good they'd never need it again. It wears off, and you need more.

Jesus is the absolute opposite. He alone can supply a satisfaction that nothing else on this earth will ever come close to supplying—one that fills us not just for this life but for all eternity. Not only does it not wear off, but it actually gets better with time, continuing to unfold and unfurl until it's a whole museum of joys and pleasures. The creek of peace becomes a river of delight.

Jesus made it clear that, like bread, he must be looked to for survival (John 6:25–59). Many of his followers walked away after hearing this (v. 66). Jesus asked his disciples if they wanted to bail also. Peter asked, "Where would we go? You have the words of eternal life" (v. 68 CEB). He wasn't saying there was no one else offering guidance or words of wisdom. There was. There were plenty of wannabe messiahs in Judaism, which he had been raised in, emperor worship was spreading throughout the Roman Empire, and pagan perspectives abounded. Peter scanned the options in his mind and realized there was no one coming close to the kind of supernatural meat and potatoes that Jesus offered up freely to anyone who would believe in him.

But there's more.

Bread abounds in the life of Jesus. From the bread Satan tried to get him to make out of rocks (Matt. 4:3) to the bread of the Last Supper, this wonderful food features prominently in Jesus' three years of public ministry.

We know that one of Jesus' favorite things to do was to break bread with people—any people. Sinners, saints, prostitutes, publicans. He was nicknamed the friend of sinners (Matt. 11:19) and was willing to eat with anyone who wanted to hang. Jesus, the party animal. Probably not the image Sunday school drove home for you. But the popular image of the Son of Man's ministry is summarized by the three words *eating and drinking*.

It can also be hazardous to eat with people. One time in Seattle, my daughter Daisy hit me in the head with a hammer while we were cracking open crab claws. In Jesus' day it was viewed as hazardous to eat with the *wrong* people.

To eat with someone brings about a great unity. If we share an appetizer, some of it goes into and becomes a part of me, and some of it goes into you, and so in that way we are bonded. So it was in the ancient perspective as well. With contempt Jesus' enemies spit through their teeth while talking about the riffraff who Jesus sat around the table with. Matthew 9:11 explains, "When the Pharisees saw it, they said to His disciples, 'Why does your Teacher eat with tax collectors and sinners?'" They simply couldn't fathom why he would allow himself to be united with people who would defile him with their super sinful cooties.

If I were Jesus, when I got wind of their criticisms I would have shot back, "Because otherwise I would have to eat all my meals alone. Your question insinuates that you are not sinners, but the reality is none of you would be able to sit with me and I would have to eat every meal at a table for one."

Jesus wasn't, and isn't, surprised or worried about being defiled by our dirty, sinful condition. That's the reason he came. "When Jesus heard that, He said to them, 'Those who are well have no need of a physician, but those who are sick. But go and learn what this means: "I desire mercy and not sacrifice." For I did not come to call the righteous, but sinners, to repentance'" (Matt. 9:12–13).

I am the bread of life.

I see much, much more in this than just, "At long last we can find something to satisfy our desires, unlike the things of this world that leave us wanting more." It is not just about stomachs not being empty but about hearts being full. And nothing causes a heart to be full like being invited to a party—which is ultimately what living bread is all about.

This living bread was on Jesus' mind when he told us to pray for it in the Lord's Prayer. "Give us this day our daily bread" (Matt. 6:11). N. T. Wright observed:

> The prayer to the Father for daily bread was part of [Jesus'] wider and deeper agenda.
>
> At the heart of it stood a central biblical symbol of the kingdom: the great festive banquet which God has prepared for his people. . . .
>
> The banquet, the party, is a sign that God is acting at last, to rescue his people and wipe away all tears from all eyes. . . .
>
> "Give us this day our daily bread" means, in this setting, "Let the party continue."

The church is a table held up by a cross. It is nineteen people crammed into a booth for dinner. And judging by the smells coming from the kitchen, Mexican food is on the menu.

Speaking of parties, Jesus frequently compared the kingdom of God to the festivities that accompanied a wedding. Check out Matthew 22:1–10 (MSG), which tells the story.

Jesus responded by telling still more stories. "God's kingdom," he said, "is like a king who threw a wedding banquet for his son. He sent out servants to call in all the invited guests. And they wouldn't come!

"He sent out another round of servants, instructing them to tell the guests, 'Look, everything is on the table, the prime rib is ready for carving. Come to the feast!'

"They only shrugged their shoulders and went off, one to weed his garden, another to work in his shop. The rest, with nothing better to do, beat up on the messengers and then killed them. The king was outraged and sent his soldiers to destroy those thugs and level their city.

"Then he told his servants, 'We have a wedding banquet all prepared but no guests. The ones I invited weren't up to it. Go out into the busiest intersections in town and invite anyone you find to the banquet.' The servants went out on the streets and rounded up everyone they laid eyes on, good and bad, regardless. And so the banquet was on—every place filled."

Weddings, and the parties surrounding them, are often the pinnacle of human joy. You put your best clothes on and come ready to laugh until it hurts. And eat lots of dessert. Dancing. Crying. Watching the ring bearer veer off course or the flower girl get stage fright. It's all the best parts of being alive. And that is God's kingdom.

All of history points forward to, and is explained by, feasting and celebration—a party. This is why turning water into wine at a wedding was Jesus' first miracle. He was pointing to life on the other side of Gethsemane. It was in that garden, as he prayed, that he contemplated taking up the cup. In the end, despite everything in him telling him not to do it, he was willing to go through with the plan for the joy that was set before him (Heb. 12:2). Whose joy? Yours!

HE SWALLOWED YOUR SORROW SO YOU COULD SIP ON JOY!

He swallowed your sorrow so you could sip on joy! That is why I want you to wonder at the majesty of space. To explore the moon and be willing to explore your own soul. It can awaken your curiosity so God's goodness can free you from the somber seriousness of religion and allow the vitality of friendship to lead you down a path of celebration and excitement and newness. Heaven is not just out there; the kingdom of God is within, in your inner space.

The party God is focused on is also seen in the parable of the prodigal son, which Jesus tells in Luke 15:22–28.

> The father said to his servants, "Bring out the best robe and put it on him, and put a ring on his hand and sandals on his feet. And bring the fatted calf here and kill it, and let us eat and be merry; for this my son was dead and is alive again; he was lost and is found." And they began to be merry.
>
> Now his older son was in the field. And as he came and drew near to the house, he heard music and dancing. So he called one of the servants and asked what these things meant. And he said to him, "Your brother has come, and because he has received him safe and sound, your father has killed the fatted calf."
>
> But he was angry and would not go in. Therefore his father came out and pleaded with him.

The father's decision to throw a party for his son makes so much sense. Having a son he feared dead returned to life must have been like having a lost limb suddenly restored. The pent-up energy of all the lost birthdays and holidays now had an outlet. I can't wait to make up for lost time and catch up on what we have missed when we finally get to see our daughter Lenya in heaven. This is how your Father feels about you and what he wants you to know life in his kingdom is all about. The first thing on heaven's agenda when we get there is the marriage supper of the Lamb. A wedding party!

But parties aren't just in our future; Jesus made sure of that. Before he ascended to heaven, triggering the detonation of the firework show called Pentecost, he left us a meal to point us to the massive eternal feast that is coming: the Last Supper. But do we allow it the same kind of party atmosphere?

I think that we so often get Communion wrong. Don't you normally feel like you are supposed to be sad when holding those tiny little cups and terrible-tasting crackers? I do. That's not what Jesus wants. It is not supposed to be overly serious, like a funeral. It should be celebratory. It is certainly reverent because of Jesus' suffering, but it is meant to be full of joy, not sadness. "For as often as you eat this bread and drink this cup, you *proclaim* [this Greek word literally means "celebrate"] the Lord's death till He comes" (1 Cor. 11:26).

If you visit the tomb of George Washington at Mount Vernon, you can't help but feel somber and sad. Washington was a seemingly invincible leader cut down after getting a sore throat. What? Here lie the bones of a man the British couldn't kill—but some wet weather took him out. Jesus' tomb, on the other hand, is empty! He faced the cross, absorbed the wrath of God and all the devil could throw at him, and came through on top. Next time you hold the bread and wine, remember they point not to his defeat but to his victory.

The gathering of the church to eat and drink is meant to get us ready for the day when the full-blown joy of life in the new heaven and new earth kicks into high gear. Isaiah 25:6–8 describes it as a "feast of rich food for all peoples, a banquet of aged wine—the best of meats and the finest of wines. . . . He will swallow up death forever. The Sovereign Lord will wipe away the tears from all faces; he will remove the disgrace of his people from all the earth. The Lord has spoken" (NIV).

Matthew 8:11 says of this day, "Many will come from the east and the west, and will take their places at the feast with Abraham, Isaac and Jacob in the kingdom of heaven" (NIV). C. S. Lewis imagined we will exclaim when we arrive, "I have come home at last! This is my real country! I belong here. This is the land I have been looking for all my life, though I never knew it till now." Because all the adventures we have ever had will end up being only "the cover and the title page." Finally we will begin "Chapter One of the Great Story which no one on earth has read: which goes on forever: in which every chapter is better than the one before."

That's there and then. What about here and now?

We live in the already but not yet. We get glimpses of that joy. Moments of it. Tastes of it.

Charles Spurgeon said, "Some of us know at times what it is to be almost too happy to live! The love of God has been so overpoweringly experienced by us on some occasions, that we almost had to ask for a stay of the delight because we could not endure any more. If the glory had not been veiled a little, we should have died of excess of rapture, or happiness."

That is what church can and should be.

That is what your small group can and should be; what you are inviting friends and family, strangers and coworkers to: a table propped up by a cross.

The Bible insists on using sensory language about salvation. It calls us to "taste and see" that the Lord is good, not only to agree and believe it (Ps. 34:8). In this we live, serve, love, and rejoice. We build. We sing. We worship. And we invite everybody to come in and participate in this kingdom. We beg for God to make it below as it is above, for heaven's touch to be felt on earth.

In the tension of the already but not yet, we feed hungry people. We clothe homeless people.

You and I are all on the party-planning committee. We don't have to be the life of the party, but it is our job to bring life to the party.

THE INVITATION STARTS NOW

I see time as a crucial connection between space exploration and what Jesus did for us on the cross. Not just because of the coincidental connections—it takes three days to get to the moon, and Jesus was in the tomb three days; there were twelve manned Apollo missions and twelve years from the launch of Sputnik until we landed on the moon, and there were twelve apostles and twelve tribes of Israel. But also, time shows us how vividly we can see enthusiasm and passion drain out of a project that was originally invigorating.

When the original Mercury Seven astronauts were announced, they were heralded as heroes. This happened to an extreme extent when the Apollo 11 astronauts returned from the moon. They were celebrated with parades and seen as role models for children. But attention quickly faded. The impossible morphed into the commonplace. The final shuttle mission was completed with the landing of Atlantis on July 21, 2011, closing the thirty-year Space Shuttle Program. For the next decade, American astronauts had to hitch rides on a Russian Uber to get to the space station.

That all changed with Elon Musk's SpaceX. It is such a big deal to have Americans going to space from American soil again. That new program, along with Richard Branson's Virgin Galactic, Jeff Bezos' Blue Origin, and other commercial enterprises, have started to restore interest in space in the public at large. They seem to be reawakening the American imagination. The prospect of going to Mars for the first time, and potentially seeing humans live there, has provoked interest. A collective fascination for the stars is stirring once again in a way that hasn't happened in my lifetime.

Don't mishear what I am saying. The men and women of NASA have been working just as hard; we just haven't seen the same support for or awareness of their programs since the early days of the space race. The entropy began as early as Apollo 12, later on in 1969, after the famous first moon landing. Viewers plummeted in comparison. And then Apollo 13 wasn't even broadcast. There was no public energy surrounding what was going to be the third landing on the moon. Until a tank ruptured, endangering all three astronauts' lives, they weren't even getting television coverage. They were sending videos from space, but none of the networks were airing them. The CapCom (capsule communicator) on the ground didn't have the heart to tell them this, though, so the astronauts went through their updates oblivious to the fact that no one was watching. America had grown bored. It took a crisis for people to stop yawning and care.

Passion can deteriorate in the same way spiritually.

We who have been invited to the party can forget how amazing grace is, forget that our tables are held up by a cross, and turn our faith inward instead of outward.

In a sermon, I heard Tim Keller retell a story about an experience Tony Campolo had one night. The story reminded me of how winsome our faith can be when we, like Jesus, are the friend of sinners.

Tony flew to Honolulu for a meeting. Because of the time difference between the East Coast and Hawaii, he woke up at 3:00 in the morning. Wide awake and super hungry, he decided to go find some food. He ended up sitting at the counter of the only place open—a filthy, seedy diner. A server with a greasy apron—presumably the owner of this mom-and-pop establishment—came over to ask what he wanted. Seeing a display of donuts under a glass dome, the pastor said, "I'll take a donut and a cup of coffee"—what seemed to be the least awful option. The server walked over, filled his cup with coffee, and with his grimy hand grabbed a donut and set it down in front of the pastor.

As the pastor was trying to decide whether he should eat the donut, the door opened and three women walked in. They were prostitutes who had just finished working in the red-light district of Honolulu; he couldn't help but hear their conversation as they talked about their night's work, the clients, and so on.

As they finished eating, one of the girls said to the other, "It's my birthday today. Can you believe it?" The other girls started ragging on her, asking her what she wanted for her birthday. She scoffed, "Present? I haven't had a birthday gift in my entire life." With that, they paid their bill and walked out.

Stunned at what he'd just heard, the pastor called the server over and said, "Those three girls who just came in—do you know who they are?"

The server said, "Oh yeah, they come in every night about this time."

The pastor continued, "That one girl who was sitting on the far left—it's her birthday today. What do you think if tomorrow night we had a party for her?"

The server called to his wife in the back, "Honey, come out here. You've got to hear this. This guy here wants to throw a birthday party for Agnes."

She barreled out and, looking the pastor directly in his eyes, she said, "Mister, whatever you want, you can do. Agnes is a really good girl. She's in here every night. I'm telling you, I know she's in a rough line of work, but this girl really is sweet. She's one of the good ones."

The pastor asked, "Can we use your restaurant?"

"Anything," she insisted, "anything you want."

"Okay," he said, "I'm going to get streamers. I'm going to get balloons. I'm going to get a cake."

The server slammed his hand on the counter as he said, "Not the cake. I'll get the cake." The pastor agreed, all while thinking, *Just don't touch it with your hands, bud, and that'll be good.*

Then he said, "And invite anyone who knows Agnes, anybody who's ever met her. Ask them to come as well."

At 3:00 the next morning, he went back to the diner and decorated the place. And like clockwork, at 3:27, Agnes and her friends came through the door. She froze like a deer in headlights. Tears streamed down her thickly made-up face as everyone began singing "Happy Birthday." Agnes looked down and saw her name written in the icing. Begging them to please not cut the cake, she explained that she'd never seen her name on a cake before and wanted to show it to her mom, who lived two blocks away. Agnes ran out the door with the cake.

Standing there in a crappy diner, at 3:30 in the morning, in a room full of prostitutes waiting to cut a cake, the pastor decided to make conversation. He said, "Hey, would anyone object if I said a quick prayer for Agnes?"

No one knew what to say, so he prayed God would bless her, that she would know his love for her, and that she would find her value and purpose in him.

As the prayer ended, Agnes returned with her mother (and the cake), and the party continued. The server grabbed the pastor by the shirt and said, "You didn't tell me you were a preacher. What kind of a church do you lead anyway?"

Thinking on his feet, the pastor responded, "Well, I guess the kind of church where we would throw a party for a prostitute at 3:30 in the morning."

The server said, "No way. That kind of a church doesn't exist. But you know what? If it did, I would go to it."

Touché.

When we lose our why, we lose our way.

As Jesus' followers, we aren't meant to ever lose sight of what the living bread points to—the feast that is coming, what an honor it is to have been invited, and the obligation that is ours to let others know they, too, are on the guest list.

Warning: this will be messy. Does this message of grace mean truth doesn't matter? Of course not. It would be unloving to not speak the truth. This incredible tension has caused casualties on both the side of compromising what the Bible teaches and the side of becoming "sinner"-condemning Pharisees. In the middle is Jesus. Grace and truth. Unyielding fidelity and unceasing kindness. Remember, our goal isn't to win arguments but to win souls—and that is worth the tension and the mess.

CONQUER YOUR INNER SPACE

My vision for these Conquer Your Inner Space moments is twofold. First, I know this book is a lot to digest. We are going fast and covering a lot. So I want these moments for us to check in and give you a chance to breathe and process, and for your stomach to settle so you don't get informational acid reflux. And second, to give you a hot second to apply what we've discussed so you don't just learn and move on without taking action. The learning is worthless if it doesn't lead to living differently. Sometimes there will be a prayer, other times some questions. I'll mix it up depending on what is needed, and there won't be a formula, so you won't be tempted to just skip over them.

Who is one person in your life who is far from God that you can invite to church with you on Sunday, even if it means getting out of your comfort zone? List a few people in your life who you are praying will come to know Jesus. Are there any new people who have moved to your street or building you can introduce yourself to and welcome to the neighborhood?

Don't let the party stop with you!

CHAPTER 6

HOUSTON, WE HAVE A PROBLEM

WHEN NEIL AND BUZZ BEGAN THEIR DEPARTURE FROM THE MOON to return to the command module, Columbia, and ultimately to the earth, a massive number of switches needed to be thrown inside their spacecraft. They had trained for up to fourteen hours a day, six, sometimes seven days a week, to know which switch did what and when it would need to be flipped—all while strapped into simulators. Then they had to read and write reports and memorize mission rules after these training sessions.

In the three modules that comprised the Apollo 11 spacecraft, there were a total of 678 switches and 410 circuit breakers. They were all organized into the spacecraft's subsystems. *Subsystem* is basically 1960s NASA jargon for a device that performs a specific function, such as providing air to breathe, power for the engines, or a toilet—because even in space you have to go somewhere. The CSM (command

and service module) and LM (lunar module) had the same subsystems but were tailored in their design for where they would operate. The LM would only ever operate in a vacuum, so aerodynamics wasn't a factor. But the CM would have to survive reentry to earth. And both needed toilets.

Everything was ridiculously efficient. Let me nerd out for a second while I tell you *just* how efficient it all was. A scuba diver's underwater tank of oxygen, which would last sixty minutes while scuba diving in Cancun, would last for fifteen hours in Apollo. How? We don't absorb all the oxygen we take in when we breathe before we let that breath go. When we breathe out, we exhale good oxygen along with carbon dioxide. The life-support system on Apollo scrubbed exhaled air to eliminate CO_2, then recycled the O_2 to be breathed again. The same machine also removed moisture, eliminated odor, monitored the cabin's pressure, dispensed cold and hot water, and circulated coolant to keep all their gear at the right temperature. The entire unit was about the size of an air conditioner you would mount in the window of an apartment. Not bad for 1960s tech, huh?

The fuel cells that generated electrical power by mixing hydrogen and oxygen also did double duty, because this process created drinking water as a by-product. The beta versions of the machine used on Gemini missions led to water carbonated with hydrogen bubbles, which posed no health risk for the astronauts but made for uncomfortable gassiness during missions. Uncontrollable flatulence while locked in a tiny capsule with other humans for days on end and spending time trapped in a space suit is brutal. On Gemini 7, the capsule splashed down in the ocean after a fourteen-day-long mission to test rendezvous procedures that would be crucial before Apollo could be attempted. Two of the three frogmen swimmers who were there when the hatch opened vomited in response to the fragrance that came from the tiny cabin and its inhabitants.

The fact that the oxygen and hydrogen that made both power and water had to be stored at super cold temperatures led to incredible breakthroughs in leakproof, insulated storage devices. Consider this. "If an Apollo hydrogen tank were filled with ice and placed in a room at 70°F, it would take 8.5 years for the ice to melt. If an automobile tire leaked at the same rate as these tanks, it would take 30 million years to go flat." Take that, Yeti.

The ingenious design of Apollo also utilized the landing gear of the lunar module as a makeshift launch platform. What they landed on, they departed from. The spidery legs of the Eagle lunar module from Apollo 11 (as well as five other LMs used in NASA missions: 10 [which crashed into the moon after being released from lunar orbit at 50,000 feet], 12, and 14–17) still reside on the moon today. But before anyone could take off, a number of those 678 switches needed to be flipped.

Both astronauts in the Eagle had charts duct-taped to the walls on either side of the lunar module to indicate the desired switch positions for the various circuit-breaker panels. They hardly needed these, since they had trained so much they could literally find the right switch blindfolded. As they went through the checklist, they were horrified to realize one of the switches had broken off. Apparently after the space walk, when they were getting back in the cabin wearing their bulky space suits and backpacks (a rather unceremonious affair not unlike being born backward), Buzz had accidentally hit one of the breakers with a corner of his space suit and broken it off.

Buzz was born to be an astronaut. Not only was his mother's maiden name Moon but he also shared a name with a very famous, yet-to-be-created spaceman with the last name of Lightyear. (In childhood, Buzz's sister called him "Buzzer" since she was unable to pronounce "brother," and it stuck.) "He was the first astronaut to have a PhD," and he was the first person to take a selfie in space (which he did on an extravehicular activity on a Gemini mission). But now he had potentially jeopardized their safe return home with a clumsy mistake.

In his book *Men from Earth*, Buzz elaborated: "The little plastic pin (or knob) simply wasn't there. This circuit would send electrical power to the engine that would lift us off the moon. . . . We looked around for something to punch in this circuit breaker. Luckily, a felt-tipped pen fit into the slot."

How amazing to think that a potentially mission-threatening problem was solved through such a simple improvisation. It was very much a "hold my beer while I try this" type of thing. Luckily they had a pen in their vehicle—but think how many times you have been rooting around in your glove compartment for one and come up empty-handed. How incredible that they not only had a pen but their makeshift switch worked.

THE CHURCH IN EPHESUS

Jesus has a similar style. When he sees issues in the hearts of his followers, his solution is to pick up a pen. Communication is not his last resort but his first. And with his words he both builds us up and cuts away that which would hold us back. As William Shakespeare said, the pen is mightier than the sword. In this case, he keeps us soaring through the air and keeps us from missing out on our mission with his words that are as sharp as a sword (Heb. 4:12). Just as Buzz's pen got the astronauts off the ground, the words of Jesus will address the broken and damaged areas of your life.

Buckle your seat belt. What you are about to read could change your life.

Jesus' first letter of the seven he wrote to seven churches, recorded in the book of Revelation, is to the church at Ephesus. I know you might be intimidated by the book of Revelation, but don't worry. We are in this together, and this is going to help you. I promise. John—the same John who was a disciple of Jesus and wrote the gospel bearing his name—was a prisoner on an island called Patmos, where he had been banished during a time of intense persecution of Christians. While incarcerated he was given the apocalyptic vision that is contained in the book of Revelation, and among other things, there were seven messages he was to deliver. Here is the first one:

> "To the angel of the church of Ephesus write,
>
> 'These things says He who holds the seven stars in His right hand, who walks in the midst of the seven golden lampstands: "I know your works, your labor, your patience, and that you cannot bear those who are evil. And you have tested those who say they are apostles and are not, and have found them liars; and you have persevered and have patience, and have labored for My name's sake and have not become weary. Nevertheless, I have this against you, that you have left your first love. Remember therefore from where you have fallen; repent and do the first works, or else I will come to you quickly and remove your lampstand from its place—unless you repent. But this you have, that you hate the deeds of the Nicolaitans, which I also hate."'" (Rev. 2:1–6)

The seven churches were all real churches in real cities. They had pastors. They had congregations. And they had problems.

They are also you and me.

Symbolic of seven types of churches that exist throughout history, the seven churches exhibit the different characteristics represented in just about every congregation of people who call themselves Jesus followers. In them we find ourselves at our best and at our worst. Jesus wrote to these seven churches because he cared enough to commend and to correct them.

In the previous chapter we learned that every time the Last Supper is eaten it points forward, rejoicing in the second coming. The return of the King. When we read the words Jesus spoke to these seven churches, we can determine if we are on the right track and where we might be missing the mark, as we hold fast until he returns.

EVERY TIME THE LAST SUPPER IS EATEN IT POINTS FORWARD, REJOICING IN THE SECOND COMING. THE RETURN OF THE KING.

Just like Jesus' "I Am" declarations and words spoken from the cross, these words are life and light. In them we find overlapping themes and the chance to overcome. As we study them, you and I will see the things these people dealt with that needed exposing. By the way—where it stings, lean in and don't push back. Think of it this way: If you had a tumor growing inside you, wouldn't you want to know? It's not just big things that cause breakdown; often it's the small, hidden things that go unchecked.

ALL THE SMALL THINGS

The two deadly incidents in the Shuttle Program (1981–2011) were caused by relatively minor issues.

- 1986: A faulty O-ring caused Challenger to explode seventy-three seconds after liftoff.
- 2003: A large piece of foam that fell off a fuel tank on takeoff damaged vital carbon panels on Columbia's left wing. These damaged tiles were unable to function as a heat shield, creating vulnerable spots that caused the shuttle to burn up on reentry.

And then there's the near-tragedy of Apollo 13, in which the fateful words we now know as "Houston, we have a problem" were spoken after tanks ruptured, losing oxygen and causing the crew to have to abandon the landing and instead use the lunar module as a life raft. It was a chain of events that began with a rather small issue. One of the oxygen tanks "had been accidentally dropped during maintenance . . . causing slight internal damage that didn't show up in later inspections."

At about nine o'clock at night, the crew performed a routine cryo-stir with a fan to churn up the super cold oxygen that was not only their source of air but also water and fuel. Then, at 9:08 p.m., there was an explosion, which prompted astronaut Jack Swigert to say, "Houston, we've had a problem here." Over time his words have morphed into the famous line we know today.

How do you blow up your life? Slowly. And there are always warnings. When you are on the path to sin, God sends agents of rescue. Ways of escape. No temptation can overpower you without you having the opportunity to do the right thing.

The conviction of the Holy Spirit is a gift. A tender conscience is a precious thing. I once saw a sign in my dental office that said, "Ignore your teeth and they will go away." Little rumbles of conviction from the Holy Spirit are like that. The more you ignore him, the quieter the rumbling becomes. And soon you can lose your bearings.

According to Eugene Peterson, "David didn't feel like a sinner when he sent for Bathsheba; he felt like a lover. . . . David didn't feel like a sinner when he sent for Uriah; he felt like a king" (2 Sam. 11). Peterson also noted that "the subtlety of sin is that it doesn't feel like sin when we're doing it."

We get to let biblical figures' hindsight be our foresight.

After the massive success of the Mercury and the Gemini programs, NASA set its sights on Apollo, thinking they could do no wrong. But death followed the overconfidence. During Apollo 1, Gus

Grissom, Ed White, and Roger Chaffee burned to death strapped in the cockpit of their command module. A fire broke out while they were belted in on the launchpad for a dress rehearsal. Everyone knew space could be deadly, but no one thought such a horrific tragedy would happen on the ground.

Michael Collins, who would go on to be a part of Apollo 11, was friends with the astronauts who died. He had the impossible job of notifying one of the widows of her husband's death. He later described how it caught them off guard and why it shouldn't have:

> We worried about engines that wouldn't start or wouldn't stop; we worried about leaks; we even worried about how a flame front might propagate in weightlessness and how cabin pressure might be reduced to stop a fire in space. But right here on the ground, when we should have been most alert, we put three guys inside an untried spacecraft, strapped them into couches, locked two cumbersome hatches behind them, and left them no way of escaping a fire.

It was a tense time, and many thought Apollo 1 spelled the end for the space agency. In the days that followed, NASA's director of flight operations, Gene Kranz, gave an incredible message to his team that has been called the speech that saved NASA. All of the men in the Mission Operations Control Room (MOCR, pronounced *moker*) listened in stunned silence as he pulled no punches but instead assumed responsibility and called everyone to a higher standard. It galvanized the team and caused everyone to shoulder their share of the blame, learn from their mistakes, and redouble their efforts.

> Spaceflight will never tolerate carelessness, incapacity, and neglect. Somewhere, somehow, we screwed up. It could have been in design, build, or test. Whatever it was, we should have caught it.

> We were too gung ho about the schedule and we locked out all of the problems we saw each day in our work. Every element of the program was in trouble and so were we. The simulators were not working, Mission Control was behind in virtually every area, and the flight and test procedures changed daily. Nothing we did had any shelf life. Not one of us stood up and said, "Dammit, stop!"
>
> I don't know what Thompson's committee will find as the cause, but I know what I find. We are the cause! We were not ready! We did not do our job! We were rolling the dice, hoping that things would come together by launch day, when in our hearts we knew it would take a miracle. We were pushing the schedule and betting that the Cape would slip before we did. . . .
>
> From this day forward, Flight Control will be known by two words: "Tough and Competent." *Tough* means we are forever accountable for what we do or what we fail to do. We will never again compromise our responsibilities. Every time we walk into Mission Control we will know what we stand for.
>
> *Competent* means we will never take anything for granted. We will never be found short in our knowledge and in our skills. Mission Control will be perfect.
>
> When you leave this meeting today you will go to your office and the first thing you will do there is to write "Tough and Competent" on your blackboards. It will *never* be erased. Each day when you enter the room these words will remind you of the price paid by Grissom, White, and Chaffee. These words are the price of admission to the ranks of Mission Control.

This speech was critical not only in filling those at NASA with energy but also in providing the right kind of energy—a resolve to not cut corners or compromise. The sentiment outlined the core values that would shape the American space program.

With this mindset, and after a nearly two-year hiatus, NASA skipped all the way from Apollo 1 to Apollo 7 (there's that seven again!), which was the next manned flight. (Apollo 2 and 3 were canceled, and Apollo 4, 5, and 6 were unmanned.) The Apollo 1 astronauts would have been pleased. Before he died Gus Grissom had said, "We hope that if anything happens to us it will not delay the program. The conquest of space is worth the risk of life."

THE SEVEN LETTERS

Tough and competent.

That's how we should approach Jesus' words to his church. He is giving us the chance to get better. He's rethinking. Coaching.

Many people today have a low tolerance for difficult conversations. We quit jobs the moment we no longer have all the good feels. Even faster when someone presents a "growth opportunity."

Troy Aikman said, "I never once threw a pass in my time in the NFL that wasn't critiqued." The higher you go in a sport, the more you receive coaching, not less.

Wounds that hurt cleanse evil and stripes clean out the inside of the heart (Prov. 20:30). Our response to correction should be gratitude. You can't get better without it.

We were eating pizza in Jackson, Mississippi. My friend Louie looked me in the eyes and told me some things he saw as concerns in my life. I have eaten a lot of pizza with him, but that memory stands out because he cared enough to chastise me.

It stung in the moment, and my mind filled up with a litany of excuses. When I was a child, my dad told me he thought I would make a great lawyer because of my ability to argue. But something inside me told me that if I pushed back, I would miss out. Instead, I listened and took heart and made changes based on his perspective.

And I became better for it.

I don't remember anything about how the pizza tasted, but I will never forget how much I appreciated the opportunity to improve as a human.

That is what these seven letters represent. Jesus is leaning in, speaking honestly. It's an opportunity for you to get better, faster, stronger.

Tough and competent.

LETTER #1 (EPHESUS)

"You have left your first love."

Revelation 2:4

Ephesus was no doubt the most important city of Asia Minor, much like New York City or Los Angeles in the United States. Located on the east side of the Aegean Sea, on the mouth of the Cayster River, it was important politically, commercially, and educationally.

The population of Ephesus at the time of this letter was upward of 250,000 people. It had a 25,000-seat theater that hosted athletic events rivaling the Olympics. It was home to the primary harbor in Asia. It was located at the center of the highway that ran from Rome to the east—the backbone of the Roman Empire.

The library of Ephesus—built around AD 100—was massive and impressive, but the city was best known for its temple to the goddess Diana (known to the Greeks as Artemis), which was one of the seven wonders of the ancient world. It was four times bigger than the Parthenon in Athens. Diana was worshiped through sex with temple priestesses. The city actually derived much of its wealth from the manufacture and sale of images of the goddess (a figure with multiple rows of breasts on her chest and a club in her hand).

In spite of its unspeakable immorality, Ephesus also had a rich heritage of Christian ministry. It's wild to think that this city hosted

Paul, Timothy, and John, "the disciple Jesus loved," as its pastors. All three taught there at one time or another. It's hard to imagine any church in history with such bragging rights.

According to Acts 19:9–10, the Ephesian church was the mother church out of whose ministry the other six were founded. So it makes sense that Jesus addressed it first.

Jesus gave his first letter through the apostle John forty years after the church was planted. Many of those in the church would have been second-generation Christians. Passing the baton of belief to the next generation is always the most complex and important leadership responsibility anyone faces.

The letter begins with a tone of high praise—for their labor, patience, doctrinal integrity, and ceaseless activity. They even hated the deeds of the Nicolaitans. That's good, right? Yes, it is good. It means that their pastors didn't lord their authority over the people (*nico* means "over" and *lateans* means "laity," or church members who aren't clergy).

Nicolaitans were clergy who practiced sexual immorality and perversion under the guise of ministry. They were abusing the designation of "priest" (1 Peter 2:5). And the Ephesians weren't into it. Theirs was a church that was busy making plans—the program was full and the calendar was packed. They had a legacy of significant pastors in the past who had served there and were proud of it. And they were very worried about being defiled, so they made sure that none of their members lived in sin or believed false doctrine.

But sadly, there was a problem for these once on-fire saints.

Jesus said, "You have left your first love" (v. 4).

Near my family's favorite breakfast restaurant is a church with a beautiful statue of Jesus in its front yard. One day I dropped my family off to get on the queue for a table, and as I was parking the car, I noticed that there were people going into the building for a service. It struck me that Jesus was still outside. It was a snowy day and flakes had collected over his brow and arms. I stood there looking, and it made me sad. It made me think of these words and the times they have been true in my heart. It's like Jesus was saying, *Everything you are doing for me is amazing, Levi. I see the church, and there's the steeple, and look at all the people. There is this one small thing, though. You have left me out in the cold.*

Ouch.

Besides that, how was the play, Mrs. Lincoln?

It's like a birthday party with decorations, cake, and presents, but the birthday boy isn't invited.

How did a once-vibrant church planted by Paul and pastored by Timothy end up here? It's called drift. And drift happens. That's how you lose your first love.

You used to hold her hand, open every door, scatter rose petals. Now you hardly grunt to acknowledge she has arrived. Or think about your first car. There was a day when you drove carefully with hands at ten and two, checking mirrors and blind spots before every turn. Now you eat a cheeseburger with one hand, draft an email with the other, and drive with one knee.

We never drift in a positive direction.

You don't doze off while heading to work and accidentally find yourself in the company parking lot. You end up in a ditch or an ambulance—or worse. Drifting doesn't take you to where you want to go; it takes you away from it.

I find myself drifting away from God so easily because of small decisions like these:

- rushing through my devotions to check a box instead of worshiping a Savior
- nursing my injured ego when snubbed by someone, either in reality or in my mind, instead of talking to God

- gorging on too much entertainment
- indulging too much in food and drink
- caring too much about what people think

This is as obvious as it is predictable. As is my go-to reaction when my soul is not as healthy as it should be.

- I pour myself even more into the things I do for God.
- I spend longer hours at the office and give myself more work.
- I don't pull away from service; I throw myself into it, full tilt.
- I think, *Surely this activity will make up for the disconnect inside me. The deadness. The distance.*

Whether this is all driven by a sense of penance or pragmatism, I don't know. But I know this for sure: I can relate to the Ephesians. It's my go-to home church when things are not well inside the house of Levi.

This church is full of type-A alpha females and males, and we are productive, efficient, and hoping the God we have been ignoring notices all we are doing for him. Does he care if I'm snippy with my wife? Preoccupied when my kids are trying to talk to me? Seething over my collection of snubs and wounds? Toying with little lusts and not-so-little jealousies? Emotionally unavailable to my staff? Surely not. Look at all I am getting done.

I feel like Martha, the sister of Lazarus, can relate. She was annoyed by her sister, Mary, who was sitting at Jesus' feet while she was doing all the hard work of putting on a banquet for him (Luke 10:38–42). Must be nice to get to just sit there all *la-di-freaking-da*, but the guacamole isn't exactly going to make itself.

Talk about burying the lead. Not to say what Martha was doing wasn't important, but when you have Jesus in your house giving a message . . . don't miss it.

The biggest ouch for me is the fact that she interrupts Jesus' sermon and demands that Jesus tell Mary to help her work.

Oof.

Spoken like an Ephesian. Takes one to know one.

Here's the takeaway: when you are doing things for God but not spending time with God, you begin to act like you *are* God.

I know your works, but you have left your first love.

Everything is going well but this: your walk with God is missing the walking-with-God part.

Jesus said that when we visit the prisoner, give cold water to the orphan, or are kind to the afflicted, we are doing those acts of kindness to him (Matt. 25:40). I wonder if this is not a clue to undoing what the Ephesians omitted. In their quest to be doctrinally sound they had become emotionally cold. But in what way? How do we measure this?

If we grow cold in our love for Jesus, it will show up in our love for people. When I am close to Jesus, I care more about what he cares about: *people, people, people.*

If you asked the Ephesians if they loved Jesus, they certainly would have thought you were crazy. "Of course we do. Have you not seen how big our Bibles are?"

But in his criticism of the church, Jesus was saying, in effect, "How can you say you love me if you don't show love for the poor, the outcast, the foreigner? When you are more focused on what you are doing for God than what he has done for you, it is easy to lose the beating heart of compassion for others in need." You can easily degenerate into a skewed perspective that tells you other people have what is coming to them.

In Jesus' day it was commonly believed that someone who experienced disease must have sinned to deserve their condition. Thus, to help someone out would be to interfere with God's justice being doled out.

Convenient. I don't have to raise a hand to help my neighbor *and* I get to feel good about not doing it.

Eventually this self-rewarding and self-congratulating way of looking at the world reached the point that if someone was born with a serious medical condition, people shrugged their shoulders and guessed that the baby must have sinned in the womb or that the parents must have done something horrible. When you turn from grace, it doesn't just hurt you; it hurts those God wants to touch through you.

REMEMBER, REPENT, RETURN

What's the solution? How do we remember our first love?

There is a surefire, three-step way to get back that tender feeling: *remember, repent, return*. The same knob that turns the water off turns it on.

Remember

Remember what? What was it like to be newly saved? To be so broken over your sin you couldn't believe God could change you? That there could be hope?

Remember what God went through to reach you. That the Spirit hunted you. That your grandmother prayed for you. That your sister invited you. Most of all, remember that Jesus died for you. There's a reason the Last Supper is a call to remember. To go back. To feel it all over again.

Unless you intentionally remember, you will automatically forget. How cocky you become when you stop dancing with the one who brought you.

Remembering also gets you out of the "God, I thank you that I am not like other men" kind of prayers Jesus referred to in the Gospels. In the parable in Luke 18, Jesus told about a Pharisee who was smug, inflated, and full of death on the inside. He actually prayed, *I thank God that I am not like other men; I fast, and pray, and give* (vv. 11–12 paraphrase). He was basically the biblical version of Gaston from *Beauty and the Beast*, so full of himself it is nauseating. Yet if I am honest, I can often relate to this sort of self-centered spirituality. The other man, a tax collector, saw himself clearly. He fell to his knees, beat his breast, and refused to even lift his eyes to heaven. His prayer was not eloquent or put together. He didn't know how to do it. He put it simply: "God, be merciful to me a sinner!" (v. 13). Yet Jesus said the latter man, not the former, was "doing it right" in God's sight.

The real trick is to be able to pray a prayer like that years and decades after you got saved. All God has done in your life and through your life makes you no less in need of his mercy and grace—and no less deserving of his favor. Just because you were born again onto third base doesn't mean you hit a triple.

Remembering helps you shrink down to your proper size and puts things back into perspective. It helps you mend your Ephesian ways and fosters gratitude in your heart.

Repent

Change your mind. Do a U-turn. Flip the switch. Make the conscious choice to not leave Jesus out of the things you do for God.

To put it in the terminology of Martha's story, this doesn't mean you don't cook for the banquet; you just don't miss out on the opportunity to sit at Jesus' feet before you start serving.

Return

Go back to what you did at the beginning, when you were amazed by grace. When you spent time with Jesus every day and talked to him like he was right there in the room. Time spent together is powerful medicine in any relationship. Much like when you were first dating your spouse, you likely stayed on the phone all the time and refused to hang up—even to brush your teeth.

When you have left your first love, Jesus, you can return to him by crossing the street and going out of your way to bless someone who is unlike you. By raising your voice and extending your hand to the forgotten, the incarcerated, the distressed.

This is pure and undefiled religion (James 1:27). It's not only an action that helps make things right in the world but also helps make things right inside you.

It's clear that the stakes were high for the church in Ephesus. They are equally high for us now: in essence, Jesus said, "Do these things or I will come to you quickly and remove your lampstand from its place—unless you repent."

Opportunity, favor, blessing—a glorious life being a city on a hill, with new doors opening up to you—this is the ever-expanding living thing that your relationship with God, and thus any ministry you do, should flow out of. But when this vibrant faith turns into religious obligation—when you live by a quid pro quo of "if I do this, God will surely do that"—you will shrivel up on the inside. The passion will gradually slow down, and one day you will wake up, seeing nothing of the grand, glorious way God intends for you to live—and you'll wonder why. Soon someone is eating ramen noodles off reclaimed wood that used to be the side of your church.

They say the body keeps score. Football players see the effects of concussions, boxers experience speech issues, and mixed martial arts (MMA) competitors and wrestlers end up with cauliflower ears. Eventually spiritual dullness starts to show in the same way. And, like Uzzah, one day you will put the ark of the covenant on a cart and steady it—and find that God has reached the terminal end of his patience (2 Sam. 6:3–7). Uzzah was the Israelite who had God in a box and officiously assumed "responsibility for keeping [God] safe from the mud and dust of the world," according to Eugene Peterson. "Men and women who take it upon themselves to protect God from the vulgarity of sinners and the ignorance of commoners keep showing up in the religious precincts. . . . Uzzah's death wasn't sudden; it was years in the making."

Or like Michal in 2 Samuel 6:16, you will judge King David's uninhibited dancing in worship of the Lord. Alexander Whyte said, "Those who are deaf always despise those who dance."

When you first got saved and started out serving God, you were small in your own eyes, just happy to be on the team. That kind of spirit makes it easy for God to bless you. The problem is that it is easy to get used to being saved, to having a spot on the team. That's why legalism is so attractive.

In the drama of grace, you are offered an opportunity to be a different character. The gospel portrays you as the damsel in distress. The trap of religion is that it entices you into thinking you're the hero.

The gospel only has one hero, and you are not it. Neither am I. That is good news. It's a relief. It means that tomorrow the weight of the world will not be on your shoulders. It is your job to bask in the glow of the one whose problem it is to keep things ticking. You and I are along for the ride.

Jesus' last words to the Ephesians do not end with condemnation, but invitation: "He who has an ear, let him hear what the Spirit says to the churches. To him who overcomes I will give to eat from the tree of life, which is in the midst of the Paradise of God" (Rev. 2:7).

The Ephesians' story didn't have to end with them mired in loveless religion, and neither does yours. You can come alive in a genuine relationship with God that will turn all the dull and dead spiritual activity into on-fire reality. Awaken your wonder. Your relationship with Jesus can be real. Pulsating, heartfelt, raw, vibrant, and full of joy. Leaping and whirling like David before the ark, totally alive.

I love the footage of Apollo 16 astronaut John Young saluting the flag on the moon. Watch it on YouTube. It's amazing. He doesn't just stand there and salute; he leaps and jumps and is so full of zeal, clearly thrilled to be experiencing one-sixth of the gravity that normally holds you down on earth.

C. S. Lewis called the Trinity "a kind of dance," an idea that Tim Keller took one step further when he called it "the dance of God." Your relationship with God is meant to be you joining that dance. Jumping to salute. Hopping for joy. Happy to be alive and not dead. You were not a person of God; now you *are* a person of God! You were not a child of the King; now you *are* a child of the King!

CONQUER YOUR INNER SPACE

Once while I was at Space Center Houston watching a video that covered the Columbia disaster, my astronaut friend Shane leaned over to me and said, "Ever since that tragedy we did a backflip in the space shuttle upon arrival at the International Space Station, because doing this would have spotted the damaged heat shield." During this Rendezvous Pitch Maneuver (RPM), those still in the space station would perform a visual inspection on the ship looking through the windows. If there were problems, they would know it before attempting reentry to earth.

You can't see your own back. You need someone else to do that.

Who can you ask to tell you where you are weak and keep you from disintegrating? The good news is that if you have left your first love, you can return to it. As the old adage goes, "No success is final, and no failure is fatal." Apollo 13 has been described as "a successful failure"—a failure because they didn't accomplish the original mission, but a success because of how they responded to that failure and kept it from becoming catastrophic. You might not be where you want to be, but it doesn't have to get any worse! To quote Andy Grove, former CEO of Intel, "Bad companies are destroyed by crisis. Good companies survive them. Great companies are improved by them."

Today is the day to remember, repent, and return.

CHAPTER 7

FIFTEEN SECONDS TO PARADISE

THIS IS MORE THAN I CAN HANDLE.

I'm barely hanging on by a thread right now.

I don't have the bandwidth for one more thing.

How can I be behind in every single area of my life at the same time?

The throbbing ache of being overwhelmed could be coming from the pressure of school. The challenge of parenting. Your aging parents and their medical situations. Juggling multiple jobs. Things that need to be done around the house. And now you have a cold. Who gets a cold in the summer? And is that a canker sore on your tongue? Often it is a mixture of multiple things happening all at the same time.

In that moment, just the thought of trying to add anything to your plate seems all but impossible. It feels like it's all more than you can bear, and you don't know how you are going to cope. You are being asked to do more than your system can process. So you get into the shower because you don't know the last time you washed your hair, and as you attempt to catch your breath while standing under the stream, you glance over to the towel bar and realize there's no towel hanging there. And you think to yourself, *This is the end. I have reached the end. This is the straw that broke the camel's back. Who exactly is this sicko who is loading all this straw on this poor camel's back?* And then there's the horrible, cold, wet future that awaits you after your shower as you streak to the linen closet, preferably without scarring any of your children for life.

The screen flashed to life—1202.

It was a program alarm. Neil's eyes darted from the window, where he saw the moon coming closer, over to his right to the abort button. He and Buzz had traveled a quarter of a million miles from earth to the lunar surface and from there they were in the final sixty-mile-long, twelve-minute descent to the moon in the lunar module Eagle after disconnecting from the command and service module, named Columbia. This was by far the most complicated part of the mission, and so much could go wrong.

The third man, Michael, was now alone in the CSM Columbia. This is where he would remain while Buzz and Neil landed the Eagle and then returned to Michael and the CSM for the trip home. But now it looked like the Apollo 11 mission would be a failure.

During the planning phase, NASA had considered three options when trying to decide how to put a man on the moon.

Option one was called direct ascent. It involved just one rocket and one ship. The single giant booster would shed spent stages along the way, and the single ship would land on the moon and return.

Option two was called earth orbit rendezvous (EOR), and it

involved two rockets and two ships. After separate rocket launches brought the various components needed, everything would be assembled in earth's orbit and then taken to the moon. This was the plan favored by famed German rocket scientist Wernher von Braun.

Option three was the lunar orbit rendezvous (LOR), and it involved one rocket and two ships. In this plan a single Saturn V launch vehicle would carry the components that would travel to the moon. It'd be two spacecraft composed of three modules: 1) the command module, which the crew would return to earth in; 2) the service module, which was basically a U-Haul to store consumables (water and oxygen) plus a propulsion system that would get them to lunar orbit (so, a storehouse crossed with a powerhouse); and 3) the lunar module. The intent was for the two spacecraft to separate from each other so the LM could land on the moon and then reconnect (rendezvous) with the CSM in orbit while going around the moon at great speed. Neil described it this way: "Bringing two spacecraft together in a rendezvous in space will be like maneuvering a boat into a moving dock in the middle of the night with only a half pint of gas." Yikes.

And even if the two ships did end up reconnecting without sailing past each other or smashing into each other, they had to achieve docking with the limited views outside the windows. Michael gave insight into the difficulty of docking: "The docking process begins when the two vehicles touch and the probe slides into the drogue. They're held together then by three tiny capture latches, and it's almost like tiny little paper clips holding together two vehicles, one of which weighs 30,000 pounds and the other 5,000. It's a tenuous grasp."

LOR was originally ridiculed as outlandish but eventually won out after being championed by engineer John Houbolt, who risked his reputation and career by aggressively promoting the concept. He passionately believed it would work and took on "the task of convincing the naysayers—within Langley, at the highest levels of NASA,

Congress, and industry—that LOR was the only technique that would enable meeting the goal set by President Kennedy of landing on the moon before the end of the decade of the 60s."

Despite its challenges getting off the ground in time, LOR was deemed the only feasible option and proved to be the vital link in the nation's successful effort to place humans on the moon.

According to NASA, "Houbolt's supervisor, Ed Garrick, congratulated him by commenting, 'I can safely say I'm shaking hands with the man who singlehandedly saved the government $20 billion.'" His persistence had resulted in the critical decision that ultimately ensured the success of the Apollo program and won him the NASA Exceptional Scientific Achievement Medal for his work. Option three it would be.

Having made the 240,000-mile, three-day journey from Florida, all that was left was for Buzz and Neil to descend in their tiny, spider-legged contraption the sixty miles to the lunar surface. But at a point 40,000 feet above the moon, this alarm threatened everything. Imagine—to have gone so far only to go home without having taken one small step.

The lunar module, Eagle, was so light it bordered on rickety. It had to be. For every pound of the spacecraft, they had to bring three pounds of propellant if they were to get back to the command module orbiting above, a.k.a. their only ride home. And there are no gas stations anywhere on the moon yet.

One of the compromises was extremely thin exterior walls. There is not much pressure on the moon, and none in space, so it didn't need to be thick. As a result, the lander ended up with aluminum walls a hundredth of an inch thick—so thin they could be punctured with a butter knife. And easy on the goods, fellas, because if there were to be a breach in one of those paper-thin walls while they were out of their suits, they would die, not just from the lack of oxygen but also the temperature. Keep in mind, the moon reaches 250 degrees in sunlight and -250 degrees in the shade, so your blood would either boil or freeze. Oh, and you would suffocate too. There are a lot of things that can kill you in space.

The lunar lander was difficult to fly and easy to crash. Neil once said that on a difficulty scale of one to ten, walking around on the moon's surface was a one; descending in the LM was a thirteen.

Later on Neil said, "There were just a thousand things to worry about while descending." Knowing that the descent had no margin for error, NASA created a simulator for the landing, as they did for every aspect of the mission.

Through dress rehearsals the prime and backup crews had "landed on the moon" two thousand times and crashed one thousand times by the time they *actually* landed on it. Some simulations were done underwater, which mirrors space because of the way neutral buoyancy mimics being in low gravity.

How do you get experience for the final landing on the moon? For this NASA created something quite possibly more dangerous than going to space: a machine called the Lunar Landing Research Vehicle (LLRV), nicknamed the "flying bedstead," and the later Lunar Landing Training Vehicle (LLTV). Its General Electric CF700-2v turbofan engine could support only five-sixths of the vehicle's weight. Remaining lift was provided by a pair of hydrogen peroxide rockets, which simulated operation of the LM's descent engine in one-sixth gravity and gave the pilot a flying-by-the-seat-of-your-pants feeling.

(I know, I know. You are wondering why you need to know all this. It was probably the hydrogen peroxide that pushed you over the edge, huh? But you gotta admit—it's pretty cool that it has uses outside of toothpaste, hair dye, and first aid.)

Basically this training vehicle was like a super wobbly, upside-down grasshopper-helicopter thing that flew straight up from the ground hundreds of feet into the air. Coming back down to earth you had to keep it stable so it didn't tip over. Easier said than done.

In a simulation exercise on May 6, 1968, Neil lost control of the LLRV-1, ejecting to safety seconds before it crashed to the ground in a fiery explosion that would have killed him. When he pulled the ejection handle, he was launched into the air at two hundred miles an hour, sustaining a force of 15 Gs. Pilots will tell you that any landing you can walk away from is a good one. Neil walked away from that one with only a bloody tongue—he'd bitten it when he ejected—and bruised buttocks from the ejection seat, but other than that he was fine.

Neil's cool, unflappable nature prevailed. He dusted himself off and returned to work to finish out the office day at his desk, not thinking the incident warranted leaving work early.

By the end of the Apollo program, two other pilots had ejected safely from crashes, and NASA considered scrapping the LLRV altogether. But astronauts who had made moon landings vetoed the idea, insisting it provided vital experience and accurately forecasted how the lunar module handled. Neil adamantly said that, its danger notwithstanding, the experience gained from his working with it allowed him to remain composed while flying the real thing.

Hurtling toward the moon's surface, Neil knew it was now no longer a test. This was the real thing, and there would be no second attempt at a landing on this mission.

The so-called fourth astronaut was churning away calculations. It was the world's first portable computer and the father of all laptops that would follow. People love to point out that the modern cell phone has more computing power in it than NASA had when we landed on the moon. Supposedly this statement irks to no end the people who were involved in the moonshot.

Consider this: to leave the earth and arrive at the moon is staggeringly complicated in and of itself. But the astronauts couldn't just go *to* the moon or its gravity would grab ahold of them and slam them into its surface. Game over. They needed to arrive at precisely sixty-nine miles above the moon and enter into an orbit around it. No less, no more. The precision required is equivalent to tossing a dart at a peach from twenty-eight feet away, not hitting the peach itself, but instead grazing the peach fuzz without touching the peach skin.

It gets worse. Both bodies are in motion. The moon travels at 2,300 miles per hour, so you can't go to where the moon is; you have to go to where it's going to be. This means you throw a peach into the air, and then when it's twenty-eight feet away, you chuck a dart at it and hit the fuzz but not the skin.

Welcome to orbital mechanics. The entire process, from the precise moment of liftoff to the burns (carefully choreographed ignitions of the thrusters to finesse their direction) that take place at each stage along the way, must be precisely synchronized to perfection. It is a masterpiece symphony of moving pieces and controlled explosions. And the most daunting part, yet to be attempted, was the final descent from orbit to landing. There is no atmosphere to slow your fall as you brake with thrusters, slowing from thousands of miles per hour to just a couple of feet per second, so you can land in an area where it is safe to do so.

Fortunately, very little of this was left to human error. A computer had been built to manage these astounding calculations and keep them on course. Larger and more primitive than our cell phones today? Absolutely. But as we discussed earlier, much of the technology pioneered in the moonshot paved the way for what we enjoy today. Walter Isaacson observed, "While trying to solve the problems of manned space flight, scientists laid the foundations for satellite television, global positioning systems, microchips, solar panels, carbon monoxide detectors and even the Dustbuster."

Do you have more power in your cell phone than they did then?

Technically, yes, but it's kind of like bragging that you can bench press more than your great-grandpa can. Even if you could, would you really take pleasure in being stronger than someone you wouldn't even exist without?

In my opinion, this is the same reason you should treat history with humility and respect and not historical snobbery. It's easy to armchair quarterback those who have gone before us and point out how they should have handled things differently, or even cancel them for their sins. These people were mere mortals. You and I can easily slip into the assumption that we would do things differently if we were in their shoes, but we might have done far worse. We should seek to learn from the mistakes of those who have gone before us, but a better use of our energy would be to be inspired by their bright moments, and then ask the more difficult question: What will future generations make of the decisions we are making? Especially the ones we make with our fancy, soul-destroying phones we are so proud of?

The computer had brought Neil and Buzz so far and done so well, but then, with just 40,000 feet to go until the lunar surface, the sound of an alarm filled their ears and made their hearts jump.

"Program alarm," Neil said with more intensity than his usually flat voice ever contained.

Buzz punched a button to see what caused the computer to make this sound, and both of them read it simultaneously.

"It's a 1202," Neil said.

CapCom Charlie Duke's response revealed the consternation everyone at Mission Control felt. "Twelve . . . twelve-oh-two," he acknowledged.

There were 650 million people watching live when program error code 1202 flashed across the screen, jeopardizing the entire mission. If Mission Control couldn't determine what it meant, and fast, the mission would be scrubbed and the astronauts would have to return to earth in defeat. The problem was that neither the astronauts nor Mission Control had ever seen this particular code before. They had no idea what it meant.

With some urgency now, Neil insisted on feedback: "Give us a reading on the 1202 program alarm."

Fortunately, one man knew what to do. He was a part of back-room support, an assistant to one of the men in Houston at Mission Control who you have seen in the space movies wearing short-sleeved, white button-up shirts with ties, and chunky black glasses. They hunched over their terminals, chain-smoked with headsets on, and occasionally glanced up at the enormous screens in the front of the room.

Norman Mailer referred to Mission Control as "the astronauts' brain on earth." What is less known is that each of those men had a corresponding room full of nerds feeding them data and information in real time. For the room you see, there are six rooms of people in the back you can't see. Each room, including Mission Control, smelled like stale cigarette butts, cold coffee, and day-old pizza.

Steve Bales was the guidance officer, a young expert responsible for telling the flight director, "We're go on that alarm," a.k.a., "We don't have to abort." Steve was twenty-six. He sat "in the trench," the front row of Mission Control, and when the alarm unexpectedly went off on that all-important day, it fell to him to make the billion-dollar decision on whether to call off the landing. Fortunately, Steve and his team were ready.

After a different but similar alarm had triggered an abort in a simulation eleven days before the launch, Gene Kranz had ordered Steve and his team to study "impossible alarm codes" and to know which ones they could still fly through and which ones they could not. They didn't have to understand why they were triggered, just whether they were mission critical or not.

Twenty-four-year-old Jack Garman was the member of Steve's

back-room support who knew these codes cold. He had prepared for this moment. It was his time to shine. He had analyzed and inventoried the error codes that could potentially be triggered by the computer. On his desk he had created a cheat sheet of twenty-nine codes he thought could show up, however improbably, and what his response to them would be if asked.

When the 1202 code showed up, Jack was prepared to say that it was merely an executive overflow caused by the stress of the system. It was the computer's way of letting everyone know it had been asked to do one too many things and was overloaded, so it was going to restart itself without unnecessary tasks. Once cleared, it should be okay. Had Jack not known the meaning, the astronauts might not have landed on the moon in 1969, failing to meet President Kennedy's goal and giving the Russians time to beat us to the punch. Engineers at MIT who coded the computer said they would have probably called an abort, but Jack knew the computer could handle it.

In his memoir *Magnificent Desolation* (so titled because those were the two words that came out of his mouth to describe the lunar surface in response to Neil calling it "magnificent"), Buzz wrote that it was his mistake that caused the overload. A rendezvous radar should have been turned off after disconnecting from Columbia and not turned back on until redocking. But Buzz left it on in case they had to make a quick getaway from the moon, and he didn't think the computer's memory would be unable to handle that extra task. It ended up overloading the system and chewing into precious RAM.

Since the computer knew to prioritize vital functions, it was going to ignore the rendezvous radar and focus on the landing. Slowly and calmly, Jack reassured Steve on their private loop, "It's executive overflow. If it does not occur again, we're fine . . . that has not occurred again—okay, we're go. Continue."

Steve let Gene know: "We're go on that, flight." And CapCom

Charlie, knowing Neil would be considering the abort button, quickly relayed it to the astronauts. Neil replied, "Roger. We got you—we're go on that alarm."

The code was cleared. In fact, it was only twenty-one seconds from the moment the code sounded to when it was cleared.

And the descent continued. Because the radar was still on, the code came back on multiple times. Each time it was cleared.

But soon they had a new problem. They were running out of fuel.

They hadn't accounted for the extra boost in speed they'd get when separating with Columbia; a burst of air came from undocking (think opening a carbonated drink), and Eagle had reached a speed of twenty feet per second too fast on descent—over halfway to abort limits. If it increased to thirty-five feet per second above projected, they would have to abort.

The excess speed caused the astronauts to overshoot the landing site the computer was originally trying to take them to. It was now bringing them down on rocky terrain covered with boulders and craters, which would be impossible to land on. Neil described the scene as they neared the surface of the moon: "There were boulders big as Volkswagens strewn all around. The rocks seemed to be coming up at us awfully fast, although of course the clock runs about triple speed in a situation like that. My attention now was directed almost completely out the window, and Buzz was informing me of the important computer and instrument readings. At about 400 feet it became clear that I would have to take over [from the computer]."

Neil shifted from autopilot to manual and steered over the unlandable terrain, the extra flight time causing them to burn through excess fuel. As they came in for the final approach, an alarm tripped, letting them know they only had sixty seconds' worth of fuel left.

Just imagining this pressure makes my heart beat faster and moves me to the edge of my seat. But Neil and Buzz didn't even have seats to move to the edge of. Seats had been removed from the Eagle to save weight. To fly the module, the astronauts had to stand at the controls and peer through the tiny windows. Tethers were installed to keep them from floating away, and Velcro strips were attached to the floor with the corresponding Velcro on the bottoms of their boots.

While Armstrong flew, Aldrin gave him readings:

> **ALDRIN:** [Eagle] 540 feet, down at 30 [feet per second]. Down at 15 . . . 400 feet, down at 9 . . . 54 forward . . . 350 feet, down at 4 . . . 300 feet, down 3½ . . . 47 forward . . . 1½ down . . . 13 forward. 11 forward. Coming down nicely . . . 200 feet, 4½ down. 5½ down . . . 5 percent. Quantity light . . . 75 feet . . . 6 forward . . . 40 feet, down 2½. Picking up some dust. 30 feet, 2½ down. [Faint] shadow. 4 forward. 4 forward. Drifting to right a little . . .
>
> **HOUSTON:** 30 seconds [fuel remaining].
>
> **ALDRIN:** Contact light. [This means that at least one of the three 67-inch probes beneath the spacecraft's four footpads had touched surface, flashing a light on the instrument panel.]

Buzz later said that in his estimation they had fifteen seconds of fuel left in the tank before they would have had no choice but to abort.

Fifteen seconds of fuel was all that separated the mission from being a success or a failure.

Had they taken just fifteen seconds longer to avoid one more crater, puzzle over the program alarm, deliberate over the right spot to land, or slow themselves down just a bit more, they would have had to cancel the mission, and world history would have turned out completely differently.

It's amazing how much can change in fifteen seconds.

EVERYTHING CHANGES

Eternity can turn on a dime.

Jesus was crucified between two thieves who were being punished for the charges brought against them.

The cross Jesus died on was originally reserved for the thieves' friend, and the probable ringleader of their little gang—a man the Bible identifies as Barabbas (Matt. 27:16–18). He was a murderer, the kind of guy your mother warned you about, and the two thieves clearly made the mistake many of us have made of not heeding Mom's advice.

These two men were implicated in capital crimes they did not commit. They were not murderers, but they were thieves, and so weren't without blame; they were accessories after the fact, and by one of the two's own admission, guilty as charged. But now they were dying for a crime their friend committed, and the worst part was that he went free.

In an attempt to get out of condemning a man he knew to be innocent, Pilate had picked Barabbas to be a contestant alongside Jesus in *Israeli Idol.* Pilate had instituted a new annual tradition to curry favor with the Jews: releasing one person condemned to die (Matt. 27:15). He figured by putting Barabbas alongside Jesus that the Galilean would be an obvious choice. He saw Barabbas as the worst possible option to go free—no one would want him walking the streets. Pilate was convinced they would select Jesus as the lesser of two evils.

Pilate underestimated the power of jealousy. The religious leaders smugly chose Barabbas and convinced the people to cry out, "Give us Barabbas!" (John 18:40 NIV) and then to chant "Crucify him!" when asked what was to be done with Jesus (19:15). Just the week before, these same fickle people had called out to Jesus, "Hosanna! Blessed is He who comes in the name of the LORD!" (12:13).

If you live by the approval of others, you will die by it. It is essential to simply be who God has called you to be, to appreciate affirmation but not need it. In modern terms, the goal is to be hungry but not thirsty. It's good to have drive but not an excessive need for validation.

We know from a parallel gospel record (Mark 15:32) that at the outset of Jesus' crucifixion, the two thieves had mocked him, crying out for him to prove himself by saving himself and them. Jesus would prove that he was who he said he was, but not by coming off the cross; he would reveal his identity by remaining on it.

IF YOU LIVE BY THE APPROVAL OF OTHERS, YOU WILL DIE BY IT.

Another of the Gospels shows that one of the two men later had a change of heart. Seeing his friend begin to berate Jesus yet again, he told him to knock it off: "We receive the due reward of our deeds; but this Man has done nothing wrong." Then he turned to Jesus and spoke these familiar words: "Lord, remember me when You come into Your kingdom" (Luke 23:41–42).

Jesus met his gaze and, for the second time, spoke words from the cross.

LAST WORDS #2

"Assuredly, I say to you, today you will be with Me in Paradise."
Luke 23:43

The words came quickly. Around fifteen seconds was all it took for a man whose life had become a living nightmare to be promised paradise.

Before we carefully examine Jesus' words and what they mean to us, we must first ask how the thief knew to call Jesus a king.

The answer is simple. It was right there in front of him, in three languages: "JESUS OF NAZARETH, THE KING OF THE JEWS" (John 19:19).

It was customary for the Romans to put the crime of the condemned on a sign over their heads. Much like the films you watch in driver's ed of mangled cars and red asphalt, meant to educate you about the perils of drunk or distracted driving, this was all meant to serve as a deterrent to crime. People would see the crucified victims and the signs as an unmistakable message: We don't mess around. Do this, and here is what will happen to you.

In Jesus' case, Pilate chose to poke some fun at the Jewish leaders who had forced his hand in Jesus' execution. They hated the insinuation and wanted the writing changed to "*He said,* 'I am the King of the Jews,'" but Pilate rejoined, "What I have written, I have written" (John 19:21–22).

This sign sat as a gospel tract in this thief's hour of need. They were the final words he would read before dying.

Those dying of crucifixion would often scream and shriek at those below, spitting and even peeing on those passing through the soak zone. The two thieves' original barbs and insults in Jesus' direction speak more of their pain in the moment and unhappiness with themselves than any actual animosity toward him. I think that what began to chip away at the penitent thief, in addition to the words he might have been reading, was Jesus' complete and utter calm.

"As a sheep before its shearers is silent, so He opened not His mouth" (Isa. 53:7).

This was a man in control of himself. He showed up on top of Skull Hill depleted—that is for sure—but not panicking. As he was nailed to the cross he was composed.

Who dies like this?

When Jesus finally did speak, it wasn't to beg for mercy, protest the unjustness of his sentence, or declare his innocence. He had actually prayed for the forgiveness of those who were putting him to death (Luke 23:34).

It must have felt like a live wire had run down the thief's back when he made out the words on the sign. All this, combined with what he had no doubt heard through the grapevine about Jesus' miracles and teachings, must have swirled together in the thief's mind as he suffered. Soon his pride and bravado began to melt, and all that was left was the incomprehensible pain he was in and the awareness that he would soon leave this world.

It has been said that the common denominator of all people everywhere is that we are all guilty, empty, and scared to die. So he must have made the connection: *If this is a king, then he must have a kingdom. If he has a kingdom, perhaps there is a place for me.*

"Remember me," he said (Luke 23:42).

We all have a fear that we will be forgotten, that when we die it will be like our lives never happened, that we weren't here, and we didn't matter. In an amazing way, what we witness in the salvation of this man in Jesus' second statement from the cross is actually an answer to the prayer he prayed in his first: *Forgive them*. Jesus was just a few feet away from the thief, but at that moment he was paying for the sins he had committed. Not only would he be forgiven; he would not be forgotten.

Jesus died first.

The Romans brought out sledgehammers and broke the legs of the two thieves (John 19:32). But Jesus, who had already died, did not experience this cruci-fracture that was so common. A soldier pierced his side, and blood and water ran out of his heart, proving the pericardium (the watery sac surrounding the heart) had already been ruptured (v. 34). He was dead.

The thief had watched Jesus die. As he saw the soldier approach his cross to break his legs, ending his ability to raise himself up on tiptoe, he would have known he would soon suffocate and die too. It still hurt, and was scary and painful, but he now had hope.

Ironically, the most peace-filled moments of his life were also the most painful. He knew his suffering would soon be over, and he was confident about what was going to happen next.

Jesus' promise to the thief—"Most assuredly you will be with me in paradise"—was only ten words in the original Greek, but those few words speak volumes. Starting with the first two: "Most assuredly." From these words we learn that heaven is real.

The function of these words in the original language was essentially a pledge of confirmation, like what you get when you book a hotel or a car online. It gives you confidence to have a confirmation.

DESTINATION: JESUS

I don't like traveling without knowing exactly what I will drive and where I will stay. A confirmation email provides a position of certainty. That is what Jesus was offering. It also tells us that heaven is near, today.

As mentioned, it took Michael, Buzz, and Neil three days to get to the moon. Even with a top speed of 25,000 miles per hour, the moon was still far, far away. Heaven, it seems, is nearer than the moon. But for sure, God is not far from any one of us.

In the Jewish understanding of time, the new day begins at sunset. Jesus died at noon, so the most "this very day" could be was six hours away.

I happen to think it is even closer than that.

Stephen, in Acts 7, sees it right before his eyes. It's just presently invisible to us.

Paradise. What a wonderful word! It is my favorite New Testament description of heaven. It feels more tropical than ethereal. And it sounds more like surf and infinity hot tub than stained glass and angelic cherubs.

Heaven is being with Jesus: "You will be with Me."

It is not just a place; it is a person. Whenever and wherever we get to be with Jesus, that is heaven. For in his presence is fullness of joy, and at his right hand are pleasures evermore (Ps. 16:11).

The gold streets (Rev. 21:21) or the sea of glass (Rev. 4:6; 15:2) or even being able to fly (Phil. 3:21; Acts 1:9) aren't what will make heaven so great; but rather it's the ability to bask in the presence of one so wonderful and glorious that it makes the glow of the sun unnecessary.

In *The Lion, the Witch and the Wardrobe*, when Lucy and Susan got to wrestle and hug and ride on the golden lion, Aslan, after he rose from the dead and caused flowers to bloom, they couldn't tell if being with him was more like playing with a kitten or a thunderstorm.

This is wonderful tension. Not safe, but good.

The thief next to Jesus knew not only where he was going (paradise) and when he would arrive (today) but also who would be there to greet him (Jesus).

The afterlife is mysterious and so shadowy and unknown. That's why we fear it. But for this man on this day, he knew all he needed to know. His King would be there to pick him up on the other side. I can just imagine the settled look that must have come over him. The last face he spoke to before he died was also the first person he spoke to in eternity.

But the mysteries of this passage go deeper.

Ephesians gives us some insight into what Jesus did in between dying and being there at the gates to greet the thief. He apparently descended and freed the captives held in the grave (4:9).

This is wild, and understanding this is way above my pay grade, but followers of God in Old Testament times were saved based on what Jesus *would do* on the cross. That's instead of the way we are saved today, by faith, believing what Jesus *did do* on the cross.

The cross is still the only means of salvation because *Jesus* is the only name that can save. But for those in Old Testament times, the crucifixion hadn't happened yet, so their salvation was given on credit. Until the payment was made, their souls were not taken to be in God's presence immediately. They went to a place called Sheol.

Stay with me.

Sheol means "the grave" and is used in a generic sense all throughout the Old Testament as the place all the dead—righteous and wicked—go. It seems there were compartments to Sheol. One side was called Gehenna and is basically what we understand as hell. This is described as a place of weeping and gnashing of teeth and sorrow (Matt. 25:30). This is ultimately not the final format of hell, for there is a day of final judgment coming that the book of Revelation describes as the great white throne judgment (Rev. 20:11). It will be followed by sentencing to the lake of fire—the final destination for the devil, his demons, and all those who rejected Jesus' salvation (v. 15).

THE CROSS IS STILL THE ONLY MEANS OF SALVATION BECAUSE *JESUS* IS THE ONLY NAME THAT CAN SAVE.

Let me stop right here and tell you that hell is not a subject I love talking about. I resonate with a quote often attributed to D. L. Moody, which says that no one should be able to talk about hell without tears forming in their eyes.

Yet let's remember that no one will end up in hell who didn't choose to go there.

There are two kinds of people: those who say, "Thy will be done" on their way to heaven and those to whom God says, "Thy will be done" on their way to hell. You practically have to crawl over the cross of Jesus, which God planted in your way, in order to get to hell, since he has given us the means to be saved at such a precious price. C. S. Lewis said, "It is hardly complimentary to God that we should choose Him as an alternative to Hell: yet even this He accepts."

If there is no hell, the cross makes no sense. Why would God have allowed his Son to suffocate and die such an awful, bloody death if there were not such a heavy consequence we were being rescued from?

I read once of a police officer who pulled a man over for an expired registration. The man acknowledged he was in the wrong but had fallen on tough times financially and had no option. "There's no explanation for why I haven't done it, except I don't have the money. It was either feed my kids or get my registration done," the man pleaded.

The police officer listened but, in the end, gave the father of two young children a ticket, deepening the pit the man was in financially. Now just two weeks away from Christmas and further behind than ever, desperation filled his heart. But when the police officer had driven away, he realized there was not just a ticket in his hand—it was wrapped around a hundred dollar bill. He was able to use the money to pay the fine and update the registrations on his car and his wife's car. The experience not only helped this man out in a tough moment, but he also later said the experience of grace restored his faith in God.

This story is a snapshot of the gospel. That police officer allowed the full consequences of the law to be felt, but then he absorbed them himself. Only through the reality of hell and the romance of the cross can God be both just and the Justifier of those who believe. God didn't just give humanity a warning; he threw the book against our breaking of his law, but then at the cross he also paid the fine personally.

God also never displays one of his attributes to the exclusion of any others. For instance, he is loving *and* he is just, both at the same time.

Another compartment of Sheol is often referred to as Abraham's bosom. Yes, bosom.

Abraham believed God when he made promises to him, and God accounted it to him for righteousness (Gen. 12:2–3; Gal. 3:6). In that moment he became the prototype of what it looks like to be saved by faith in Christ. Therefore, all those who followed after him joined him in this place of rest for the righteous dead. Even Jesus referred to a beggar named Lazarus who had died and was taken to be with Abraham (Luke 16:19–31). Many people believe Jesus' activity after dying on the cross included a trip to Abraham's bosom to set free all those who were there and bring them with him to heaven.

It's so epic to think about.

Esther and David, Ruth and Rahab, Jochebed and Jonah, Samson and Seth, Manoah and Noah. What a spectacle they made, whooping and hollering on a grand victory parade making its way from Sheol to heaven. I bet they didn't even try to keep their voices down.

Colossians 2:15 tells us, "Having disarmed the powers and authorities, he made a public spectacle of them, triumphing over them by the cross" (NIV). So what the devil thought would be a death blow against Jesus actually pounded the final nail in his own coffin. And then Jesus arrived in heaven with all these saints in tow!

What about the thief on the cross, the one who asked Jesus to remember him? First of all, his death by suffocation would have come quickly after his legs were broken. Unable to stand, he would have died in minutes.

As I write these words, I find it traumatic and difficult to think about someone struggling to breathe and eventually losing consciousness.

It hits me as a father. I think about the Thursday in 2012 when Jennie and I helplessly watched our daughter Lenya lose consciousness after an asthma attack and never wake up again.

But it gives me hope for my daughter, for the thief, for you and me, that Jesus endured just such pain and agony. And that by his sacrifice he began to knock down the dominoes that will end in all things being as they should be. Everything will be restored to God's original design and perfection: "Not only that, but all the broken and dislocated pieces of the universe—people and things, animals and atoms—get properly fixed and fit together in vibrant harmonies, all because of his death, his blood that poured down from the Cross" (Col. 1:20 MSG).

Heaven is not your final destination. This earth is. Heaven comes down. As N. T. Wright, inspired by David M. D. Lawrence, said, "Heaven is important, but it's not the end of the world."

For the Jesus person, there is life after death, and life after that too. As the thief closed his eyes here and came to in heaven, it's possible that he got to be the very first person to die and go straight there without first visiting Abraham's bosom.

What an honor.

What a historic opportunity.

YOUR OWN JOURNEY WITH JESUS

What conclusions can we draw from all of this—especially if you still have questions about your own standing before God?

1. No one is so messed up that they can't be saved.

This man hanging on the cross next to Jesus? In his life he did wrong and deserved what he was experiencing. That's from his own lips. He didn't have good church attendance. He never wore a cool Christian T-shirt. He didn't even get baptized. All he ever did was trust in Jesus as King.

That was enough.

Colossians 2:14 says, "Christ has utterly wiped out the damning evidence of broken laws and commandments which always hung over our heads, and has completely annulled it by nailing it over his own head on the cross" (PHILLIPS).

Jesus broke the code.

In God's eyes this thief became completely and totally perfect.

2. It is dangerous to delay.

You might have less time than you think.

A minister once faithfully warned his people about the danger of putting off accepting Christ. A man asked him, "Preacher, what about the thief on the cross?" The minister replied, "Which thief?"

That is a frightening question because there were two thieves, not one. And only one made it to paradise.

Doubt and delay are dangerous. Both thieves saw and heard the same things as they hung on their crosses. Both had the same opportunity to accept Jesus as Savior. And however long they thought they had before their deaths, it was actually less, because the leg-breaker was on his way.

Had the other thief planned to get right in his soul later on in the day? The computer never read error 1202; the mission was aborted by a Roman sledgehammer. Don't delay to tomorrow what God is calling you to do right now! Talk to God in prayer before you end this chapter. It doesn't have to be polished or pretty. I think prayers are most powerful when they are raw and gritty.

You might feel you are too far from your heavenly Father, that he is an eternity away, but this interaction proves that the one small step of calling out to Jesus is actually a giant leap in reconnecting with your Savior. Calling out to Jesus turned the thief's cross into a launchpad. And as the final seconds of his life ticked away, the countdown stopped measuring time remaining until death and started measuring time remaining until eternal life.

It's only seconds to paradise.

CONQUER YOUR INNER SPACE

What does this mean to those of us who feel overloaded? To those who struggle with anxiety and worry that makes us feel like we are behind and failing and drowning? Remember that the 1202 alarm came about because Buzz left the radar switch on when it should have been off. The computer ended up overloaded because its finite abilities were exceeded. It was trying to do more than it was capable of doing or needed to do.

It is easy to make the mistake of misinterpreting panic as a sign that there is something wrong with you when it might just be something wrong with your schedule. My counselor has helped me see that I often put pressure on myself to be and do more than any mere mortal can, and then I freak out and worry that I am broken *because* I am freaking out. The spiral continues.

The thieves who were lashing out at Christ weren't actually mad at him; they were frustrated at themselves and projecting it on him. Often the people we treat the worst are actually a reflection of how we feel about ourselves. There is always a thing under the thing.

When you end up overloaded, it is often because you haven't left appropriate margin. Your Creator built you to need daily rest, quiet times, sabbaths, vacations, and retreats. They are meant to be tastes of paradise to keep you going. What are you trying to do that God never asked you to?

When you feel a 1202 coming on, see that as a light on the dash alerting you to a prioritization malfunction. In seasons of busyness it's hard to justify prioritizing emotional, spiritual, physical, and mental health, but it is vital that you do what the computer did: prioritize vital tasks and do first things first.

If you grow your no, God will bless your yes.

CHAPTER 8

IF NOT ME, ANOTHER

YOU CAN'T STARE AT THE SUN WITHOUT DAMAGING YOUR EYES, but you can stare at the moon for hours. And when you do, the moon is actually allowing you to enjoy the sun's light that's not safe to look at directly. In fact, all the moon's light is borrowed.

Buzz Aldrin's father was really frustrated by the fact that Neil Armstrong was put on a postage stamp and the caption read: FIRST MAN ON THE MOON. As he saw it, the stamp should have read, FIRST MEN ON THE MOON, acknowledging that two men stepped foot on the lunar surface that day. Neil went first, and then twenty minutes later he was joined by Buzz. The case could have been made that Buzz should have egressed first because NASA's protocol dictated that the most senior of the pair would remain in the ship in case anything went wrong, and Neil was the commander. But because the door was positioned on Neil's side, Buzz would have had to crawl over him to get out, and in tests it was proven to be impossible to do without somehow damaging the fragile lunar module in their bulky space suits. So, because of the orientation of the door, it was determined that Neil would take the first small step and be the one to make history.

Buzz's official statement on the subject was that he was happy for Neil and happy to get to go on the mission at all. He was completely honored to get to be the second man in a very small club of humans (at the time of this writing, twelve) to step on the moon. His first wife, Joan, said that privately he was "devastated" by it. Early on he had launched an unsuccessful campaign to anyone who would listen, trying to lobby his way into being the first out. What he didn't know was that there had been a clandestine meeting of top NASA brass, including Chris Kraft, Deke Slayton, and others, where they decided that in large part because Buzz wanted it so badly, it should be Neil, who had never asked for the opportunity. In their eyes his lack of personal ambition qualified him as the right choice.

By Buzz's own admission, even the lesser glory that accompanied being the second man did not enhance his life. He noted in his memoir that "the instant celebrity brought to us by the lunar landing took a toll on everyone in my immediate family" and that "it was worse than any of the effects of space travel, on me and my children."

The kickoff was a parade down New York City's "Canyon of Heroes" route from Broadway to Park Avenue to City Hall to the United Nations headquarters. An estimated four million people packed the streets as ticker tape rained from the skyscrapers. Later that night President Nixon presented all three men with the Presidential Medal of Freedom at a state dinner attended by forty-four governors and ambassadors from eighty-three nations.

What followed was travel to twenty-nine cities in twenty-four countries in forty-five days on the Giant Leap Tour, where they had the unusual luxury of traveling around on the president's backup plane, Air Force Two. In every city they visited around the world, they received "exhausting acclaim" with a fervor rivaling Beatlemania at its apex.

The night the tour ended they were invited to sleep in the White House. Over the course of the tour one hundred million people saw the astronauts in person, and as many as twenty-five thousand had shaken hands with at least one of the astronauts.

It is difficult to imagine how blisteringly disorienting this would be.

Afterward, Buzz pointed out that NASA has a plan to train its astronauts for every scenario for leaving the world but gives no help as far as what to do when they get back.

Indeed. Ironic that they spent hundreds and hundreds of hours in a simulator to get ready for the mission to the moon that they knew could kill them. They even did survival training in the Grand Canyon, jungles, and deserts in case they missed the ocean on reentry. But there was no preparation for the thing waiting to destroy their marriages, their sanity, and the lives of their children no matter where they landed on the planet—fame.

We often spend so much time fretting over what we will do if our dreams don't come true. Perhaps we should ask the question, What if they do? Will you have the character to sustain yourself wherever your gifts, talents, and blind luck take you?

Michael Collins seemed to agree with Buzz's assessment of the shadow side of celebrity. In his book *Carrying the Fire*, he wrote, "Fame has not worn well on Buzz. I think he resents not being first on the moon more than he appreciates being second."

I include this not to shame Buzz for what so obviously was in his heart but because it shines a light on what I know is in mine. Deep down in all of us there is a suspicion that Ricky Bobby is right: "If you ain't first, you're last." This is nothing new. Two of Jesus' disciples sent their mother to Jesus asking that they might get special treatment (Matt. 20:20–21). And an entire clandestine fight broke out as to which disciple was the greatest immediately *after* Jesus disclosed to them the breaking news that he had come to serve and not be served by painfully dying on the cross. The irony.

Dark forces are at work inside all of us, and they tell us we have to be separate, to be special, to be superior, or else we don't matter. These ego-driven whispers never take us to a healthy place. They drive us on an insatiable, alienating, and isolating quest to prove ourselves. To be validated. To be enough. To matter.

THE BRIGHTER LIGHT

In John 1 we read about the men who came to Jesus' cousin John asking him who he was. They wanted to know, "Are you the sun?"

"No," he said. "I am the moon. The sun is coming."

And the Son did come.

And John the Baptist rejoiced.

Remember, the moon gives off no light of its own. Its glow belongs to another. We see it shine in the night sky, but it reflects the star that is the light at the center of our solar system.

So many of the mistakes I have made in life have come from not having the clarity of identity that John the Baptist modeled. Refusing to humble myself, I want the attention that belongs to another.

Not long after Jesus' baptism in the Jordan River, after attention grew to a fever pitch around Jesus and then the crowds began to wane, some of the disciples came to John and said to him, "Teacher, that Galilean you baptized is now baptizing and everyone is going out to him" (John 3:26 paraphrase).

They wanted John to be outraged. To be indignant. Self-righteous. To be as small of soul and myopic in vision as they were.

Baptism is our *thing.*

John taught them and us all a master class.

"He must increase, but I must decrease" (John 3:30 ESV). Translation: This was always the plan.

No doubt they were shocked. But then again, their Obi-Wan Kenobi ate bugs, so perhaps they were a little beyond being surprised by him. Regardless, he gave a picture in John 3:19–21 to explain what he meant, but also to show what it means to be a Jesus follower. It is to wake up every day hoping that in your actions a brighter light than you would catch people's eyes and captivate their attention.

"Christ in you, the hope of glory" (Col. 1:27).

When all the attention is on Jesus, things are going right.

BORN TO BE A MOON

Neil handled intense fame spectacularly. He was unquestionably the man for the job—a man of incredible discipline and willpower who, in an age where smoking was ubiquitous, only allowed himself one cigarette per year. He refused to see himself as someone special. You can't detect ego in his spirit or the way he carried himself before or after the mission, at least not in any interview I have ever seen or listened to. If anything, he was almost superhuman in how unromantic his language was about the whole thing.

In an interview with Douglas Brinkley and Stephen Ambrose for NASA's oral history project, they pushed him to admit he was dazzled by the historic significance of it all, even just a little bit. "Would you every night, or most nights, just go out quietly and look at the Moon? I mean did it become something like 'my goodness.'" Completely stoic, he deadpanned, "No, I never did that," and explained that he and the team mostly just went over plans and checklists again and again.

Before the Apollo 11 mission, when press reporters tried to compare him to Christopher Columbus, he demurred and pointed to the team of astronauts who were not selected but could have just as easily done the job if called upon. Basically he was saying, "If not me, it would have been another." Meaning, there had been only one Columbus, but "there were ten astronauts at least who could do the job, and hundreds of men to back them up." He was the representative of a collective will. When reporters asked if he would at least recognize that his endeavor was equal in magnitude to Columbus's adventure, he replied, "Our concern has been directed mainly to doing the job."

Neil refused to be made into someone who was elite, or special, or a big deal. He saw himself as one more part of a very big and dedicated team. It's possible that his aversion to grandiosity led him to refuse to take Communion along with Buzz.

At least part of Buzz's motivation for his Last Supper on the moon was the fact that Christopher Columbus had observed Communion upon landing in the New World. He liked that this ceremony recognized the immense import of the moment and his part in it.

Buzz wanted to mark the moment in a big way. That simply was not how Neil rolled. Mr. Matter of Fact and Captain by the Book. He was there to do a job and was somewhat allergic to anything that went beyond the punch list.

I have to pause here and acknowledge how relatable and convicting all this is. I feel for Buzz having to deal with the turmoil associated with being the second man on the moon and having a hard time with that. I can so easily get wrapped up in my ego even if I am trying to do something God-honoring. But it's pretty difficult to appreciate what God is doing in your life when you are fixated on what someone else gets to do.

The quest to conquer inner space so often amounts to leaning into the grace in *your* life and to stop wishing God gave you someone else's calling. Easier said than done, of course. And social media doesn't help because it's often edited, and everyone else's curated feed looks better than the real-life situation in front of you. But leaning into grace is the gateway to peace. I promise. Victory over the insidious, peace-robbing cancers of pride and jealousy, in whatever forms they take, comes when you become grateful for who you are and what God has called you to do. Then, and only then, can you say with David, "The boundary lines have fallen for me in pleasant places; surely I have a delightful inheritance" (Ps. 16:6 NIV). And you can shine the light of the sun like the moon you were born to be.

THE QUEST TO CONQUER INNER SPACE SO OFTEN AMOUNTS TO LEANING INTO THE GRACE IN *YOUR* LIFE AND TO STOP WISHING GOD GAVE YOU SOMEONE ELSE'S CALLING.

LIGHT TURNS OFF THE DARK

It is impossible to overstate how important a light was in Jesus' day. There were no streetlights, no flashlights, and no iPhone lights you could turn on. When the sun went down, you couldn't flip a switch and bathe your home with light.

The world has changed much since the invention of the incandescent light bulb. For one thing, we sleep less. Ten hours of sleep used to be normal. Today, the average American adult gets 6.7 hours of sleep.

If the sun went down and you didn't have a light with you, well, you were in very real danger. A lit lamp gave you safety and the ability to travel, to read, to talk with a friend, to cook food. Light turns off the dark because darkness is the absence of light.

When a child has a bad dream, besides wanting their parents and a glass of water, nothing will comfort them more than having a light turned on.

"I AM" #2

"I am the light of the world."

John 8:12

Jesus is God's light come into our world to show us the way, comfort us, and illuminate our hearts and lives with his glory. When we have his light we can see God clearly, we can see ourselves clearly, and we can see each other clearly. When the light dims and we begin to stray from him, everything gets dim. We start to think we look better than we do.

I always think I look pretty good in airplane bathrooms. It's the dim light. Unhealthy things grow in our hearts when we are in the dark. That is why we need constant exposure to his light, which sanitizes, purifies, and disinfects.

One of the reasons we read, memorize, meditate on, and apply God's Word is because it gives us access to the light we need. Psalm 119:105 reminds us, "Your word is a lamp to my feet and a light to my path."

The light you find in God's Word will reset the things that are out of whack in your heart and mind.

I am a performance-obsessed human being. Productivity is my heroin. When I am not in a healthy place, I can only feel good when I am being successful and efficient and getting things done. But that is to put myself at the center of the universe. It places me in a seat I am unable to occupy, and so I start to fall apart.

Idolatry leads to anxiety. When I am having my mind transformed, my motives inspected, and my values constantly recalibrated by God's light, I remember I am a son—and that is my identity. My value doesn't come from what I do, but who I am.

It's true for you too. You are a human being, not a human doing.

God is the sun; we are the moon. God made two great lights—the greater light to govern the day and the lesser light to govern the night (Gen. 1:16). We don't produce any light, but as we follow his orbit round and round, we get to glow.

Jesus didn't stop with his assertion that he was the light of the world. He said in the Sermon on the Mount in Matthew 5:14–16:

> "You are the light of the world. A city that is set on a hill cannot be hidden. Nor do they light a lamp and put it under a basket, but on a lampstand, and it gives light to all who are in the house. Let your light so shine before men, that they may see your good works and glorify your Father in heaven."

What's beautiful about this arrangement is that it takes all the pressure off. You don't have to conjure up the light; that's the sun's job. The moon doesn't have to strain; it just needs to reflect.

At the moment of our salvation, Jesus puts his life and light inside us. Our job is to keep from obscuring and hiding the light.

A TRUE REFLECTION

During a challenging time emotionally for me, Debra Fileta, a friend and counselor, shared an insight that has stuck with me. God the Father spoke, "This is My beloved Son, in whom I am well pleased" (Matt. 17:5). He said this at his baptism. Before any miracles had been performed. No demons had been cast out. No sermons preached. His Father was pleased with who he was, not what he did. From that standing he was anchored and able to go and do all he did. Not to earn anything, but because he already had everything.

As I heard her say these words, I felt a weight shift from my soul. I so easily connect my worth with my work. I am only as good as my most out-of-the-box idea. My most over-the-top sermon illustration. Consistent meteoric growth. The converse is also true. It's easy to downshift into self-loathing and despair when creativity stalls, growth lags, or I feel like I have plateaued, as though the only question God is up there in heaven asking is, "What have you done for me lately?"

Is it a coincidence that Satan's biggest agenda was to get Jesus to try to do something to prove he was the Son of God—turning rocks to bread, jumping off the temple, and things like that? Jesus' secret sauce, and the key to conquering inner space for me and you, is resting content in the fact that we don't have to do *anything* to prove ourselves. We are loved. That is enough.

With our flaws.

In our brokenness.

We don't have to perform to be loved.

We are not merely our contributions. If I never preached or wrote again, I would be loved. If you never memorized another verse, or shared your faith, or got a promotion at work, or if your kid's birthday isn't a Pinterest-worthy affair, you would be loved exactly the same as if you did all you are capable of doing.

That love is a safeguard against the devil, who desperately wants to trick us into forgetting. If he can do that, he will have us perpetually auditioning for what is eternally ours.

Your Father in heaven is incapable of loving you more or less than he does right now.

This just in: the tryouts have been canceled. You can rest in the exquisite comfort that comes from being comfortable in your skin. Your regular-old, blemished, unphotoshopped, wearing-sweats-at-two-in-the-afternoon-on-a-Saturday, hair-in-a-messy-bun-for-the-third-day-in-a-row skin. For in that skin you've got something much better than a part in a play or a spot on the team. You are part of the family. And Jesus saved you your fave seat on the couch.

I like the analogy of us being the moon because sometimes the moon is a tiny sliver and other times it is a supermoon. When Buzz and Neil landed on the moon, from the earth's perspective, it was a waxing crescent moon. It is also marvelous to think that from the moon Neil, Michael, and Buzz looked back toward home and saw a crescent earth. The following illustrations show where they landed. The image on the left shows what the moon looked like from earth on July 16, 1969.

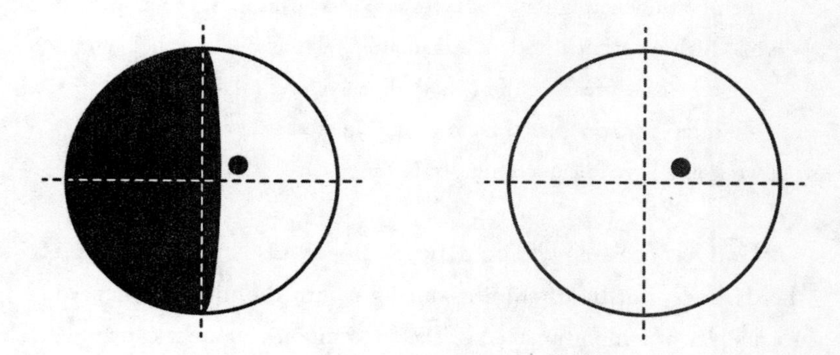

This had been determined because the angle of the sun optimized their view of the landing site. The difference depends entirely on how close the moon is to the sun and how much of it is eclipsed by the earth. Michael included this highly complex math in his book, *Carrying the Fire*:

> Since Neil and Buzz would be approaching from the east, and since they didn't want the sun's glare in their eyes as they were trying to measure crater depths, etc., the sun at the time of landing should be behind them. . . . If too high, i.e., overhead, the craters and boulders would not cast shadows, and depth perception and obstacle avoidance prior to touchdown would be a real problem. Too high also meant the surface would be too hot. Too low, and the shadows could

> get so elongated that they would obscure other useful details and again make a visibility problem for the crew. A sun angle of about 10 degrees was deemed perfect. . . . Since it takes roughly thirty days for the moon to go through this cycle once (i.e., 360 degrees), it turns out (dividing 360 by 30) that at any one spot on the moon the sun angle changes by 12 degrees per day. Therefore, if you want the sun to be approximately 10 degrees above the horizon, behind you as you touch down at a particular spot, there is only one day in the month which is good for landing. . . . During July, the month in which we wanted to land, Tranquility Base would reach a sun angle of 10 degrees on the twentieth; therefore, to allow time for the three-day trip, plus time to get all squared away in lunar orbit, we would have to launch July 16. *Voilà!*

God has declared war on darkness not by his church shouting at the darkness but by us shining our light into the night. When you are a sliver of a moon instead of the supermoon you are meant to be, you limit the hope he can give through you. Not because the sun isn't bright but because there is less of you that you are willing to allow him to use.

I have a glow-in-the-dark flashlight mounted in the garage right by the fuse box. It absorbs light when the lights are on, and during the night it shines out what was put into it. If I hid it all day under a cloth, it would have nothing to give when the time came.

Your closeness to God by day is directly connected to your ability to shine by night. The world is full of hurting people who are enveloped in the darkness of idolatry, anxiety, and lust. God loves the people in your school, your city, and at your job so much he sent you there to work and learn and live.

Yesterday I had a vivid reminder of what it's like to be the moon I was born to be. My favorite place to have quiet time is in our backyard by a gas firepit. I spend all but the coldest mornings there with my journal, Bible and devotionals, and books I am reading. Between sips of piping hot coffee, I sing and sigh, read and repent, reflect and absorb the light that surrounds the throne. Sometimes I really feel God; other times I don't.

GOD HAS DECLARED WAR ON DARKNESS NOT BY HIS CHURCH SHOUTING AT THE DARKNESS BUT BY US SHINING OUR LIGHT INTO THE NIGHT.

Yesterday I woke up at 4:30 a.m. with a headache. My alarm was still two hours away, so I took an ibuprofen and headed into the shower with a LaCroix. Often if I take a shower in the dark, and alternate between hot and cold water, it can help me deal with a headache. By 5:00 I hopped back into bed and hoped that when 6:30 came I would be rested and reset. No such luck. I crawled from bed feeling worse. In my time by the firepit I felt like I was clawing through mud to get anything from the Bible. Even the birds chirping did little to cheer me up, I dropped to my knees to pray and ask God for strength as though I felt him right next to me. I took a screenshot of one of the verses I read in Psalm 103 so I could read it again later.

The part that stuck out to me was about your youth being "renewed like the eagle's" (v. 5). *That's good,* I thought, *and man do I need it.* Near the end of the day, I would discover that more light had absorbed into me than I thought when a longtime friend called to let me know terrible news: he had been diagnosed with prostate cancer and the doctors wanted to rush him into emergency surgery. He asked if I would pray for him.

Hearing desperation in his voice, I felt the Spirit of God prompt me to pull up the screenshot and read the psalm to him. I put the phone on speaker and found my way to the camera roll. Reading it to him, I noticed something that had been there all along, only it hadn't stuck out to me at the time: he "heals all your diseases" (v. 3). I heard his voice break up as I proclaimed faith in a God who is greater than cancer and who would be victorious in this situation, whether through divine intervention, medical precision, or welcoming him to heaven.

After we prayed and hung up, I felt myself glowing with God's love and power in this situation. I actually muttered to myself, "Wow, I just came alive in battle." I had felt tired and been dragging all day, but now I felt strong and brave. It wasn't my strength I was feeling; my youth had been renewed like the eagle's when God shone on me, but it wasn't until I stepped out to use his light that it was activated. This is what David meant when he said God makes his hands strong for war (Ps. 144:1).

I wonder how different that moment would have been—that moment I needed to be a city on a hill and a light on a lampstand—if I had neglected my time with God or put a basket over my heart. Because the sun had shone on me, I was able to reflect it in the dark. Talk about magnificent desolation!

CONQUER YOUR INNER SPACE

I don't write this to make you feel guilty for yesterday but to help you feel inspired for tomorrow. When the sun rises in the morning, don't miss it. But also take heart knowing that you are not the sun; the responsibility to shine is not yours. It is God's, and he loves you. Think of it: the most important person in the universe loves you!

So rest well.

Good night, moon.

- Take some time to reflect on your desire to have what others have been given and what impact it has had on you.

- Write a few sentences in a journal about some things that bring envy into your heart or give you jealous thoughts you are struggling with.
- Write out John's challenging sentence: "He must increase, and I must decrease." Then reflect on the way focusing on Jesus curbs selfish, self-destructive thoughts.

CHAPTER 9

OUT OF THIS WORLD

WE ARE ALL EITHER CHASING OUR CHILDHOOD OR RUNNING from it.

At the age of six, Neil Armstrong took his first ride in a small airplane with his father. He never looked back. Consumed by science, reading, and making model planes throughout elementary school, he thought one day he would become an aircraft designer. Even as a teenager he would hitchhike miles away to the closest airfield to take flying lessons.

As a child I played lots of different sports. My parents didn't make us kids play any one sport in particular but encouraged us to try those we were interested in. There was one rule: no quitting. We didn't have to do it again the next year, but anything we started we had to finish. Consequently I played lots of sports for one season, which in most cases was enough to show me that it wasn't my thing. Often I would realize this after the third or fourth practice and then be disillusioned, only to have to gut it out.

I played basketball for one season. T-ball the next. I ran cross-country until I realized I don't really like running. I played soccer for a season and it taught me that I still didn't like running, even when kicking a ball was added to the equation. Skateboarding came and went; of all the sports it likely lasted longer because I identified with skater culture more than being a jock. This, of course, is complicated, and brings up deep-seated problems I will spare you of and save for my counselor.

The only sports I really enjoyed as a kid, and that stuck with me for more than a minute, were snowboarding and skiing, riding bikes, and tennis. Getting a Huffy bike as a child meant freedom. I'll never forget how I felt when my dad, wearing a Pink Panther costume, rolled it into my birthday party at Tickles Fun Factory. I rode that thing into the ground. When I got a Diamondback mountain bike in fifth grade, it became my trusty steed. Hopping curbs and riding down dirt embankments and arroyos was life. Riding to Circle K with neighbor-hood friends to buy a Slush Puppie and put quarters into the Mortal Kombat video game machine was a weekly pilgrimage.

More than anything, tennis and skiing give me feelings of nos-talgia for my childhood. They represent memories of both summer and winter. My mom worked as a ski instructor and a volunteer ski-patroller during winters when I was a young child in Colorado. Today, stepping into a ski lodge smelling of chili, hot cocoa, wood, and sweat brings back powerful memories, transporting me to those nascent years. Even the hard memories—a broken collarbone snowboard-ing in seventh grade and being taken off the mountain in a sleigh attached to a snowmobile—make me smile, though with a little bit of a grimace. And skiing with my kids today gives me a profound joy that is difficult to really understand. When I see them riding on the chairlift, or licking snow off their gloves, I want to cry, laugh, and hug my parents.

I resonate with the words of Honoré de Balzac: "Some day you will find out that there is far more happiness in another's happiness than in your own. It is something that I cannot explain, something within that spreads a glow of warmth all through you."

When summer came we played tennis. My mom and dad were both avid tennis players and card-carrying members of the United States Tennis Association (USTA). Summers were for tennis. We would watch their matches and take lessons in the afternoons. I idolized Andre Agassi and his bold look and aggressive style of play. I grew my hair into a spiky-on-top, mullet-in-the-back ode to his style. I imagined myself playing like him when I grew up. To this day I haven't read very many books I enjoyed more than his autobiography, *Open*. I remember watching televised professional matches with my family. And our Colorado tennis club had big watch parties for the Grand Slams.

It was during a third-grade tennis lesson that I first began wheezing, which led to the diagnosis of exercise-induced asthma. It is a horrible feeling to have an asthma attack. If you have never had one, know that it is difficult to express how scary it is. Gym teachers should have sympathy and never tell a young person having one to "walk it off," "suck it up," or anything horrible like that. The first time I had one, it felt like all of a sudden a gorilla was sitting on my chest. I couldn't get air in. I was pulling at the skin at the base of my throat trying to let more oxygen in, and I just couldn't. It's a dreadful thing.

But the asthma didn't deter me, and the next four years I dutifully puffed my inhaler before tennis practice. But then my eyesight betrayed me, putting my tennis playing on the rocks. I remember distinctly everything getting fuzzy when I was a little sixth grader, and I didn't have a clue what was going on. In my book *I Declare War*, I talk about how it affected my grades. It doesn't take a genius to know it would spell disaster for depth perception and the ability to see a ball flying at your face. So I got glasses. The first time I played while wearing them it was weird to see the reflection of my face glaring back at me in the edge of my lenses. My peripheral vision was nonexistent, and when I got sweaty, the glasses slid down my nose. I put the racket down. It wasn't fun anymore.

It was a turbulent time made even more difficult because I was beginning a challenging identity crisis and the era of "Ratboy" that would last several years. Again, I've written the soul-crushing details of this time in my life in *I Declare War*.

The tennis court that used to be a sanctuary for me was no longer a safe space. I stopped showing up. I wish I had pushed through, or gotten contacts, but I didn't know how to put into words all that I was feeling. They also didn't have a tennis team in my high school, otherwise I might have kept it up.

I didn't touch a tennis racket for the next twenty years.

When we moved to Montana, we discovered that right outside our home was a neighborhood tennis court. On a whim, I bought Jennie, myself, and Alivia rackets. But they began gathering dust. I blinked, and another five years went by.

Then, one day, as I was driving by a tennis court, a wave of happy feelings washed over me and I had the impulse to play. I texted a friend who played and asked if he wanted to hit balls. We did. And all the love instantly came surging back. We played all summer. My kids all expressed interest in it as well because I was spending time playing and loving it. (Note to parents: your kids hear what you say but have a much easier time doing what you do. For instance, if you want your kids to have a passion for the Lord or serve humanity, lectures will never be as powerful as a simple demonstration.) Soon, our whole family was in lessons and spending a lot of time at the tennis court. My oldest is on her high school tennis team, and we have become a tennis family.

I have found that having an outlet like tennis has been incredibly helpful for conquering my own inner space. All you can think about is the next serve or return, or whether the ball was in or out. And something about sweating is so good for the soul.

The love of playing tennis has extended to watching it. The past three years we have been watching it on TV as a family. The rowdy US Open, the stuffy French Open played on a clay court, the down-under Australian Open, and the granddaddy of them all—Wimbledon. We get really into it, throwing "Breakfast at Wimbledon" parties where we eat strawberries and cream and pop the corks on some bubbly bottles (Martinelli's for the kiddos). We love cheering on our household favorites, Serena Williams, Rafael Nadal, and Roger Federer. The young ones get a kick out of Nadal's little routine he goes through before each serve. My daughter Liv roots for Djokovic, though I have warned her that a house divided cannot stand.

Last year my fascination for space and tennis came to a head when just prior to the coin toss before the final match of Wimbledon, they played a video explaining the backstory on why this coin toss was special.

The coin had been to space.

In the video they showed the gold coins being etched and prepared and then flown with an astronaut to the International Space Station, where the coins floated around in zero gravity. Next they showed them tossed into the air, rotating around and around, at the All England Lawn Tennis Club on center court where the men's and women's singles finals take place during Wimbledon. The narrator was describing the coin toss while they intercut footage of rocket launches. NASA imagery with tennis footage. Epic.

When the video ended, the same coin that had spent six months on the football field–sized International Space Station was now in the hand of the person about to toss it. It took my breath away. What foresight, planning, and insight. Long after the match ended, and my disappointment over Federer's loss had faded, that moment stuck with me. The value and wonder of that coin did not come from what it was, but where it had been.

THE CHURCH IN SMYRNA

Keep that in your mind as you read what Jesus said to the church in Smyrna in the letter to them recorded in Revelation 2:8–11:

> "To the angel of the church in Smyrna write,
>
> 'These things says the First and the Last, who was dead, and came to life: "I know your works, tribulation, and poverty (but you are rich); and I know the blasphemy of those who say they are Jews and are not, but are a synagogue of Satan. Do not fear any of those things which you are about to suffer. Indeed, the devil is about to throw some of you into prison, that you may be tested, and you will have tribulation ten days. Be faithful until death, and I will give you the crown of life.
>
> "He who has an ear, let him hear what the Spirit says to the churches. He who overcomes shall not be hurt by the second death."'"

In January 1962, Neil and Jan Armstrong's daughter, Karen—nicknamed Muffie—died on their sixth wedding anniversary. No parents should have to endure such a loss, and in a way that tarnishes such a special day. Her death came after a six-month-long fight with cancer that was discovered after she fell while playing on a playground. In September of that same year, Neil was selected to be an astronaut. At the time he was a former navy pilot who flew and had a hand in the design of experimental planes for the National Advisory Committee for Aeronautics (NACA), the precursor to NASA. He piloted planes like the X-15, which flew six times past the sound barrier and up to 350,000 feet above the earth. It was all as he dreamed as a child. And now he would get to go even higher as an astronaut, but it would be while simultaneously walking through the greatest of griefs. It is important to remember details like this because it humanizes a larger-than-life individual and helps you remember that, in addition to doing something so epic, he and his family were carrying a heavy emotional load.

Yes, he had dreams come true. But also nightmares.

So it often is.

The church in Smyrna received a pretty glowing report card from Jesus in this letter in Revelation. Jesus had nothing but positive things to say to them. That is noteworthy.

How did they do it? I believe it can be attributed not to who they were but to where they had been. They had experienced light and shadow, breathtaking beauty and agonizing heartbreak.

It is not coincidental that of the seven churches, they had endured more pain and hardship than any of the others. This is not to say that you should seek out suffering, but rather that you should rethink how you view the sufferings you face. Properly understood hardships and fiery trials are not occasion for despair, but rather an opportunity to be developed.

LETTER #2 (SMYRNA)

"Be faithful until death, and I will give you the crown of life."

Revelation 2:10

The second church Jesus wrote a letter to was Smyrna, which, today, is a city in Turkey called Izmir. At the time it was a beautiful city. The name *Smyrna* means "myrrh," which was a resinous perfume made from rare spices and an important commodity, since it was used as perfume for the living and the dead. The city's port was known as the Port of Fragrance. Smyrna was a center of science, medicine, and athletic competition. Homer, the author of *The Iliad* and *The Odyssey*, was born in the area. And because of the way it was built, the city appeared to have a crown rising above its hilltops, which Jesus referenced in his letter as he promised a crown to his followers who overcame hardships.

This educated coastal city named after a beautiful fragrance became an anything-but-lovely place to be a Christian.

The church at Smyrna was persecuted greatly.

It became a city wholly given over to idolatry as the emperor, Domitian, had demanded to be worshiped as God. The church reacted just like Shadrach, Meshach, and Abednego had in the book of Daniel, refusing to bow to the king or his image (Dan. 3). They knew that Jesus was Lord and the emperor was not. Polycarp, a disciple of the apostle John and the bishop of the church in Smyrna, was burned at the stake and became a famous Christian martyr.

The city had a large Jewish population that had turned on the Christians and added to their suffering. The Jews were exempt from emperor worship because their religion had been legally accepted by Rome, and the Jews hated the Christians and did whatever they could to get them in trouble. Jesus wrote to tell his followers in the city that they hadn't made a mistake by following him. They had held on to him and he had noticed—and through their pain he was producing something powerful.

History bears witness that the persecuted church is the purified church.

An interesting fact about myrrh is that it has to be crushed before it gives forth any fragrance. The more it is crushed, the more fragrant it becomes. The same thing was true of the church in Smyrna. The more the people were crushed, the sweeter the aroma of their faith became. The more they were stomped on, the more the world caught the fragrance of their love. The crushed church is the fragrant church. God permitted Satan to bruise the church in Smyrna; the harder he bruised it, the more he released the fragrance of its grace, and the more devastating its testimony became to darkness.

The same is true in your life. That is why the most extraordinary miracles often come from the most difficult places. As Tertullian put it, "The oftener we are mowed down by you, the more in number we grow; the blood of Christians is seed."

So many verses and passages in the Bible tell you to react to pain in the exact opposite way you feel in the moment. Translation: Your mind will scream at you that all has gone wrong, but with faith you must believe that things are going to be better than alright.

Peter said it like this in 1 Peter 1:6–9:

> In this you greatly rejoice, though now for a little while, if need be, you have been grieved by various trials, that the genuineness of your faith, being much more precious than gold that perishes, though it is tested by fire, may be found to praise, honor, and glory at the revelation of Jesus Christ, whom having not seen you love. Though now you do not see Him, yet believing, you rejoice with joy inexpressible and full of glory, receiving the end of your faith—the salvation of your souls.

Paul said it, too, in Romans 5:3–5:

> Not only that, but we also glory in tribulations, knowing that tribulation produces perseverance; and perseverance, character; and character, hope. Now hope does not disappoint, because the love of God has been poured out in our hearts by the Holy Spirit who was given to us.

And in James 1:2–5, we read that James, the brother of Jesus, said it like this:

> My brethren, count it all joy when you fall into various trials, knowing that the testing of your faith produces patience. But let patience have its perfect work, that you may be perfect and complete, lacking nothing. If any of you lacks wisdom, let him ask of God, who gives to all liberally and without reproach, and it will be given to him.

Please remember that none of these three were strangers to suffering. Peter's life ended when he was crucified upside down, Paul was beheaded outside Rome, and James was killed with a sword in Jerusalem. That was just how they died. While they lived, they experienced even more hardship and difficulty.

Paul took a minute to detail some of the hard things he had been through in a letter to the Corinthian church you can find in 2 Corinthians 11:25–27:

> Three times I was beaten with sticks. Once they tried to kill me by throwing stones at me. Three times I was shipwrecked. I spent a night and a day in the open sea. I have had to keep on the move. I have been in danger from rivers. I have been in danger from robbers. I have been in danger from my fellow Jews and in danger from Gentiles. I have been in danger in the city, in the country, and at sea. I have been in danger from people who pretended they were believers. I have worked very hard. Often I have gone without sleep. I have been hungry and thirsty. Often I have gone without food. I have been cold and naked. (NIRV)

That would be enough to make anyone throw their hands up in despair.

NEW WEIGHT, NEW DEPTH

Maybe you can relate. Not to the exact sequence of events Paul faced but perhaps because one thing after another after another has hit you. Life can feel like a Russian nesting doll of calamity.

Sometimes it just feels like too much. In those moments it is normal to question God's goodness or his sovereignty, and even his existence.

Paul saw things differently. He chose to view the sufferings he experienced in life through an eternal lens to put them into proper perspective. I love how he turns mountains into molehills in 2 Corinthians 4:16–18.

> Therefore we do not lose heart. Even though our outward man is perishing, yet the inward man is being renewed day by day. For our light affliction, which is but for a moment, is working for us a far more exceeding and eternal weight of glory, while we do not look at the things which are seen, but at the things which are not seen. For the things which are seen are temporary, but the things which are not seen are eternal.

I appreciate how Eugene Peterson translates verse 17 in *The Message*: "These hard times are small potatoes compared to the coming good times, the lavish celebration prepared for us."

It's not like Paul didn't truly suffer. So how could he say such things? Paul knew what the trials were doing in him and what was awaiting him when he left this world.

You can do the impossible when you can see the invisible. It taps you into secret power.

This kind of anticipation applies the weight of glory. Weight has everything to do with gravity. Gravity attracts objects together. The greater the mass of an object, the stronger the force of gravity. The smaller the mass, the lesser its gravitational pull.

The moon is smaller than the earth, so on the earth things weigh more, and on the moon they weigh less. A hundred-pound weight here would only weigh about sixteen pounds on the moon, which has one-sixth gravity.

YOU CAN DO THE IMPOSSIBLE WHEN YOU CAN SEE THE INVISIBLE. IT TAPS YOU INTO SECRET POWER.

When you are standing on the promises of God and not on your circumstances as they presently appear, the mass of your trial gets smaller and the power of its pull gets weaker. By applying the weight of glory, you put it in perspective.

In C. S. Lewis's *The Screwtape Letters*, written from the perspective of a demon, we read, "Our cause is never more in danger than when a human, no longer desiring, but still intending, to do our Enemy's will, looks round upon a universe from which every trace of Him seems to have vanished, and asks why he has been forsaken, and still obeys."

If your worship is to be unconditional—where you can give it to God not just when the sun is shining and the birds are chirping and traffic lights are green, but in the midst of searing pain when all around your soul gives way—you must fix your eyes on what can't be taken away. You will become like Job, who lost everything one can lose on this earth and yet said, "Slay me and I will still follow you" (Job 13:15 paraphrase). And on an ash heap he worshiped.

In time you will see something precious developing in your soul, not in spite of the hardship but because of it. Worship that hurts like hell heals like heaven. Like the coins that went to space, you will gain a new depth, sweetness, and mystery that is not based on what is in you but what you have been through and where you have been.

CONQUER YOUR INNER SPACE

At the beginning of every tennis set, the score is love-love. For whatever reason, that is how they say zero-zero. But it is also what I want you to remember when you look at your life and you see a big fat zero—the spouse you don't have, the job you just lost, the pants you can't fit into, the friend who isn't there. That, dear heart, is a space God wants to flood into and fill with his affection.

Take a deep breath. Hear your Father promising you the crown of life and reassuring you of his unending, unfaltering, never-giving-up-on-you love. The one who hung the moon loves you! And he likes you too. When you understand how God is working through your trials, through the zero-zero feelings of life, you will be like the church in Smyrna and not pray for a weaker burden but instead a stronger back. You will foster a faith that is out of this world.

Love-love.

CHAPTER 10

NINE AND A HALF FINGERS

WHEN HUMANS FIRST FLEW TO THE MOON AT THE END OF THE 1960s, billboards all across the country displayed the words "Beautify America. Get a haircut." The nation was deeply divided over the controversial Vietnam War, and you could generally tell where someone stood on the subject simply by the length of their hair. Some young people favoring peace were swept up in the hippie vibe and rebelled by refusing the clean-cut aesthetic favored by their parents. The old people, defined as anyone over thirty, tended to trust the government, while those under thirty tended to openly question everything "the man" did and took nothing for granted.

During the Apollo 11 mission, Houston read news stories to the astronauts on their way to the moon as they woke up and rehydrated their coffee. One of the stories was a headline piece about how Mexican immigration officials began refusing entry to American hippies—unless they first bathed and cut their hair.

I was sitting in the chair at the barbershop one Saturday morning when a man walked in and plopped down. He had long, shoulder-length, silvery gray hair. "I'm cutting it all off," he announced. "My girlfriend is not going to be happy. But it's getting in my way when I'm working." He was in his fifties, short, stout, and had a friendly disposition.

"How long have you been growing it?" I asked, resisting the impulse to ask whether he had considered a hair tie. After all, *if Momma ain't happy . . .*

Barbershop conversations are one of the simple pleasures of life. The way men so freely dialogue with perfect strangers in that environment is a thing of beauty. Cars. Projects. Adventures. Kids. Work. Conspiracy theories. It's pretty predictable and often mindless, but it is enjoyable. As long as it doesn't get political.

My barber and this newcomer, who had introduced himself as Tony, began talking. He had worked twenty years himself in the hair industry, working for Vidal Sassoon, so they had lots to talk about. But my interest was piqued when in passing Tony mentioned that he had just spent a year living in an Airstream trailer seeing the country. This was a curiosity gold mine for me, and my new goal became getting Tony with the long hair to tell all on his cross-country trailer adventure.

I took a risk, interrupting their barber fraternity moment to say, "I want to hear about your year in the Airstream. What was that like?"

He lit up. "Oh, it was great. We had no plan—just picked up when we felt like it and stopped as long as we liked." I asked about the highlights. He thought for a moment and said, "Driving the Pacific Coast Highway down from Oregon to California, and probably Crested Butte, Colorado. We didn't have reservations or anywhere we needed to be, so it took the pressure off, and we were really able to enjoy ourselves. Everyone at some point in their life should do that."

I said, "It sounds amazing, but I would probably need to wait for the kids to move out of the house."

He piped back, "A divorce worked for me."

There you go.

We were deep into his favorite things about his Airstream when the door opened and another man with shoulder-length hair burst in. He headed straight for the bar (yes, there was one) and poured himself a whiskey (I know, but this is Montana) then plopped down two seats from Tony. There is a code of men and it goes like this: You never take the urinal next to one that's being used if there is another option. And you always space yourself out unless you have no choice in seats. These are inviolable and sacred.

New guy was about the same height as Tony and built like a bulldog. His frame was compact but brawny, and my money was that he was in the construction trade, judging from a build clearly forged from hard work with his hands. Speaking of his hands, I noticed immediately that the pointer finger on his left hand was missing at the middle knuckle. There was medical tape over the nub that was now the end of that digit. It seemed fresh too. I was done with the Airstream stories and really hoped that by the end of my haircut I could probe the depths of the mystery of the missing finger. But I had to play it cool. Barbershop etiquette. The convo is freewheeling and must not be forced.

Tony asked this new man if he was also there to have his locks shorn. "Hell no, just a trim," he immediately responded, evidently very proud of his follicular flex.

New guy and Tony started up a little get-to-know-you banter, but I wasn't really paying attention; I was too distracted by his finger as he held his whiskey glass. I heard him say his name was Spike at one point and noticed he described himself as a punk-rock skater rat. If I had to guess, I would place Spike in his late forties, maybe fifties. Now the long hair made sense. He definitely had a *Dogtown and Z-Boys* vibe, and I could see him thirty years ago in an abandoned pool on his Powell-Peralta.

I sensed my opportunity for a good finger-amputation story passing as Tony and Spike made small talk, but then out of nowhere hope revived when Spike said something about the last year being rough because he got some sort of bacterial or viral infection in his hand that ended up being really serious. My ears perked up. Maybe his finger was taken by some invisible, flesh-eating attack. Now that his hand injuries were on the table, I seized the opportunity. Careful to not sound too eager, I asked if that was what had caused the injury on his left hand.

"Oh, this?" He pointed at me with the stump of his left index finger. "Nah. I cut this off with a table saw a few weeks ago while at work."

"Oh man," I said. "That's awful."

"Yeah," he said as he gestured with a jerking motion. "I sneezed while making a cut and cut it right off. It was one hell of a sneeze," he explained with a shrug of his shoulders.

Jackpot. This was now one of the best barbershop stories I have ever heard.

He added, "Crazy thing is, I heard it before I felt it. After the sneeze I realized the saw made a funny noise, like it was struggling to cut through something, but I didn't feel anything right away. But when I looked down, the finger was cut off and hanging on by the skin."

I didn't want any more details like that or I would get woozy. Still, I asked, "What did you do? Put it in milk?"

"I just grabbed it, held it tight, and yelled for the homeowner whose house I was working on. He came in but I think it was worse for him than me. He got me a big bag of ice; I put both hands in, holding the wound and finger tight, and he drove me to the hospital. They couldn't get it back on because the body protected itself so well when it was cut off, all the blood vessels closed themselves down immediately and started to heal. So, they just sewed it up. They take out the stitches next week."

My barber and I both nodded in amazement. With my eyes open huge, I looked over to Tony as though to commiserate: *What a crazy story.* He apparently was not able to resist the urge to attempt to equal this tale. He nodded, then shrugged and offered, "My girlfriend had something similar happen. She dropped a circular saw on her lap and it cut her thigh open."

I shook my head at Tony's clear attempt to top with a terrible story about his poor mutilated girlfriend. Then I turned my attention to Spike. "How has the adjustment been with a shorter finger? Do you have phantom sensation and all that comes with that?"

"Yes!" he said. "Like poking the buttons on keypads and stuff, I keep missing—my depth perception is off. And when I go to grab something, like the ATM card coming out of the machine, I drop it 'cause I try to grab it between my thumb and the part that's no longer there." The best was when he said every night before bed he gently and affectionately touches the tip of his wife's nose, but the first time he tried to do it after the injury he misjudged the distance and ended up poking her in both eyes with his thumb and middle finger.

"Your muscle memory was trained. You had that finger for a long time," I reasoned. "Give yourself time."

"Yeah," he said, accepting this. "I had that thing for fifty years."

Bingo on judging Spike's age, I thought.

Just then my haircut ended. I said goodbye to my barber, Tony, and Spike, and I went on my way. I thought of them both as I walked toward my car. And they both have come to mind this week. Tony, because I have seen multiple Airstreams on the road, and Spike, because it's spring and with all the pollen in the air I have been sneezy. I don't often use power tools, but it definitely put the fear of God in me about ever coming close to one without first having taken my Claritin.

TWO FOR ONE

In John 10 Jesus gave two different "I Am" declarations that are meant to be understood together.

"I AM" #3 AND #4

"I am the door . . .

I am the good shepherd."

John 10:7, 11

First, he compared himself to a door, and then he said that he is a shepherd. We get two "I Am"s for the price of one!

Here it is in context in John 10:1–18:

> "Very truly I tell you . . . anyone who does not enter the sheep pen by the gate, but climbs in by some other way, is a thief and a robber. The one who enters by the gate is the shepherd of his sheep. The gatekeeper opens the gate for him, and the sheep listen to his voice. He calls his own sheep by name and leads them out. When he has brought out all his own, he goes on ahead of them, and his sheep follow him because they know his voice. But they will never follow a stranger; in fact, they will run away from him because they do not

> recognize a stranger's voice." Jesus used this figure of speech, but they did not understand what he was telling them.
>
> Therefore Jesus said again, "Very truly I tell you, I am the gate for the sheep. All who have come before me are thieves and robbers, but the sheep have not listened to them. I am the gate; whoever enters through me will be saved. They will come in and go out, and find pasture. The thief comes only to steal and kill and destroy; I have come that they may have life, and have it to the full.
>
> "I am the good shepherd. The good shepherd lays down his life for the sheep. The hired hand is not the shepherd and does not own the sheep. So when he sees the wolf coming, he abandons the sheep and runs away. Then the wolf attacks the flock and scatters it. The man runs away because he is a hired hand and cares nothing for the sheep.
>
> "I am the good shepherd; I know my sheep and my sheep know me—just as the Father knows me and I know the Father—and I lay down my life for the sheep. I have other sheep that are not of this sheep pen. I must bring them also. They too will listen to my voice, and there shall be one flock and one shepherd. The reason my Father loves me is that I lay down my life—only to take it up again. No one takes it from me, but I lay it down of my own accord. I have authority to lay it down and authority to take it up again. This command I received from my Father." (NIV)

This is the Psalm 23 of the New Testament. In calling himself a shepherd, Jesus was telling us that he is going to take care of us. To have a shepherd looking after you is to have someone completely devoted to you. Endless attention and meticulous care—that is what shepherds give to their sheep. Night and day, shepherds are on call. David chased down a bear and a lion to protect his sheep (1 Sam. 17:34–36). That's what Jesus will do for you.

It should also comfort you to know that you are not a number to your heavenly Father. Not just one of trillions who have lived. He knows your name. And he wants to have a relationship with you. He cares about Tony and his divorce. Spike and his finger injury. Every good and bad thing you have faced matters to him.

You have a Shepherd, and he cares for you.

But what about the fact that he called himself a gate?

Middle Eastern sheep pens in those days would have been circular enclosures with a gap used for entry and no door. The shepherd would use these communal enclosures for the night, and after feeding and watering the flock, would lead them in. He would be the last to enter and would lie down in the opening.

Jesus is a shepherd and a door because in this context the shepherd *is* the door. He would lay down his life for the sheep, metaphorically and literally. To the wolf, and every other predator out there, the message was clear: *To get these sheep you first have to go through me.*

THAT IS WHAT YOUR SAVIOR DID FOR YOU ON THE CROSS. NO ONE TOOK HIS LIFE FROM HIM. HE LAID IT DOWN FREELY.

That is what your Savior did for you on the cross. No one took his life from him. He laid it down freely.

You are free to enter through the door. You have in-and-out privileges. Come in and out and have good pasture. Listen to his voice. Call his name when you are in trouble. Even if it's been a while and you are ashamed and feel bad. He will put you on his shoulders and sing and rejoice that you have been found (Luke 15:5). He loves you.

You've heard that before, haven't you? That God loves you?

You might push back on that: "Yeah, yeah, Levi. He loves the whole world."

He does. But it's not just the world. It's you. He knows and loves you.

KNOWN AND NAMED

The first time anyone ever saw the earth from the moon was on Apollo 8. There was earth, a kaleidoscope of color turning in a black sky. Swirling, cotton-white clouds revealed brilliant blue oceans beneath their breaks, while brown and green stretches of forest and jungle covered entire countries. Thin bands of blur followed the curvature of earth, an incandescent skin of atmosphere come alive in a sea of darkness. In his retelling of the Apollo 8 mission, Robert Kurson described Frank Borman's reaction to seeing earthrise: "The planet just hung there, a jewel on black velvet, and it struck him that everything he loved . . . was on that tiny sphere."

God doesn't see just the whole world; he sees you personally. He knows your name.

I love how at NASA everything is given a name, from Alan Shepard's Freedom 7 flight to John Glenn's Friendship 7. (All Mercury flights had the number seven in them to pay tribute to the original Mercury Seven.) When they landed the Eagle on the moon, they named their location Tranquility Base. After President Kennedy's death, the launch site at Cape Canaveral was renamed Kennedy Space Center, and the site of Mission Control in Houston (the first word ever spoken from the moon) is called Johnson Space Center. In 2016 NASA named a computer building after Katherine Johnson, a veritable human computer who performed calculations for Apollo 11 and the Space Shuttle Program.

Katherine and the others involved in those precise calculations deserve every bit of attention they can receive. Imagine firing a bullet across Los Angeles and hitting the edge of a sheet of paper. That is the accuracy it took to return to the earth from the moon and strike the atmosphere at exactly the right angle to not bounce off it or be burned up while traveling at 25,000 miles an hour and reaching exterior temperatures of 5,000 degrees Fahrenheit, half of the temperature at the surface of the sun.

Unbelievable.

If you haven't seen the movie *Hidden Figures*, you really should get on that.

My favorite names are from Apollo 10, the dress rehearsal for what would be the big shebang, the eleventh Apollo mission. They called the lunar and command modules Snoopy and Charlie Brown. Because the lunar module was going to go down and "snoop around," the nickname made sense.

But there's more. Snoopy had become the symbol of NASA's quality assurance program after the Apollo 1 fire. Astronauts referred to the cap they wore in flight under their helmets as the "Snoopy cap" because it looked like the one Snoopy wore in the cartoon when he was pretending his doghouse was an airplane. So when the ship's official names were announced as Charlie Brown and Snoopy, the long-standing link between Charles Schulz's *Peanuts* and the space program became official.

It's not just another mission. Everything has a name.

So do you. No one is just a number to God. He knows Spike and Tony, Neil Armstrong, and Alan Shepard. And your Good Shepherd knows you. He knows every hair on your head. He sees when the sparrow and the skill saw fall. He knows everything you have ever done—good, bad, and in between—but he still cares for you, calls out to you, wants to walk with you, provides for you, and listens to you. He also knows what names have been spoken over you—for good or for evil. He was there when you were told you were worthless or made to feel like an object. He speaks a better word over you at this moment and always. There is no more important sound for you to tune your ears to than his voice saying your name.

CONQUER YOUR INNER SPACE

We all carry pain and shame with us into life from the harsh words that have been spoken over us and can become self-fulfilling prophecies. The Old Testament character Jacob caught his twin brother's heel at birth and so was named Heel-Catcher, which would basically be like having the nickname Deceiver or Cheater today. His first mistake ended up as his name. When God gets ahold of a life, one of the first things he does is speak a better word. He changes a name to reflect the way he changed a life: Abram became Abraham, and Simon became Peter. God changed Jacob's name to Israel, which means "governed by God." If you will allow yourself to be governed by God, he will speak a better word over you too.

On a separate sheet of paper, make a list of the harsh names that have been spoken over you in the past. (For example, when my wife, Jennie, was in fourth grade, cruel girls called her "four-eyes with a mustache.") Don't hold back. Get it out.

Next, light that paper on fire. Do it over a sink maybe.

Then write down some words God speaks over your life—for example, *loved*, *confident*, *brave*. This is your new identity.

CHAPTER 11

RITE OF PASSAGE

SUPPOSEDLY WE KNOW MORE ABOUT OUTER SPACE THAN WE DO about the deep, dark mysteries of the ocean on our own planet earth. That's crazy. And it's humbling because it makes us realize how much there is we don't know.

Parenting is like that. By the time you have kids, you think you have learned a thing or two about life. That is, until all of a sudden you realize, despite whatever wisdom or knowledge you think you have accumulated, that you are an absolute perpetual beginner at managing the little humans right under your nose in your home. The list of things you know is a lot smaller than the list of what you don't know.

Alivia is our firstborn daughter, and she's very bright. Now taller than her mom, she has a beautiful smile that makes me light up like a Christmas tree every time I see it. Liv has a dark sense of humor that she got from me and a sweetness that she got from her mom. She is really good at reading people and is the first one to want to step up or stick up for anyone who needs defending or help.

As the first of five to be born into our family, Alivia has received the good and bad of being the oldest. She stays up later than the littles, gets more freedom, and has a phone instead of an iPod Touch. She is also expected to help out on a different level by babysitting her siblings, helping me move furniture, and other big-kid chores. When people ask me what it's like having multiple kids, I like to tell them that the best thing is how compound interest kicks in. And though you are outnumbered, ultimately the older ones become more and more helpful with the younger ones.

Alivia is the only one of our four kids to have the experience of being an only child. All the rest entered into ever-escalating levels of chaos. Lennox, our fifth child and the only boy, was basically born into a movie set of *Little Women* set in the year 2017.

Then there is the issue of being wet behind the ears. Liv got us when we were green as parents, which is really just one of those "six of one, half a dozen of the other" scenarios, because when you are the firstborn your parents are inexperienced, but by the final child they are exhausted. Not sure which is better or worse. I would like to think that we have gotten better child by child, but sometimes it feels like it's *Groundhog Day* when it comes to making the same mistakes over and over. The ocean of parenting is deep and often murky.

Doing a thing repeatedly is no guarantee of progress. Contrary to popular opinion, practice doesn't make perfect; it just makes it permanent. If you practice one thousand layup shots a day you won't become perfect at it if you're practicing it the wrong way. Your muscle memory will soak it up, all right; it will just be permanently ingrained incorrectly. Stephen Covey expands on this idea that runs somewhat glaringly in the face of the "get your ten thousand hours and you'll be fine" flawed understanding of the Malcolm Gladwell concept:

> On the individual level, the problem is that many people aren't into the idea of continuous improvement. So they're working in a

> company—maybe they've been there for ten or fifteen years—but instead of having fifteen years of experience, they really only have one year of experience repeated fifteen times! They're not adapting to the changes required by the new global economy. As a result, they don't develop the credibility that would inspire greater trust and opportunity. Often they become obsolete. Their company and/or the external markets outgrow them.

The point is, practice isn't enough because you can be cursed by knowledge. (Google it. It's a thing. And there is an amazing *Harvard Business Review* article you should read about it.)

You are also limited by the blind spots you are oblivious to. A deliberate attempt must be consciously, constantly, and systematically made to improve, or all the practice in the world will just bake in bad habits, and you'll be no better off in a decade than you are today. This is as important to consider in business and in ministry as it is in relationships.

RELATIONSHIP MOMENTS

Speaking of relationships, being married a lot of years is not enough to ensure a great one with your spouse. You must continue to do the hard yards of showing up in intimacy, vulnerability, and repentance. Quantity is a poor substitute for quality. You have to study your partner, study yourself, and continue to be present and accounted for in more ways than just physically if you are going to grow. Your heart, soul, mind, and strength must all be equally employed if you want to thrive.

In the seventeen years I have been married to Jennie, I feel like I have been married to ten different women, and I am sure she feels like there have been as many or more Levis involved. Without giving each other the space to grow and putting the emotional effort into pursuing each new iteration of the other, the relationship will stay stuck in a previous evolution. It's like a snake that hasn't shed a skin that has outlived its usefulness. If you aren't intentional, you will be married to someone different in your mind than they are in reality.

Jennie and I like to go paddleboarding and canoeing together. Invariably at some point we end up floating in the middle of a lake, and unless we tether the boards together, or continuously dip our oars here and there, we will drift. There are two essential rules of drifting:

1. It is as inevitable as gravity.
2. As I stated back in chapter 6, you never drift in the right direction.

In a relationship, drifting means distance. And since nature abhors a vacuum, relational distance always leads to disaster. You must remain vigilant to close the gap because the devil will send someone in to pursue your spouse if you won't. This is why in 1 Corinthians 7:5 Paul says that couples must not let too much time pass without enjoying the gift of sex.

Okay, marriage moment over.

The same mechanics are at play when it comes to parenting. Humility is essential to guard against the incorrect assumption that lots of experience must surely make you better at it. You must be willing to grow, adapt, pivot, learn, admit you are wrong, and change tactics on the fly. You also need to study your children and give them space to grow too. Turn your back for a moment, and drift happens. It should also give you comfort to know that great parents are made, not born. It's not binary; you are not either "good at parenting" or "bad at parenting" and therefore stuck where you started. You can grow if you are willing to.

One of the most freeing things to admit to your children is that you don't know what you are doing. Trust me, they have probably already figured that out, and it will not come as a huge shock to them. But they will be relieved by your honesty. Jennie and I tell our kids all the time that we have never been here before, and neither have they, so we will all do our best to figure this out. It is counterintuitive, but when you, as a leader, admit this to those under your authority, it really freshens the air.

We often feel a self-imposed pressure to set the course of our families or organizations. But in seeking to project confidence when there is a lack of certainty, we miss out on the power that comes from the win of transparency. We fear transparency will lead to diminished respect—that's understandable, but unfounded. If you're open, it will humanize you in the eyes of your offspring and your direct reports. The key is, regardless of the ambiguity, you must still plot a course to move foward even though you are not entirely certain it's the right choice. General George Patton had it right: "A good plan violently executed *now* is better than a perfect plan next week." Passion covers a multitude of sins.

We often tell our kids, and our ministry team, that even though we may have never been somewhere before, even if we aren't entirely certain about an opportunity or choice or decision, we are going to swing for the fences anyway. And if along the way we realize we have made the wrong decision, we will admit it as we jettison what isn't working. It's frightening but freeing to lead and parent this way. My friend Pastor Craig Groeschel likes to say at the end of many of his leadership podcasts, "People would rather follow a leader who is always real than a leader who is always right."

Remember this: at home and at the office, you are trying to work yourself out of a job. Banish job security out of your heart and mind. Replace yourself. In your field of business or ministry, this means you raise up leaders who aren't just given a task to do but the authority and space to develop grit and become strong decision-makers. At home this means bearing in mind the end of your kid's childhood, when you are no longer the adult in their life, and making sure in that moment that there will still be an adult in their life—them!

We've all seen the consequences of overparenting or, just as disastrous, underparenting. The same is true of discipline. Too much or too little discipline will discourage your children and stunt their growth. There is a reason the Bible compares rearing kids to archery (Ps. 127:4). Both holding on to the arrow too long or letting go of it too soon will keep it from hitting the target.

It is so easy to err in either direction. We might be domineering and make all their decisions for them or wrap them up in Bubble Wrap and require them to wear their helmets indoors. Or we might give no direction or protection. Either will prove disastrous to you and your kids. Doing too much for them too long keeps them dependent on you in a way that will send them into the lion's den of life unprepared. Your job is to get them ready for the lion's den. The moment they were born, a clock started ticking—and that clock represents the moment they will leave your home able to make all decisions for themselves. There are only about 940 Saturdays from when your child is born through their high school graduation.

When it comes to being a parent, I am Jekyll and Hyde. Having kids exposed a temper in me that I wasn't totally prepared for. It is so easy to lose my cool and snap at them. Then, out of guilt, I attempt to compensate by coddling and erring on the side of being the cool parent.

Raising kids is much like driving a stick shift. Too much gas or too much clutch can cause things to go wrong. You're after that constant finesse, a dance on the pedals. The analogy isn't perfect, but it does work to the extent that every car, like every child, must be handled differently and uniquely, depending on the moment.

Jennie and I still aren't quite sure what we're doing most days, but we are eager to learn and hungry to improve. So much of parenting is trial and error, though some days it feels like error and error.

TURNING THIRTEEN

One of the big epiphanies we had when Alivia turned thirteen was that we needed something to mark the occasion with more weight than a typical birthday celebration could carry. As we mulled it over, Jennie and I both felt like some sort of ritual or ceremony was in order. She was on her way out of childhood and entering a new epoch of development, and it begged for pomp and circumstance. We had made the "sex talk" a priority, not as a one-off event but as an ongoing dialogue for our kids, but what was needed here was something else entirely.

One person told me that his dad allowed each child to select a trip to anywhere on the planet and he would take them, just the two of them. The only rule was that he wouldn't work on the trip. It wasn't a mixture of business and pleasure; it was all about the adventure and the relationship. I loved it. We put that on our developing plan as an element, but we still wanted some sort of memorable day that involved the whole family and our tribe.

I also began to research how other cultures around the world treat coming of age. The Jews have bar and bat mitzvahs; other cultures have piercings and tattoos and spirit quests for their budding adolescents. I came across a TEDx Talk that expressed everything I was looking for. In it the speaker, Ron Fritz, described how important rites of passage are when you are between being a kid and being an adult. You should watch it: it's called "An Exploration of Coming of Age Rituals and Rites of Passage in a Modern Era." I watched the video multiple times, and it hit me—our culture sucks at this all-important transition. The best we have is a driver's license or a high school graduation. And while a diploma or a laminated card from the DMV or MVD (depending on the part of the country you live in) is important, there is still something missing. Those pieces of paper declare you fit to drive a car or enter college, but they do nothing to confer blessings or clues as to your worth or place in the world.

No wonder young people end up searching for other things to make a man or woman of themselves. Losing virginity, getting drunk for the first time, reckless behavior or crime—these are understandably and yet so unfortunately looked to in the absence of mile markers on the highway of growing up.

In his book *Greenlights* Matthew McConaughey says that a rite of passage for him was when he beat up a bouncer in front of his dad. "That night was my rite of passage. Dad let me in," he recalled. "It was the night I became *his boy, a man* in his eyes."

Perhaps this is why there are so many boys and girls running around in adult bodies confused about who they are, what they are supposed to be, and what really matters. I asked one friend if anyone had ever told him that he was becoming a man, and he said, "No, not until I was in prison." There, a Christian man took him under his wing and explained to him who he was meant to become.

That was all I needed to hear.

The real problem with never being set up for success in the transitions of life is that you miss out on the opportunity to receive blessings that are meant to be passed down from one generation to another. You can't very well give a blessing you never knew you had.

As we were working on our plan for Alivia, our friends Bill and Jessica Cornelius came to Montana on vacation. While they were in town, they told us what they planned for their daughter Sophie to commemorate her coming of age. It involved fire and water and being alone in the woods. Sort of a cross between "Sweet Sixteen" and *Survivor.* Perfect inspiration.

Fully inspired, we marked the day of Alivia's celebration on the calendar so we would keep it open. When the day came, everything was in order. At least as much as was possible.

It began in the cemetery, only Liv didn't know that was where we were going. Our little girl was thirteen and all we told her was it was an important day and she needed to wear her Chacos. I teased her that my goal was for her to have a Chacos tan by day's end. (Google it. It's a thing, and she has now embraced it.)

We had family breakfast together at Chick-fil-A (her favorite restaurant on earth), and then we handed her the first of a series of envelopes. She would have to decipher the clue and instruct us where to take her in order to fulfill a mission that was also on the card, written in the form of a riddle. Only then could she receive the next card.

The first card pointed her to the cemetery. We realized that, in all the family trips we'd taken to her sister's grave over the years, Alivia had never had the chance to be there alone. As her next-closest sibling, Lenya shared a special bond with Liv, and, in several different homes, they shared a room as well. Until Daisy was born, it was just the two of them.

We drove to the opposite side of the graveyard from Lenya's plot, put a bouquet of flowers into Alivia's hands, and told her to take her time and spend as long as she wished. When she arrived, she found a picnic blanket and a present wrapped up and addressed to her in Lenya's honor. The card told her that we wanted her bestie to be involved in her day, and how she would always be a part of her life as she made this journey into adulthood.

A long time passed, and Jennie and I both eventually grew concerned that this was a lot to put on her—maybe too much. Should we go over? One hour in, and we were already crumbling. Just as we began to doubt the entire plan, she walked up with noticeable streaks down her cheeks. We all had a big hug.

The rest of that day involved cliff diving, to express courage needed for the leap into the unknown; an ice bath, to represent doing the hard things and not giving up when everything inside her told her to; sweating it out in a nearly two-hundred-degree sauna, representing the strength to stand strong in fiery trials; and, one of my favorite elements, going alone into the market and paying for a stranger's groceries. This symbolized the need to be aware of others and the power of a single act of kindness.

The final element, like the first, was one she had to face alone. The clues led her to a river where she found a paddleboard she was to use to fight her way upstream as she sailed up the river of womanhood. This had obvious overtones of going against the current of culture and its pressure to conform if it flew in the face of her inner compass and values as a child of God. She was to make her way to the head of a river, fight her way into the lake that fed it, cross the lake, touch the shore, and return.

This rite had two surprises. First, though she was told it had to be done alone, halfway up the river Jennie was waiting on a paddleboard to ride with her. This was to emphasize that her mother would help her make sense of all coming difficulties, and that she never had to be alone because God would always position the right people on the river of life if she would embrace them. They rode together to the "end," across the lake. The final surprise was that while she was expecting a return trip, at the other end of the lake we had a crew on hand to greet her and terminate the paddle.

A few weeks earlier Jennie had asked Liv to make a list of every female she looked up to and wanted to be like. The list included some staff members from Fresh Life, our church; a small group leader; her aunts; a few others from the church; Shelley Giglio, who has always had a special connection with Liv; and Jennifer Lawrence the actress. I laughed when Jennie told me she was on the list. She explained that Liv loved her strength and spirit in the Hunger Games movies. We invited all these role models to stand on the dock holding balloons, along with her sisters and brother and me, to welcome her when she arrived. Sadly, Jennifer Lawrence was not present, but the rest of us cheered for her as she fought the headwind that made the last leg of her journey difficult. She was beaming as she pulled up to the dock.

A pontoon boat outfitted with drinks and pizza was on hand, and Liv, her mom, and the leading ladies in her life all loaded up and headed out for a sunset cruise. They ate and celebrated her, officially welcoming her into the wolf pack of women.

Out on the lake, one by one, each of these women boarded an enormous inflatable slice of pizza that was tied to the boat. Each one floated out on it with Liv for a segment we called "a pizza wisdom," where they could candidly and confidentially share regrets, lessons learned, and anything else from their hindsight they wanted her to know for the coming years. Before the boat pulled back up to the dock, they had all spoken blessings and life and prayed over her. She would never have to doubt where she fit in the hierarchy of life. Her place in the community was clearly articulated. She belonged.

When we got back home, the whole family sat around the fire in the backyard and laughed and talked about her rite-of-passage day. Then we pulled out an iPad and she watched videos sent by a few women she had put on her list who lived out of state. They weren't there in person but wanted to be a part digitally. We ended as a family by putting our hands on her and praying over her, making sure she knew we were behind her and in her corner, come what may.

Just before we all turned in, Jennie handed Liv one last envelope. The return address indicated it was from Kentucky. She opened it and the blood drained from her face as she began reading the card. It was a handwritten letter from the Mockingjay herself. Jennie had reached out to Jennifer Lawrence's office and explained Liv's admiration for her and what we were trying to do that day. And Jennifer wrote Alivia a letter encouraging her in God's plan for her life.

Next, we explained to Alivia that she could choose a trip to take with me—her dad and the number-one man in her life—anywhere on earth. And I wouldn't preach, have a meeting, or write so much as an email while we were gone. She mulled it over and eventually made her choice. We were going to go on a Disney cruise to Mexico. We purchased tickets, booked flights, and began excitedly making plans. Unfortunately, COVID-19 hit just prior to our planned date of embarkation. The news was full of people stranded on cruise ships, so it didn't seem ideal, and we contemplated backing out. Mickey Mouse made the decision for us. A week before we would have left, Disney emailed us to let us know that the cruise was going to be rescheduled. Then they gave us a credit and let us know we could use the tickets whenever cruises were a thing again. We both can't wait to take that trip.

I am so glad we invested the time and energy into celebrating Alivia's rite of passage. It made it feel like not just a chapter, but a new section was written into the book that is her life.

TIME TO GROW UP

As you think through the power of family relationships, consider Jesus' next statement from the cross in John 19:26–27:

> When Jesus saw His mother and the disciple He loved standing nearby, He said to His mother, "Woman, here is your son." Then He said to the disciple, "Here is your mother." So from that hour, this disciple took her into his home. (BSB)

In this incredibly moving moment, Jesus demonstrated, right up to the end, that he did not come to abolish the law but to fulfill it (Matt. 5:17) and in so doing honored his father and mother (Ex. 20:12). He also came full circle and was taking care of the one who took care of him. He was not a boy anymore; he was a man. And as a man, he exhibited what true maturity is: selflessness and service.

During the Civil War some of the officers were standing around with General Ulysses S. Grant. One of them was about to tell an off-color joke but prefaced his vulgarity by smirking and looking from side to side, saying that, seeing as how there were "no ladies present," he had a story to tell. General Grant interrupted, "No; but there are gentlemen." That stopped him and the story was never told. In a day when *adult* is a synonym for pornographic and strip clubs use the word *gentlemen*, we need to pay a lot of attention to his example.

ONCE A MAN, TWICE A CHILD

LAST WORDS #3

"Woman, behold your son! . . . Behold your mother!"

John 19:26–27

At the cross Mary was experiencing a role reversal as well. With her loving hands she had held Jesus, fed him, and bandaged him when he scraped his knees. But now it was she who was the recipient of the loving care and protection of her son as he arranged for her to be taken care of after he was gone. Jesus knew that his mother still needed to be loved.

Mary is representative of everyone in this world who needs to be loved and tenderly cared for. You and I are John. God wants us to do for our neighbor as we would do for Jesus' mother.

What does this mean?

We need to not be so focused on how we are treated but on how we can fulfill Jesus' dying wish and treat others well. We need to grow up. As we read in 1 Corinthians 13:11, "When I was a child, I spoke as a child, I understood as a child, I thought as a child; but when I became a man, I put away childish things."

Children are selfish. It's not their fault. It's just how they are. Putting away childish things is learning to esteem others more highly than yourself. It's becoming a servant.

WE NEED TO NOT BE SO FOCUSED ON HOW WE ARE TREATED BUT ON HOW WE CAN FULFILL JESUS' DYING WISH AND TREAT OTHERS WELL. WE NEED TO GROW UP.

This is true greatness.

Maybe no one ever told you it's time to cross over from being a boy to being a man, or from being a girl to being a woman. I realize the tension in that. There is a sense in which we need to fight to *not* grow up too. Jesus said we need to have faith like a child, for such is the kingdom of heaven (Matt. 18:2–4).

C. S. Lewis acknowledged this tension when he said,

> To be concerned about being grown up, to admire the grown up because it is grown up, to blush at the suspicion of being childish; these things are the marks of childhood and adolescence . . . But to carry on into . . . early manhood this concern about being adult is a mark of really arrested development. When I was ten, I read fairy tales in secret and would have been ashamed if I had been found doing so. Now that I am fifty I read them openly. When I became a man I put away childish things, including the fear of childishness and the desire to be very grown up.

The goal, then, is to be childlike but not childish. To retain the creative, sweet, curious, and imaginative spirit of childhood but without the tantrums, selfishness, and lack of impulse control that characterize toddlers.

Childlike wonder at the stars is still welcomed.

I believe one of the reasons God had me write this book was so that you could have the rite of passage you may never have received. It's hard to live as a grown-up when you never were told you are one. This book is an excuse for you and I to stand at the foot of the cross and listen to Jesus be so completely unfocused on himself that he could use his dying words to tenderly make sure his mom was provided for.

That is being an adult. That is being a lady or a gentleman. It's a travesty that "mature" is how movies with explicit nudity or graphic sex are described. That pornographic websites are described as "adult" and that strip clubs parade under the banner of "gentlemen's clubs." If you want to see a real adult, just look at Jesus.

Tucked inside these seven "I Am" declarations is far more than a pattern for you to follow; there is power for you to harness to conquer your own inner space and explore all that is hidden inside you. Jesus is calling you to shed the selfishness of spiritual adolescence and graduate into the calling that is on your life. There is a great warrior inside you yearning to come out.

I have a feeling you know deep down what you need to do. It will be hard, but you have what it takes. Perhaps you never had a father or mother sit you down and look you in your eyes and tell you your value and invite you to take a step forward in maturity and love.

This is that. Receive it.

I BELIEVE ONE OF THE REASONS GOD HAD ME WRITE THIS BOOK WAS SO THAT YOU COULD HAVE THE RITE OF PASSAGE YOU MAY NEVER HAVE RECEIVED.

No vessel makes it to space hauling everything it left the launchpad with. It jettisons spent fuel cells and stages. In the case of the Saturn V launch vehicle, it lifts off at more than three hundred feet tall, but by the time it gets back to earth it is a fraction of that—"a cone eleven feet high and thirteen feet across the base." It sheds most of itself relatively quickly. It leaves earth tipping the scales at six million pounds but comes back a svelte eleven thousand pounds. Incredibly, it loses five million pounds in the first 150 seconds. When you see the teensy Apollo 11 command module in the Smithsonian, you might realize it could fit in your kitchen. How did it get to the moon and back? It didn't hold on to things that no longer served a purpose.

CONQUER YOUR INNER SPACE

There are some things about childhood that we abandon to our detriment, and others we hold on to far too long. Is it possible that you are hanging on to things that are holding you back? That scaffolding that was meant to simply get you to the next stage is now in your way?

Your God is a Father to the fatherless, and he will not leave you, though your own mother may forsake you (Ps. 27:10; 68:5; Isa. 49:15). It's hard to raise your kids to be strong adults when you are still a child yourself. Perhaps you don't know anything different from the childish things you should have outgrown.

If so, start here:

- Make a list of things in your life that are holding you back, and ask God for his help to jettison them. It could be a toxic friendship, a habit, a substance, or just a pattern of negativity or selfishness.

- Now make another list of the God-sized dreams that only the "grown-up," childlike (but not childish) version of you could accomplish. It could be starting a company, going back to school, or building a treehouse.

Today is the day to head out into the deep. No, it won't be easy. But it will be worth it. And you won't be alone for any of the challenges. God is with you, and he has a plan.

This is your permission slip.

This is your rite of passage.

It's time to leave the kiddie pool behind. A blue ocean of opportunity is in front of you, and the world needs the love that is inside you.

CHAPTER 12

BURN, BABY! BURN!

WHEN APOLLO 11 LEFT THE EARTH, THE ENTIRE NATION HELD ITS breath. Fitting, because at the moment of its liftoff, the rocket, Saturn V, was burning as much oxygen as a half billion people—more than the population of the United States simultaneously taking a breath. Who was in control of all that fuel, and the orientation of the five nozzles each so large you could stick the ascent stage of a lunar lander inside it with room to spare? The onboard computer—the so-called fourth astronaut.

During launch the computer was one hundred percent in charge. At that point the astronauts were just there for the ride—and to trigger the abort if it came to that.

If you have ever tried to balance a pencil vertically on a fingertip, or a broomstick on the palm of your hand, that is just a taste of what the computer was trying to manage during takeoff from earth. Wind pushing on the 363-foot-tall launch vehicle made the computer have to control the force of those five rockets as they swiveled intelligently to keep it vertically oriented and, most significantly, ensure it cleared the tower. It also had to compensate for the "pogo" vibrations, so called because of the similarity to the forces exerted upon a pogo stick, when the liftoff puts tremendous force on the long, skinny object. If this isn't dealt with, it can lead to catastrophic failure and/or shake the astronauts to death. The computer had to do all that and more. It even was able to do all that during Apollo 12 when the rocket was struck two times by lightning during liftoff, and those inside still made it to the moon and back.

The lunar module Eagle also had an onboard computer and a sequencer written into it that allowed it to multitask and prioritize tasks by importance, sharing its extremely limited memory. As we discussed earlier, the most taxed and stressed-out the computer ever got was during its powered descent to land on the moon. That's when the sequencer was asked to do more than its capacity would allow. To compensate, it rebooted the computer, giving preference to vital functions.

There was a "master ignition routine"—a piece of the code used by several of the programs to execute maneuvers during the final descent. If you watch video footage of the moon landing, you will see Buzz punching these numbers into the computer by hand. Unbeknownst to any of the astronauts, or anyone else outside those who programmed it, that master ignition sequence had a name that only the computer could see: "Burn, baby! Burn!"

No one at NASA knew this name at the time. Years after the mission, the computer code showed up online. Computer programmers combed through it and found embedded within the code the name for the sequencer that only its creators and the computer itself would see: BURN_BABY_BURN. This was one of many Easter eggs they inserted into the thirty-six thousand words the computer had in its memory. (That, by the way, was all the computer could hold in its memory. Thirty-six thousand. A marvel in itself, as today a single email takes about seventy-five thousand words of code.) Because this code is invisible to the world, it allowed the programmers who wrote it the chance to amuse themselves and speak, even though secretly, to the issues that were going on in their day.

There were references to *The Wizard of Oz*, lines from Shakespeare, pinball references, as well as a few other things, and the programmers didn't resist the opportunity to chime in on what was happening in their culture at the time. When recalling the grandiose adventure of the space race, it's important to remember it took place during one of the most tumultuous chapters in American history: the late fifties and sixties. Vietnam, the Cold War, the Cuban Missile Crisis, the Bay of Pigs, the women's liberation movement, civil rights protests, an extremely polarizing presidential election, and the assassinations of JFK, RFK, and MLK. It seems the sixties were not a good time to be known by your initials.

This decade also produced some of the greatest music of all time. In 1968 (the first year contact was made with the moon on Apollo 8), the world first heard "Hey Jude" by The Beatles (written by Paul McCartney to help Julian Lennon process the divorce of his parents); "Revolution," also by The Beatles, as their commentary on the chaos erupting around the country that summer; "Mrs. Robinson" by Simon & Garfunkel; "Hello, I Love You" by The Doors; "Born to Be Wild" by Steppenwolf; "I Say a Little Prayer" by Aretha Franklin; and "All Along the Watchtower" by Jimi Hendrix. All in one year, people—one year. The point is, the most difficult times occasion some of the most beautiful art. This is true of the Psalms. And it can be true for you too.

Peter Adler and Don Eyles worked for MIT's Instrumentation Lab. They were tasked with developing the software for the computer in Eagle that would guide its powered descent to and ascent from the moon. America was in upheaval. Peter put it this way: "We might not have been out on the streets, but we did listen to the news." Burn, Baby! Burn! was a way of acknowledging the Black Power movement and the riots that had broken out across the country. It referred to the tagline of a popular radio DJ, Magnificent Montague, who liked to exclaim, "Burn, baby! Burn!" when he loved a song—a phrase that became a rallying cry for rioters starting in the Watts neighborhood of Los Angeles in 1965.

It is important to remember how many young people were involved in the moon landing. The average age of the engineers was twenty-eight. Don and Peter were even younger than that. These young people accomplished so much—and amid significant cultural upheaval. Their eyes were focused on the heavens, but they still lived here on a planet that was, as it is today, embroiled in turmoil.

It is precisely that tension that we as a church must constantly navigate. We are a people who should have our hearts focused on heaven, where our citizenship is, with our feet still firmly planted on planet earth. The former should dictate how we handle the latter, and not vice versa. We don't interpret our faith through the light of our physical reality but instead allow our faith to determine how we see our lives here, now.

THE CHURCH IN PERGAMUM

Jesus speaks to the struggle implicit in this split-screen reality in his comments to the church in Pergamum. Revelation 2:12–17 spells it out:

> "To the angel of the church in Pergamum write:
>
> These are the words of him who has the sharp, double-edged sword. I know where you live—where Satan has his throne. Yet you remain true to my name. You did not renounce your faith in me, not even in the days of Antipas, my faithful witness, who was put to death in your city—where Satan lives.
>
> Nevertheless, I have a few things against you: There are some

> among you who hold to the teaching of Balaam, who taught Balak to entice the Israelites to sin so that they ate food sacrificed to idols and committed sexual immorality. Likewise, you also have those who hold to the teaching of the Nicolaitans. Repent therefore! Otherwise, I will soon come to you and will fight against them with the sword of my mouth.
>
> Whoever has ears, let them hear what the Spirit says to the churches. To the one who is victorious, I will give some of the hidden manna. I will also give that person a white stone with a new name written on it, known only to the one who receives it." (NIV)

Notice that Jesus immediately acknowledges the difficulty in living as a church so close to "where Satan lives." The city of Pergamum was famous for its civilization and learning. It was home to the pagan cults of Athena, Dionysus, Asclepius, and Zeus, whose altar was located at the top of Pergamum's acropolis; the city hailed him as savior. Translation: They were Kansas City Chiefs fans wearing jerseys and war paint but surrounded by a sea of Raider Nation, and doing the tomahawk chop would be enough to get you killed (or canceled). That is exactly what happened, yet they prevailed. Jesus is saying, "Antipas got jacked, and you still held the faith."

Out of the frying pan and into the fire.

Remember, though, that the reward for progress in one area is usually an attack on a different front. "There is nothing that fails like success," G. K. Chesterton once said. It is very easy and human to go through one difficult thing and wipe your brow—*Phew!*—then let your guard down and be unprepared for the next round. But this can sap your morale.

It can be disheartening to feel like you are beginning a new crisis when the old one is barely behind you. Imagine finishing a triathlon totally depleted and being told you are actually participating in a decathlon. Lots of seasons in my ministry have felt that way, though perhaps none so emphatically as 2020.

As impossible as it can feel to navigate challenging seasons, this sensation of fatigue is nothing new. I'm encouraged while going through present challenges when I look back to the 1960s, a time when the nation seemed to be burning, and see that in those dark days we did one of the most daring and beautiful things in human history.

It can be the worst of times and the best of times at the same time.

LETTER #3 (PERGAMUM)

"To him who overcomes I will give some of the hidden manna."
Revelation 2:17

The church in Pergamum reminds us that the Christian life isn't a triathlon; it's a decathlon. And just because you are good at running doesn't mean you can ignore the shot put. Even if you endure against outright opposition and prevail, it doesn't mean much if you then give in to sexual immorality, for example.

We need vigilance in all of life. As Ronald Reagan said, freedom is "never more than one generation [or one temptation] away from extinction." To let your guard down because you are strong in one area is to invite temptation in another.

THE STRUGGLE

Let me tell you one of the best ways to overcome the tendency to put yourself in neutral: it's to never think you have arrived, but instead stay in drive and maintain your edge. Kings become fools when they stop going out to battle. Ask David.

This needs to be balanced, of course, by understanding the importance of recovery. LeBron James spends over $1 million dollars annually on recovery. There is no honor in playing hurt or pretending you are fine when you are not. Maintaining resilience and vigilance doesn't mean you don't take the time to heal, rest, and recharge. You just employ these things as a strategy, not as the goal. Lying on a beach like you are in a beer commercial is a fabulous way to spend a vacation, but if that's the carrot you are chasing your whole life, that'd be a tragedy and a recipe for disaster.

These verses about not letting your guard down are illustrated by a reference to Balaam, whose story is fascinating (Num. 22). And not just because of the presence of a talking donkey.

The Israelites had come out of Egypt but had not yet gone into the promised land. They ended up like you and I often end up—stuck in between where they were and where they were meant to be. Salvation is not meant to be a finish line but a starting line. We are supposed to go forth into growth, strength, and power.

Let's stop here for a minute so I can encourage you to think differently if you believe you've peaked or gone as far as you can go. I dare you to dream deeply and fight with all your heart against disillusionment or the spirit of lethargy. You are meant to move forward. There is more land for you to take. More of your calling for you to fulfill. More character to develop. If you are alive today it is for a purpose and not an accident.

But there will be blood.

Not everyone wanted the Israelites to make it to the promised land because if they did, they would become strong. Balak was an enemy. He wanted to oppose the children of Israel just like the devil wants to stop you. If the devil can't take you to hell, he will try to keep you from bringing anyone else to heaven with you. If you are saved, he can't take your soul, but he will try to neutralize your effectiveness.

IF YOU ARE ALIVE TODAY IT IS FOR A PURPOSE AND NOT AN ACCIDENT.

Such was Balak's approach. He tried to oppose Israel on the battlefield. That didn't work. It only caused the Israelites to rally and work together and trust in God. Next, he decided he would hire someone to cast a spell on them. The most Harry Potter–like dude in the land was Balaam, a sorcerer who lived in the next town.

Balak sent word to Balaam, but God warned him not to come. Even still, homie wanted to get paid, so he mounted up on his donkey and threw caution to the wind. This is where the *Shrek* story went down. Balaam's donkey saw in the middle of the road an angel with a sword that was invisible to Balaam's greedy eyes, so the donkey veered to the side, then crushed Balaam's foot against a wall, then simply laid down on the path and refused to move. This is when it gets entertaining.

Balaam was beating the poor donkey who just saved his life, and the donkey went off on him. Speaking in his best Eddie Murphy voice, he asked his master if he had ever been stubborn like this before. Hadn't he always been a good, compliant donkey? And didn't he think there might be a reason that he didn't want to go this way?

The crazy part about all this is not that the donkey talked, but that Balaam, who was so set on his payday, didn't even notice that the donkey was talking to him. Instead he returned fire and talked back as though it were a perfectly normal thing for donkeys to speak.

Finally, he looked up. The angel had taken off his invisibility cloak and could be seen. Only then did Balaam realize that the donkey saved his life. Now for the really crazy part. Balaam decided to go through with his mission and continued on to see Balak. If you are paying attention, that is now three times that God had done something drastic to stop him from going. Balaam was unwilling to listen.

A lot of people say they would believe in God if they saw something miraculous. But Romans 1 says that when we ignore the general revelation of creation and our conscience that points to a God, our foolish hearts are darkened and we refuse to believe—not because we can't, but because we won't (vv. 18–32).

Eventually Balak and Balaam were on a mountain and Balak said, "Okay, we had a deal. I've done my part; now it's your turn. Curse them good." Balaam opened his mouth and out came nothing but blessings for Israel. He just encouraged the stuffing out of them.

Balak was outraged. "Then Balak said to Balaam, 'What have you done to me? I took you to curse my enemies, and look, you have blessed them bountifully!'" (Num. 23:11).

So Balaam tried again. More blessings.

Balak was turning red.

It reminds me of *Liar Liar*, the movie where Jim Carrey can't tell a lie because of a wish his son makes. He tries all day to tell lies and no matter what, he can only tell the truth, no matter how devastating.

According to Eugene Peterson, this bizarre moment in Scripture illustrates something vital about God. His primary goal is to bless, not curse. Peterson said, "The widespread grassroots ignorance of this is reflected in common speech: 'God damn' is far more prevalent in everyday American speech than 'God bless.'" Peterson posits that we resort to such language when we feel displeased, inconvenienced, frustrated, or "blocked by people God has put in our lives." We think we can get God to curse them like Balak did the children of Israel, but we're way off base: "God's characteristic way is blessing. . . . 'Blessing is God's main business.'"

Before Balak could demand a refund, Balaam, the quick-thinking, donkey-striking prophet said, "You know, Balak, you technically don't need to curse them if you can get them so polluted they will hurt themselves" (Num. 31:15–16 paraphrase).

The enemy knows this too. He doesn't need to harm you if he can get you to willingly choose to walk away from God's blessings.

So that's what Balak did. He sent women into the camp to tempt and seduce the Israelite men (Num. 25).

This is the perpetual struggle of following Jesus.

While Jesus wants to light you on fire and ignite power, influence, and love in your life through the controlled burn of his Holy Spirit, your adversary, the devil, is working hard to blow you up with trials or crush you with despair. When that doesn't work, he will try to seduce you. And should that fail, he will try to trick you into a spirit of religion and pride. He will never stop trying to tear you down. So you must never let down your guard.

CONQUER YOUR INNER SPACE

Allow me to pray a prayer over you. As you read it, I hope you'll receive it. Maybe raise an open hand or two as you do.

Lord God, thank you for how you want to pour out your Spirit on us. You are so kind and generous, and I pray for my friend to receive all that you have for them. I pray that they would see the kindness in your eyes and that you not only have a plan for them, but you want to empower them to keep walking and live on purpose with your power in their veins.

Keep them from the wicked one. Refresh them. Recharge them. Remind them that they are yours and make them burn bright for your glory, for their good, and for the good of those in their lives, in Jesus' name, amen.

CHAPTER 13

GO, NO GO

THREE HOURS AND TWENTY-TWO MINUTES AFTER LAUNCH, while hurtling away from earth with Michael Collins at the helm, the command and service module separated from the third stage of the Saturn V rocket, spun around, and docked nose to nose with the lunar module, whose four-panel shroud peeled away like silver petals of a flower. Michael pulled the lunar module out of the compartment where it had been stored and for three days flew two ships, merged into one Apollo spacecraft, to the moon.

There is a diagram of this process in the front of the book, so save your place here and go back there now and look at it. You'll be amazed what all took place as they raced toward the moon. They gradually slowed down from the escape velocity of 25,000 miles per hour to a tenth of that. At the halfway point between the earth and the moon, they came under the moon's gravitational pull and began to speed up again.

One hundred hours later, the Apollo 11 crew began the complicated and risky process of leaving Michael and the CSM behind. Their mission was to cover the sixty miles from where they were orbiting the moon down to the lunar surface in the spidery ball of gold and grey that was the LM. Buzz and Neil undocked from Columbia. They performed a quick visual inspection and headed off. From the window in Columbia, Michael watched them go. Eagle looked far more like a robotic tarantula than a bird of prey, but it was on its way.

One hour and twenty minutes later, after Eagle had flown around the moon and behind it again, Neil and Buzz were given permission to begin the braking burn for the descent orbit initiation. An hour later, the final ignition for the twelve-minute powered descent to the moon began. *LIFE* magazine reporter and author Norman Mailer called this "the climax of the greatest week since Christ was born."

This was it. Permission to land had to be relayed from Houston through Michael in the ship they just left, because they momentarily lost a direct connection to the earth. Here is what they heard on the coms.

COLLINS: [Relaying] Eagle, this is Columbia. They just gave you a Go for powered descent.

Before that, a flurry of activity had taken place, unheard by Buzz and Neil. "Flight Director Gene Kranz had asked members of his White Team of flight controllers [at Mission Control] for a go/no go for powered descent." These White Team men were on shift at critical moments during the eight-day mission, most notably the descent and landing. They trained as strenuously for the mission as any astronaut and then celebrated or commiserated at their favorite haunt, The Singing Wheel, after their shifts. Kranz later reflected that, "It was, for all of us, a place of refuge where we could celebrate on the good days—and lick our wounds on the bad ones." This, by the way, is what the church can and should be when it is "a table held up by a cross."

The go/no go sounded like this:

RETRO: Go.

FIDO: Go.

GUIDANCE: Go!

CONTROL: Go.

TELMU: Go.

GNC: Go.

EECOM: Go.

SURGEON: Go.

You can listen to the recording still today. Those voices coming crackly over the intercom send chills down my spine. They are iconic and unmistakable. Especially twenty-six-year-old guidance controller Steve Bales, who yelled, "*Go!*" louder and with more excitement than anyone else. In the audio you can almost hear the usually unflappable and almost always vest-clad Gene giggle as he moves on to "Control."

Listening to it stirs something inside. Knowing the timer is counting down. That the whole point of this decade-long, billion-dollar enterprise is imminent.

"We're *go* for powered descent." The most difficult and dangerous part of the mission—and the only aspect that hadn't been able to be tested by the other Apollo missions that had built up to this moment.

Each question represents a crossroads.

Go? No go? Is it safe to continue to the next step?

MISSION ABORT?

His legs stirred again. From his slumped-down position, hanging on the weight of his hands, he began to shift. He grimaced as his back once again slid up the cross. His parched lips opened, and out of his dry throat croaked not so much words as much as a moan of agony. It wasn't so much the physical pain. That had been constant for hours. This was different. The sky turned black. It was high noon. The man on the cross had three hours to live.

My God.

There was still time for the astronauts to back out.

Some sort of launch escape system has been a vital feature of every rocket launch going back to the Mercury days. In Gemini, the astronauts sat on ejection seats and wore parachutes in the event of a rocket malfunction. The very top of the Saturn V is an escape tower composed of three solid fuel rockets weighing nine thousand pounds. When deployed they would generate 200,000 pounds of thrust—more than twice that of Alan Shepard's entire Redstone rocket that carried his Freedom 7 capsule in 1961.

Should it be deemed that the astronauts needed to quickly get away, all that Neil had to do was rotate thirty degrees the abort handle that sat to the left of his left knee, and they would be yanked away from the presumably malfunctioning Saturn V.

It is a test of nerves to be strapped to the top of a bomb so powerful it could blow a crater the size of a town in the earth and know that you have access to an abort handle that can make it all stop. (Or it could have killed them. The escape tower was never tested with humans on it. If they ever had to use it, the estimates were that they would sustain twenty times the force of gravity—which would be much better than staying connected to an exploding rocket.) This is of course only an option for the three minutes and seventeen seconds from launch until they reach orbit sixty miles above the earth—which is when the entire escape system is jettisoned.

Once the lunar module landed on the moon, Neil and Buzz had the option to instantly blast off again, firing the rockets that would send them hurtling back toward Columbia. Consider they had made it through the program alarms, the critically low fuel, the craters and boulders, and finally landed. They shut down the engine and immediately did another roll call. Now the question was stay, no stay. They had to determine if it was safe to remain for the space walk. If any of them determined it wasn't, Neil and Buzz would have been forced to abort. But they needed to decide quickly because it would take two hours for Michael to come back around. They had less than two minutes after landing to decide if they could stay. If they deemed it unsafe, they would be forced to blast back off to Columbia. Without any eating of the Last Supper, picking up of lunar rocks, planting of a flag, or walking on the moon.

Space exploration is as dangerous as it is expensive. Thirty lives have been lost in the history of space travel. Over five hundred and fifty have been to space—and more every day: *Hello Branson, hello Bezos, hello cow jumping over the moon*. So approximately one out of twenty has died. I imagine that during countdown the abort option would come into your mind a time or two.

Jesus had such an option. Twelve legions of angels at the ready. All he had to do was say the word and it would all go away. You can be sure that the devil was playing the same mind games he had in the temptation that preceded Jesus' ministry as Christ reached the halfway point of the crucifixion and spoke for the fourth time.

LAST WORDS #4

"My God, My God, why have You forsaken Me?"

Matthew 27:46

This moment of Jesus' time on the cross is also the crescendo. The climax. It's by far the most dramatic, forceful, and intense part of his death.

If you think of his seven last phrases on the cross in a picture form, it would look like a mountain—three up and three down, with this being the highest point.

There is also a mystery to it. A hopelessness and a terrible awfulness to it. Charles Spurgeon said they were by far the saddest words spoken from the cross.

Up until then Jesus had been remarkably focused—not on himself but on his enemies, his neighbor, and his mom. But after that, everything changed. Though it was high noon, the sky grew dark, and for three hours—until about 3:00 p.m.—he hung in silence.

At noon the sun should have been at its brightest. Just as Jesus, only thirty-three years of age, was being cut down in his prime.

What is the significance of the black sky? One author said he thought the sun "refused to look on such a deed of shame."

A unique star shone at his birth, lighting up the night, and now at his death a strange darkness covered the land, causing day to become black (Matt. 27:45). One of the terms in the Bible for *hell* is "outer darkness" (Matt. 8:12, 22:13, 25:30). That is what Jesus was experiencing. The outer darkness of the wrath of God, which was fitting because the reason he came was to turn *off* the dark. And the only way he could do that was by his light being snuffed out.

Theologians have called it the crucifixion within the crucifixion. It is believed that during this time the sins of the world were being placed on him. Though he was sinless and spotless, God put hands not on a goat, not on us, but on his Son, and treated Jesus as though he had committed the sins of the world.

The Old Testament prophets predicted this day would come. Daniel 9:26 tells us, "Messiah shall be cut off, but not for Himself," and Psalm 116:3 says, "The pains of death surrounded me, and the pangs of Sheol laid hold of me; I found trouble and sorrow."

Thousands of years before Christ came, the prophet Isaiah spoke of a moment when God would transfer our sin to the Messiah's shoulders. We find this in 53:4–6.

> Surely He has borne our griefs
> And carried our sorrows;
> Yet we esteemed Him stricken,
> Smitten by God, and afflicted.
> But He was wounded for our transgressions,
> He was bruised for our iniquities;
> The chastisement for our peace was upon Him,
> And by His stripes we are healed.
> All we like sheep have gone astray;
> We have turned, every one, to his own way;
> And the Lord has laid on Him the iniquity of us all.

This moment was what Jesus most dreaded.

Adam's apple, Cain's murder, Noah's drunkenness, Lot's incest, Abraham's doubting, David's adultery and murder, my pride and narcissism and lust. Every wrong thing you have done. All. Of. It.

The New Testament apostles confirmed this is exactly what happened. Galatians 3:13 says, "Christ has redeemed us from the curse of the law, having become a curse for us (for it is written, 'Cursed is everyone who hangs on a tree')." Then 1 Peter 2:24 refers to Jesus, "Who Himself bore our sins in His own body on the tree, that we, having died to sins, might live for righteousness—by whose stripes you were healed."

Would he stay or would he go?

That which he feared was coming upon him. It was all about to be placed on his spiritual account. But it's not just that he paid for it. Jesus didn't just die for you; he died as though he *were* you. Paul put it this way in 2 Corinthians 5:21: "He made Him who knew no sin to be sin for us, that we might become the righteousness of God in Him." It's been called the Great Exchange, an idea that was significantly advanced by the works of Martin Luther. In this moment Jesus was allowed to feel like and be treated as though he had personally committed every sin he was paying for.

This is why blood ran out of his sweat glands as he prayed (Luke 22:44), and he asked whether there was any other way (22:42). This is what made his chest tighten and caused him to fall headlong into the ground (Mark 14:35) and plea with his friends to wake up and pray with him (Matt. 26:41; Mark 14:38; Luke 22:46). Just as it is impossible to comprehend the vastness of space and the number of stars, so it is staggeringly, absurdly beyond understanding what distinct horrors he would face when this finally happened. While praying the night before, dreading *this* moment, he had said the words quoted in John 12:27: "Now My soul is troubled, and what shall I say? 'Father, save Me from this hour'? But for this purpose I came to this hour."

JESUS DIDN'T JUST DIE FOR YOU; HE DIED AS THOUGH HE *WERE* YOU.

FREE BIRD

In the Old Testament, two birds would be brought as a sacrifice for sin. One would be killed and one would be spared. The blood from the one that died would be placed on the wings of the other, and it would be allowed to fly away. It was spared because it was covered in the blood of the one that died (Lev. 14:1–7). Jesus is the bird that would allow us to fly free.

It is appropriate that Jesus died on a cross that should have been filled by another—Barabbas the murderer. As Barabbas lived, Jesus died in his place. Barabbas is not the only one. You and I and all those God so loves deserve the wages of sin, which is death (Rom. 6:23). Only because Jesus was willing to trade places can we go free. And when I say *free*, I don't just mean a clean pass from the sins we committed with a scolding look from God that says, *I'll let these go, but don't blow it again!* It's much better than that. Once Jesus was willing to have God see him as sin (for us), God began seeing us as clothed in his perfection and righteousness.

When the moment finally came, Jesus proceeded according to the plan. He refused to abort. He stayed on the cross and the sin of the world was transferred onto him. This was when he cried out in pain and despair: "My God, My God, why have You forsaken Me?"

Allow the sickening feeling of regret and despair and utter dread to sink in as you imagine Jesus completely abandoned and alone, experiencing the unbridled wrath poured out for sins he didn't commit. Two thousand years before this happened, near this exact spot, God tested Abraham by asking him to offer up his son Isaac. When Abraham was willing, God stopped him and in essence said, "Don't harm your son; I will provide myself for a sacrifice" (Gen. 22). On this day, that promise was kept.

God so loved the world that he gave his only Son . . . and they were separated.

It's not right.

Never in all of infinity had the dance of the Trinity—the love and beauty of community, Father, Son, and Holy Spirit—been broken. But in this moment Jesus was forsaken. Rejected. Abandoned. So you will never be. Jesus didn't appease an angry Father by dying for us. Father, Son, and Holy Spirit conspired to do something that would bring sadness to them all because it was the only way for you and I to be saved.

This is love.

FORSAKEN BUT NOT FORGOTTEN

Rejection was nothing new for Jesus. He had been rejected by the innkeeper at Bethlehem and given no room (Luke 2:7). His siblings rejected him (John 7:5). Nazareth, his own hometown, wanted nothing to do with him (Matt. 13:54–58). Israel, the nation he came to save, by and large didn't receive him. The Jewish leaders who should have supported him rejected him (John 1:11). He was abandoned by his disciples who slept when they should have prayed and then ran when he was arrested (Matt. 26:36–46). He was betrayed by Judas and denied by Peter, one of his three closest friends on earth (Luke 22).

But this was worse than anything that had come before it. He was separated from his Father as far as the east is from the west. Horrendous desolation.

That is why he cried out, "My God, My God, why have You forsaken Me?"

John Stott wrote, "Our sins blotted out the sunshine of his Father's face." And Erwin Lutzer added, "Look at these hours on the cross and you are looking into hell: darkness, loneliness, and abandonment by God."

It was literally hell on earth.

He died so that we might live.

And the most amazing thing is he paid this price with joy.

Hebrews 12:2 tells us, "For the joy that was set before Him [He] endured the cross, despising the shame."

What joy? The joy of what it accomplished—giving life to us! Author James Stalker said, "This, which was the extreme moment of suffering, was also the supreme moment of achievement. As the flower, by being crushed, yields up its fragrant essence, so He, by taking into His heart the sin of the world, brought salvation to the world."

But that's not all.

If we see this merely as Jesus suffering in life so we would not suffer after death, we miss much of what the cross means. He agonized in torment so we would view the sufferings of this life differently. He experienced:

- darkness so we could have light,
- distance so we can draw near,
- separation so nothing could separate us,
- shame so we could have honor, and
- instability so we could have security.

And because he was forsaken, you can make sense of the times you have been forsaken. The actual Greek word for *forsaken* is the same for *abandoned*. You can relate to that, can't you?

Maybe you were abandoned by a mom or dad who chose a different family or an addiction instead of you. Perhaps your spouse traded you in for a younger model. Perhaps your skin color has left you marginalized by people who view *white* as a synonym for *normal*. A friend stabbed you in the back. A company rewarded your loyalty with a layoff.

Ironically, today, while I was working on this chapter, we experienced a transition on our team with a few staff members moving on to other opportunities. It is a normal thing that happens in life, and, in the years I have led a church, it has happened many times. But it never gets easier, and I take it personally every time. *My God, my God. Why have they forsaken me?*

There is room in Jesus' wounds for all these pains. In his arms there are ten thousand charms.

Jesus can relate. And he can help you to be something much better than bitter: he can make you an overcomer.

Ask Paul. He was forsaken by all at his great hour of need, at the end of his life while on death row. Listen to the emotion when he wrote to Timothy, his son in the faith: "For Demas has forsaken me, having loved this present world, and has departed for Thessalonica—Crescens for Galatia, Titus for Dalmatia. Only Luke is with me. Get Mark and bring him with you, for he is useful to me for ministry" (2 Tim. 4:10–11).

Paul could have grown a serious chip on his shoulder and wound in his spirit. Instead we see him at the end of his life, not a grumpy old man but someone who is extending grace. I can prove it to you in two words: *Get Mark*.

John Mark had burned Paul big time, abandoning him at a critical time. This guy was basically Darth Vader to Obi-Wan Kenobi. Initially there was outrage—Paul wouldn't let Mark come with him after he dropped the ball. Paul even divided with Barnabas over it (Acts 15). But in his old age Paul had experienced healing and softened to the point where he was not dominated by the fact that he was abandoned.

This kind of healing doesn't happen in a day, but it will happen daily if you let it. As Paul got older, he got kinder. Age doesn't change you; it makes you more of what you already are. I have resolved to not thicken my skin to the point where it doesn't hurt when people leave me. The pain is proof that I loved.

In Jesus' willingness to be forsaken, you can find the secret to healing the wounds you carry from the way people have let you down. His pain can become your power. You can bring to him every injury and all the trauma and damage you have inside you and entrust it to his nail-pierced hands.

Once this process of healing begins to take the infection out of your hurts, he can open your eyes to brand-new ways of looking at what you have been through. It's called perspective. Jesus had to be crucified. It was the only way for us to be forgiven. We can now look back on it and see what it was impossible to see in the moment: that God had a plan.

After the crucifixion, it seemed to the disciples and Jesus' followers and mother that he had been taken away from them. But that's not entirely accurate. This wasn't going to end in anything taken away, but in so much more being added. God was making space. He often adds by subtracting. Persecution is a precursor to multiplication. His math doesn't really make sense on paper, but it's marvelous to watch it happen.

Consider this when someone leaves your life and you feel the sting—God is making space for someone new to come in. When he doesn't answer in silver, he answers in gold; and when he doesn't answer in gold, he answers in diamonds. If he doesn't give you what you asked for, it's because he intends to give you what you don't have the faith yet to imagine.

There is no way Saturn V could have made it to the moon if it didn't shed spent stages along the way. The first and second stages were there to get it off the ground and then into orbit, and after that they had served their purpose. If you've lost a relationship, be thankful for the season you enjoyed it, and see that person's absence from your life as perhaps a necessary ending so you can go somewhere new. Easier written than lived, of course.

GIVING: THE KEY TO RECEIVING

You should know about another angle in all of this. When you know Jesus was forsaken, it should also make you seek out those who have been forsaken. Leaving the ninety-nine, go and find the one lost sheep (Matt. 18:12).

Jesus was forsaken not just so you wouldn't be but so you would be his agent of rescue to those who feel alienated and forsaken. Wholeness is not just for you; it is also meant to flow through you. Grace that

WHOLENESS IS NOT JUST FOR YOU; IT IS ALSO MEANT TO FLOW THROUGH YOU.

is kept to yourself poisons you. And yet how quickly we become selfish and self-righteous and hoard the healing that was our only chance of getting out of captivity.

God loves you too much to let you live selfishly. His Spirit yearns for you to become free from the self-inflicted captivity of narcissism. He wants you to be a conduit for salvation. To be an ambassador of mercy. And in losing your life, you will find it (Matt. 10:39). I think the reason so many of us refuse to take opportunities to bless other people is because we are afraid—afraid that blessing for another automatically means less blessing for us. But that is a scarcity mindset that comes from applying zero-sum, earthly thinking to infinite resources. In God's economy you never have less of a thing by giving; you always have more.

It's all over the Bible.

- Only when the widow gave Elijah breakfast did she receive provision to last a famine. (1 Kings 17:10–13)
- It was when the little boy gave his five loaves and two fishes that the crowd (including him) received an all-you-can-eat buffet. (John 6:1–14)
- The world of the generous gets larger and larger; the world of the stingy gets smaller and smaller. (Prov. 11:24)
- "Remember this—a farmer who plants only a few seeds will get a small crop. But the one who plants generously will get a generous crop." (2 Cor. 9:6 NLT)

On the night Jesus held bread and wine in his hands, he gave one of the most powerful teachings on generosity anywhere in history when he was willing to do what none of his disciples dared—he washed their feet (John 13:1–17).

The disciples were all unwilling to serve each other at the original Last Supper because of the fear that if they humbled themselves and served each other they would be less. In their minds they were only important if they were more important than each other. Jesus taught them all a lesson they would never forget when he, their teacher, humbled himself and gave up the perks of being the alpha and took the place of a servant.

John's gospel tells us how he was able to be "humbled" in this moment, deprived of clout and everything we all think will make us happy. Knowing the Father had given him all things, and that he was going to the Father, Jesus laid aside his mantle and took up a basin (v. 3).

The key is that he already knew who he was, what he had been given, and where he was going. His identity wasn't up for grabs.

He had conquered his inner space.

The disciples thought, like we do, that they needed to hoard esteem and position and prominence and guard what was theirs because if anyone else's pie slice got bigger, it would make their slice get smaller.

Jesus was okay laying his life down on the cross, completely forsaken, because he knew that in so doing he would be able to take back his life from the grave and with it ours as well. He knew what you need to know: giving, not keeping, is the key to receiving.

He had the power to lay down his life and the power to take it up again. And you do too! Follow Jesus' example. Hold on to everything with a light touch. Don't cling to your image or to position or power. When, like Jesus, you know who you are and what God has promised, you don't need to be territorial, petty, or defensive. You can have a relaxed confidence and a servant's heart. As Jacob discovered, you don't need to be a foot-catcher when you know the foot-washer!

Allow Jesus' willingness to fight for you inspire you to fight for the forsaken in your world.

REMEMBER THE CROSS

There are three important takeaways from this halfway point of the crucifixion.

First, Scripture in Jesus' heart gave him traction in his grief. It is telling to me that in his great trial, his darkest hour, as he drank from the cup of suffering, his thoughts and words were guided by God's Word. Jesus' fourth statement from the cross was actually a quotation from Psalm 22:1. This is significant because this psalm contains ancient prophecies that described how Jesus would die—one thousand years before Christ and hundreds of years before crucifixion was invented. But it is also a clue pointing to what we can do when our lives are on fire. Just as Jesus did, when in pain, we can turn to God's Word to comfort us and guard our hearts.

What if you get so much of God's Word inside of you that when you are hurting, it just comes spilling out of your mouth? What if your go-to when life is hard isn't a vodka tonic or Facebook but to quote Psalm 23 to anchor yourself in hope?

All Scripture is good, but to me there is something particularly important about the Psalms. Jesus quoted the Psalms more often than any other book in the Old Testament, which tells me that praise is powerful.

Now is the time to hide God's Word in your heart.

Today.

Jesus didn't have to open a Bible on the cross—the words were hidden in his memory, there and ready to be accessed. He had made it his custom to gather with God's people, where he'd heard Scriptures read and songs sung since he was a child. If you are a parent, do not make the mistake of being an infrequent churchgoer. Take your kids. Plant them in the house of the Lord. Raise them up in the way they should go, and you can believe they will flourish in the courts of God, even in pain.

Second, Jesus overcame despair by rushing to God in it. Notice the possessive pronoun Jesus uses: "*My God, My God*, why have You forsaken Me?" And as a human in the greatest pain anyone ever faced, he asked *why*. Not just for himself, but for us. As our representative he asked the number one question humanity wants to know in the face of suffering: "*Why? Why? Why?*" We go through things we don't and can't understand. Jesus can relate. His situation was bad beyond comprehension, yet he refused to let go of his grip on God not one time but two times, saying, "My God, My God."

Jesus held on doubly strong to his Father in the fiery trial he was going through. He rushed to the arms of God in his pain, and, in so doing, he triumphed over it. So can you! When you are hurting and you don't understand, don't let go of God; hold on with both hands. Don't trade what you don't know for what you do.

Third, a word of warning that we must not reject the only means of salvation that has been offered. Jesus isn't the best way to God; he is the only way. First John 5:12 tells us, "He who has the Son has life; he who does not have the Son of God does not have life."

Salvation is a gift. But a gift must be accepted; if it's not, by default, it is rejected. So what is this ultimate gift? "As many as received Him, to them He gave the right to become children of God, to those who believe in His name" (John 1:12).

When the darkness hit and the sin was laid on him, he cried out in pain. If ever there was going to be a moment to abort, this would have been it. But he proceeded with the mission. If you keep this in your heart, it will give you strength not to abandon your post.

I read in the news about a woman in Casper, Wyoming, who received a bouquet of flowers for Valentine's Day. When she read the card, they were from her husband—which wouldn't be anything out of the ordinary, except that he died of brain cancer the year prior. She initially thought it was a mistake, or that maybe her children had sent them. But after contacting the flower shop she learned that before his death her husband had arranged for flowers to be sent to her every Valentine's Day until she, too, dies.

Commence ugly cry.

This is the gospel. Out of Jesus' death comes enough love for the rest of your life. Anytime you are tempted to doubt him, just remember the cross!

CONQUER YOUR INNER SPACE

Before you move on to the next chapter, pause for a moment and read these verses out loud. There is power in declaring Scripture over your own life. As you hear yourself speaking these words, it will change the atmosphere around you and build your faith.

"Are not two sparrows sold for a copper coin? And not one of them falls to the ground apart from your Father's will. But the very hairs of your head are all numbered. Do not fear therefore; you are of more value than many sparrows" (Matt. 10:29–31).

"You, O Lord, are a shield for me, My glory and the One who lifts up my head" (Ps. 3:3).

"Humble yourselves under the mighty hand of God, that He may exalt you in due time, casting all your care upon Him, for He cares for you" (1 Peter 5:6–7).

CHAPTER 14

HAKUNA MATATA

MY SON, LENNOX, WAS HAVING ONE OF THOSE MORNINGS. He was protesting going to snowboarding lessons, which is rare because he usually is amped about it. Jennie reminded him, "But Lennox, you love snowboarding."

"No, I hate it," he insisted.

"But you like your instructor, Michael, don't you, buddy?" his mom coaxed.

He thought about this for a moment and then said, "No. He doesn't even like *The Lion King*."

This needed no further explanation. In his world, to speak against *The Lion King* is a sin of the highest order.

There are interesting connections between Disney and NASA. Among them is the fact that when Space Mountain first opened at Disneyland in 1975, six of the original Mercury 7 astronauts (Gus Grissom had died in the Apollo 1 fire by that point) were present for the occasion. When that same ride reopened after being refurbished for the 50th anniversary of the park in 2005, Neil Armstrong attended the ribbon cutting with a space suit–clad Mickey Mouse by his side. The ride itself was inspired by Jules Vernes's 1865 book, *From the Earth to the Moon*, in which a cannon fires a bullet full of passengers to the lunar surface. Also, there is a ride at Disney World called Mission: Space, which simulates a voyage to Mars and was built with help from NASA advisers.

The Lion King was the first movie in Disney history to *not* be based on an existing story. All the others that came before it had familiarity working for them. Snow White, Cinderella, Sleeping Beauty—these were all tales as old as time, and while Walt and the gang breathed new life into them, they were not original.

It was risky to tackle the dual challenge of excellent execution and original story creation. And many at Disney doubted it could be done. Adam Grant, a prolific author, organizational psychologist, and podcaster, writes that studio head Jeffrey Katzenberg was skeptical and that *The Lion King*'s project director, Rob Minkoff, said, "No one had any confidence in it. . . . It was seen as the B movie at Disney." Katzenberg even said that it would blow him out of the water if the movie brought in as much as $50 million. It ended up as "the highest grossing film of 1994, winning two Oscars and a Golden Globe," and it had earned more than $1 billion within twenty years.

What was the secret to successfully pulling off a screenplay with the working concept of "*Bambi* in Africa" with lions?

Disney's first original movie wasn't *completely* original. They made an unfamiliar story work by overlaying it with a familiar one. They used lions as window dressing to bring to life the massively well-known plot of Shakespeare's *Hamlet*, where the jealous uncle kills his brother, and the child and true heir to the throne must face his demons and avenge his dad's death in order to fulfill his destiny. In one scene in the film, Rafiki, the wild baboon ninja sage who serves as the mentor, nods to *Hamlet* when he gives a "To be or not to be"–inspired speech.

It is the unlikely combination of lions on the savanna and an eerie channeling of the Bard's famous play structure that makes the movie so compelling.

Why do the themes of *Hamlet* and *The Lion King* resonate with us so deeply? I believe, in part, because they are true. There is a King who really did have to die in order for us, his heirs, to become who we were meant to be. And when we run from our pain and our past, we can't be positioned to face the future at full throttle. Everything the light touches is our rightful kingdom inheritance. And anything or anywhere that our Father tells us not to do or go, it's because he is motivated by our best interest. Also, he never puts rules in place except to prosper us. When he tells you not to sin it is because he doesn't want you to suffer.

This is one reason Kennedy's moonshot speech—the one about landing man on the moon and bringing him safely home by the end of the decade—resonated then, and still does today. You have to realize how gutsy it was to not just commit to the moon, but to put a date on it. It was three weeks after Alan Shepard's first flight on the first Mercury mission. He only flew fifteen minutes and only went 116 miles off the ground into space. Twenty days later President Kennedy didn't flinch as he told assembled senators, members of Congress, and the American people that we were going to send a man a quarter of a million miles to land on the moon and then get him back home again. And do this by the end of the decade. Like Disney making *The Lion King*, there was no book to base this story on. NASA was boldly going where no one had gone.

Then President Kennedy was assassinated. His death loomed large. It was so traumatizing because it was so public, so horrific, so shocking, and so senseless. He was young, beautiful, and loved. There were many consequences in its aftermath. The Camelot era had come to an end. As a result, the space program took on a whole new urgency. To many of those at NASA, and to many of the American people who funded and supported it, finishing his quest seemed to be a way to honor his legacy and not let his death be in vain.

Because the assassination was such dominating front-page news, most other things that happened on that day were forgotten. One notable example is the death of one of the greatest minds the earth has ever known: C. S. Lewis. He died on the same day as the president, one hour before JFK was shot in Dallas, after losing his battle to kidney failure.

Lewis taught at Oxford in the 1920s, and he had been an atheist for years. Intellectually, he looked at the world certain that life was pointless and all would burn up. Still he felt things that intellectually he did not believe. He felt a desire for truth, beauty, meaning, and salvation. He experienced an unexplainable longing inside, especially when reading myths and fairy tales. He called it Joy with a capital *J*. These sensations had a profound reaction on his unbelief. "I felt as if I were a man of snow at long last beginning to melt. The melting was starting in my back—drip-drip and presently trickle-trickle. I rather disliked the feeling."

One night in September of 1931 Lewis took a long walk with his friend J. R. R. Tolkien, a Christian. Tolkien and Lewis were talking about the way Lewis was moved by these myths, and Tolkien helped him understand that his reaction to what at the time Lewis called "lies breathed through silver" actually pointed to the true "myth" of the gospel.

When you are moved by a story like *Sleeping Beauty*, you know there really is an evil sorcerer who has us under his spell. When you read *Peter Pan* you know we are not meant to grow old, or just to walk; we are meant to be childlike and to fly. When you read *Beauty and the*

Beast you know love can save you from the stupidity and wrong choices of your life, from the bondage of your sins.

There really is good and evil, light and dark. There really are evil spells and there really is a handsome prince who, if he kisses you, can break the spell. These stories are "echoes of Eden" that point to Jesus Christ: his birth, life, death, and resurrection. You don't have one additional story pointing to these things; you have the reality that all these stories have been pointing to.

This conversation went on until 3:00 a.m. and was a fulcrum in Lewis's faith journey. Not long afterward, he wrote his friend Arthur Greeves and said, "I have just passed on from believing in God to definitely believing in Christ. . . . My long night talk with . . . Tolkien had a good deal to do with it."

Lewis believed that myths were not just useful in understanding the past but also in understanding the future. "If we believe that God will one day *give* us the Morning Star and cause us to *put on* the splendour of the sun, then we may surmise that both the ancient myths and the modern poetry, so false as history, may be very near the truth as prophecy."

THE CHURCH IN THYATIRA

As you read the words Jesus wrote to the church in Thyatira, see how they are all couched in themes that speak to the greater story. Revelation 2:18–29 says:

> "To the angel of the church in Thyatira write:
>
> These are the words of the Son of God, whose eyes are like blazing fire and whose feet are like burnished bronze. I know your deeds, your love and faith, your service and perseverance, and that you are now doing more than you did at first.

> Nevertheless, I have this against you: You tolerate that woman Jezebel, who calls herself a prophet. By her teaching she misleads my servants into sexual immorality and the eating of food sacrificed to idols. I have given her time to repent of her immorality, but she is unwilling. So I will cast her on a bed of suffering, and I will make those who commit adultery with her suffer intensely, unless they repent of her ways. I will strike her children dead. Then all the churches will know that I am he who searches hearts and minds, and I will repay each of you according to your deeds.
>
> Now I say to the rest of you in Thyatira, to you who do not hold to her teaching and have not learned Satan's so-called deep secrets, 'I will not impose any other burden on you, except to hold on to what you have until I come.'
>
> To the one who is victorious and does my will to the end, I will give authority over the nations—that one 'will rule them with an iron scepter and will dash them to pieces like pottery'—just as I have received authority from my Father. I will also give that one the morning star. Whoever has ears, let them hear what the Spirit says to the churches." (NIV)

With these letters, Jesus is moving counterclockwise around Asia Minor, which is modern-day Turkey. Thyatira was located forty miles southeast of Pergamum in a flat valley; it existed to guard the road that led to the much larger city of Pergamum. Thyatira was essentially like a town propped up because of an army base.

Thyatira was a small city—the smallest of the seven Jesus wrote to, yet it received the longest letter. It goes to show that no one is too small for God to care about. He hears your prayers, cares about your life and your problems, and has much to say to you.

Because it was situated on a road, it was a great place for trading and a marketplace. It was a city known for the production of its wool and dyed goods, linens, leathers, and bronze goods.

Unlike Ephesus and other cities reached during the missionary chapters of Acts, little is said of how a church was started in Thyatira. In fact, the only mention of it is in reference to a lady named Lydia who lived there but also did business in Europe and allowed her home to be a base for the ministry in Philippi. Acts 16:13–15 tells us about her:

> On the Sabbath we went a little way outside the city to a riverbank, where we thought people would be meeting for prayer, and we sat down to speak with some women who had gathered there. One of them was Lydia from Thyatira, a merchant of expensive purple cloth, who worshiped God. As she listened to us, the Lord opened her heart, and she accepted what Paul was saying. She and her household were baptized, and she asked us to be her guests. "If you agree that I am a true believer in the Lord," she said, "come and stay at my home." And she urged us until we agreed. (NLT)

Maybe, after she heard about Christ, she wanted to return home and reach out to her family and friends. Whatever the case, the church got off to a great start.

JEZEBEL'S SECRET

But by the time Jesus wrote to them, Thyatira was messed up. They were making huge mistakes sexually. That is what is meant by tolerating Jezebel. He by no means winks at any of this and directly calls them out on it. He knows that sex and romance are not peripheral aspects of your life; they are connected to your heart and will affect every aspect of it. I wrote an entire book on the life-and-death power of sex and romance called *Swipe Right*. It will point you to the urgency of walking in God's plan in this area. Straying from God's highest good will have an impact not just on you but your children after you.

In spite of Jesus' directness, notice how tender he is with them. He is intense, but it is all motivated by his desire to reward them. To give them authority. To entrust them with the morning star. Jesus is fighting for their happiness—just like he is fighting for yours.

LETTER #4 (THYATIRA)

"Hold fast what you have till I come."

Revelation 2:25

It is very interesting to me that Jesus is harder on legalism than he is on sleeping around. This is not the only time this unique emphasis is found in Scripture. Paul was ten times angrier in Galatians than he was in Corinthians.

Allow me to say something that might be shocking: God would rather have you struggle with porn than with pride.

Now, to be clear, he doesn't want you to struggle with either. Remember: salvation isn't based on you; it is placed on you. In other words, God cleans his fish after he catches them, not before—he knows there will be weakness in your life, and he will lovingly not stop convicting you to make you more and more like Jesus. He accepts you as you are but loves you too much to leave you that way. The moment you think you have "arrived" you no longer are able to receive his help. It is far better to be weak and know it, to fight and want to grow, than to be so advanced and

IT IS FAR BETTER TO BE WEAK AND KNOW IT, TO FIGHT AND WANT TO GROW, THAN TO BE SO ADVANCED AND EVOLVED THAT YOU THINK YOU ARE PERFECT.

evolved that you think you are perfect. As long as you remain humble and small in your own eyes, you can continue to become the masterpiece you are destined to be.

We landed on the moon with time to spare. Kennedy pledged it would happen by the end of the 1960s. Apollo 11 landed on July 20, 1969. Five months to the deadline. But it wasn't until four days later, when the three astronauts touched down in the Pacific Ocean, that the job was completed. Landing on the moon was only half the challenge. They had to return safely home. When that happened, the mission was truly accomplished.

Starting is not nearly as significant as finishing.

About the same time the spaceship splashed down in the Pacific, "an unknown citizen had left a lovely bouquet of flowers on Kennedy's Arlington grave with a thoughtful card that read simply, 'Mr. President, the Eagle has landed.'"

To the degree that you overcome and possess your possessions, laying hold of all that is yours in Christ (Eph. 1:3–14), you fulfill the dream the one who died for you had in his heart as he hung on the cross.

CONQUER YOUR INNER SPACE

Many at NASA, as well as the American people, found a whole new sense of urgency, audacity, and intensity in fulfilling President Kennedy's end-of-the-decade deadline. As Gene Kranz put it in his book *Failure Is Not an Option*, "Kennedy's words were to echo across the decades, and we, along with the rest of the country, found out that a noble cause brings out a nation's best qualities."

As followers of Jesus, we should take seriously, with urgency, his dying wish (John 17:20–23) that we be united, that we be holy, and that we show his love to the world. "The night is coming when no one can work" (John 9:4). You can't share your faith in heaven. There isn't a moment to lose.

As he looks at your life, what does he see that is holding you back and needs to be dealt with?

He is the real Lion King, and he wants you to rule in his kingdom (Rev. 2:26–28).

Not sure where that is?

Hint: it's everything the light touches.

CHAPTER 15

GOOD LUCK AND GODSPEED

APOLLO 11 FLEW UP FROM LAUNCHPAD 39A AT CAPE KENNEDY, Florida, at 9:32 a.m. EDT on July 16, 1969, and a decade of humanity's hope disappeared into the blinding sky. Curtains of ice slid off the Saturn V's smooth surface, evaporating into furious, swirling clouds of steam, as a cauldron of fire erupted all around. The huge spaceship groaned, heaved, and—as though under great protest—reluctantly tore loose from its moorings and broke free from the bonds of earth.

As this enormous bull bucked out of the chute, Neil, Buzz, and Michael had more things in common than their shared fate in space: the same year of their birth (1930), the same weight (165 pounds), and nearly the same height (about 5'11"). The last words these three star sailors heard from launch control before the mighty roar exploded beneath them were, "Good luck and Godspeed."

This phrase had become a NASA staple ever since John Glenn's Mercury mission where he became the first American to orbit the earth in his Friendship 7 capsule. At the time of his launch, his friend and backup astronaut Scott Carpenter had quipped, "Godspeed, John Glenn." And it stuck. For his part John indeed experienced God on his flight. From space he said, "To look out at this kind of creation out here and not believe in God is to me impossible."

I feel the same when I read the Gospel of John.

By the end of my first year at Bible college, I realized that a lot of people were going off the deep end and getting carried away with one aspect of theology to the exclusion of others, and it was making them crazy. Whether it was the doctrine of Calvinism versus Arminianism or end-time obsessions; arrogantly mocking the rapture or ardently, hyperactively, and dogmatically defending it; or seeing every blood moon or tsunami as confirmation of impending Armageddon—it was making them cold rather than on fire for Jesus. I just didn't want to end up like that.

I decided that the most important thing I could do was be madly, deeply, head-over-heels in love with Jesus and let that be the basis of any and all of my future studies. My goal was to never major in the minors but to keep the main thing the main thing and never lose sight of it.

When I returned for the next semester after summer break, my plan was to be intentionally, strategically, and devotionally anchored in the Gospel of John. And no matter what else I was reading or researching, I would let that be my ballast. John was explicit about his mission statement for writing his gospel. John 20:31 says, "These are written that you may believe that Jesus is the Christ, the Son of God, and that believing you may have life in His name." I also resonated with Paul, who said in Philippians 3:10, "I want to know Christ—yes, to know the power of his resurrection and participation in his sufferings, becoming like him in his death" (NIV).

It was one of the best decisions I ever made.

Month after month, I would work my way through the gospel every twenty-one days—one for each chapter. Each time I passed through I noticed things I hadn't noticed before. I got to know the cast of characters: Nicodemus in John 3, the woman at the well in John 4, the man healed at the pool of Bethesda in chapter 5, the man born blind in John 9. These people became my friends. I identified with them in what they each faced and appreciated what they told me about what Jesus could do for people like us.

The experience shaped me like a trellis that helps a vine grow not in an unruly way, but filled out properly and beautifully. And the power of God's grace shaped my inner space.

By the end of the semester, I became familiar with the broad strokes and basic plot twists of John and began noticing much smaller details and nuances. Far from growing tired of it or finding it stale, with each revolution I found it more fascinating and life-giving. I would spice it up with Psalms or Proverbs, and of course all the other classes I was taking had me reading widely, but this was my base camp. I felt at home among the motley crew of characters who were different in so many ways but had all had their lives changed by Jesus. Just like me.

MISSION: SAVE THE WORLD

My focused study was much like the simulator that astronauts spend thousands of hours in so that when they are at the controls of the actual spaceship, none of it feels foreign. Here is one author's description of the simulator:

> As test and fighter pilots, the astronauts had flown cutting-edge machines, but even they needed time to process the sight of the Apollo simulator. Standing about twenty feet high, it was a

> hodgepodge of sharp-cornered modules that appeared jammed together by cubist painters, jazz musicians, and mad scientists. There seemed no front or back, or even up or down, just shapes. Hundreds of cables dangled from the contraption like dreadlocks, while two narrow staircases—one circular, the other straight—led inside, or at least somewhere. Bracketing the structure were consoles of computers, instruments, and monitors for the instructors. Fluorescent white light bathed the room.

According to Michael Collins, "The first time [astronaut] John Young saw the CSM simulator, he dubbed it the Great Train Wreck." An elegant affair it might not have been, but it did the trick. Michael said his time in the CSM simulator did its job in that it prepared him for the mission. "My greatest source of reassurance came from the simulator. In the six months between January and July, I had accumulated four hundred hours in it, and it now felt like a comfortable home rather than a horror chamber."

Hundreds of hours, mornings, and evenings spent with Jesus in John's gospel have become that for me in the decades that have followed my time at Bible college: a comfortable home rather than a horror chamber. It shaped and prepared me. And no matter how many times I read it, every single time, I feel sorrow and angst as the message of the cross draws near. When the Last Supper and Upper Room Discourse begins in John 13, I can feel my pulse quicken, knowing what's coming. And without fail, the events of Skull Hill—from the trial to the torture and ultimately the crucifixion before the joy of the resurrection—rock me to my core.

I have watched the footage of the Saturn V rocket taking off hundreds of times. Buzz said he felt buoyant and got goose pimples when he stepped down on the surface.

I can only imagine. Just listening to the audio recording of the communication between Neil, Buzz, Michael, and CapCom Charlie Duke right after the landing never ceases to stop me in my tracks and make my heart thunder like a war drum.

EAGLE: Yes. Okay. Contact Light. Okay, engine stop. . . . Descent engine command override off . . .

HOUSTON: We copy you down, Eagle.

EAGLE: Houston, Tranquility Base here. The Eagle has landed!

HOUSTON: Roger, Tranquility. We copy you on the ground. You got a bunch of guys about to turn blue. We're breathing again. Thanks a lot.

TRANQUILITY: Thank you. . . .

That may have seemed like a very long final phase. The auto targeting was taking us right into a football-field-sized crater, with a large number of big boulders and rocks for about one or two crater-diameters around it, and it required us . . . flying manually over the rock field to find a reasonably good area.

HOUSTON: Roger, we copy. It was beautiful from here, Tranquility. Over.

TRANQUILITY: We'll get to the details of what's around here, but it looks like a collection of just about every variety of shape, angularity, granularity, about every variety of rock you could find. . . .

HOUSTON: Roger. Copy. Sounds good to us, Tranquility. . . . Be advised there're lots of smiling faces in this room and all over the world. . . .

TRANQUILITY: Well, there are two of them up here. . . .

COLUMBIA: And don't forget one in the command module. [Michael chimed in]

This never gets old to me.

It is impossible not to get swept away in what Walter Cronkite called "the greatest adventure in [human] history." President Richard Nixon was so excited after the moon landing that he blurted out about the "greatest week in the history of the world since the Creation." A reporter shared the quote with Billy Graham, who corrected the statement with a reminder that there were greater times, including the first Christmas, the crucifixion, and the resurrection.

Indeed.

As astounding as it is to leave the earth, Jesus did something far more impressive—he left heaven and came to earth. He didn't have to. No one took his life from him. He volunteered. And through his coming, dying, and returning to life, he has changed the world more than anyone in history.

My family and I were looking at photos from the moon landing and Lennox noticed the watch on Buzz's wrist. "Astronauts wear watches?" he asked me.

"Yes, buddy, they do. They are all issued Omega Speedmaster watches."

He thought about this for a moment then asked, "What time is it on the moon?"

That is a complicated question. Our year is how long it takes for the earth to go around the sun—365.24 days. A lunar year, on the other hand, is only twenty-seven days, seven hours, forty-three minutes, and roughly twelve seconds—the time it takes the moon to take a trip around the earth. On earth we define night as the time when our planet is between us and the sun and day as the time when the planet is not between us and the

AS ASTOUNDING AS IT IS TO LEAVE THE EARTH, JESUS DID SOMETHING FAR MORE IMPRESSIVE—HE LEFT HEAVEN AND CAME TO EARTH.

sun. Since the earth rotates as it travels around the sun, giving us one day and night every twenty-four hours, that is what we call a day. Since the moon keeps one side always facing the earth, an equivalent "day" on the moon lasts twenty-nine days, twelve hours, and forty-four minutes with approximately fourteen days of sunshine and then fourteen days of night. So basically on the moon a day is shorter than a year.

I guess being on the moon is a bit like being a parent—the days are long, and the years are short.

The astronauts keep track of time on earth while away in two ways: First, they count forward from liftoff the total elapsed time they have been on their mission. With the launch of the space shuttle, this began to be called MET—Mission Elapsed Time—but prior to the space shuttle it was called GET—Ground Elapsed Time. The Eagle landed on the moon 102 hours, 45 minutes, and 40 seconds after leaving earth. Second, they monitor the earth's clock in Houston's time. Though they launched from the East Coast, the time on Buzz's watch when they landed was the time in Houston: 3:17 p.m. CDT, July 20, 1969. In other words, the Communion wine and bread weren't the only things they brought with them on the mission that pointed to Jesus.

Jesus' birth caused time to start over. And in a majestic and eternal way, I have found freedom in the words of the one at the center of the calendar—Jesus. He came on *the* mission of all missions. His journey was to save the world and bring us back to the Father.

This is on display in the Gospel of John. And, in it, my heart learned to sing. I begrudgingly moved on to other books of the Bible but have never lost my love for John's gospel. (Also noteworthy is the fact that in addition to housing the seven "I Am" declarations, John's gospel is also home to seven signs or miracles that accompany them.)

HE DID IT FOR YOU

When Jennie and I started Fresh Life in 2007, I wanted to incorporate the same experience I enjoyed so much in that formative Bible college semester into the DNA of our church. I wanted a deep dive into Jesus, as seen through the Gospel of John, to be our foundation. Week one of our church plant I announced that the next weekend we would begin with chapter one, verse one of the book of John and stick with it as long as it took. It took a long time. Sixty-eight Sundays, to be exact. For well over a year we crawled on our bellies through the book. It was fantastic. For me, anyway. And our church survived and grew through it too. I didn't come up with a clever series title or have slick art back then. It was plain and simple: "That You May Believe: A Study of the Gospel of John." I'll smile about it until I die.

I already told you that what stuck with me most about this gospel is the people whose stories it contains. But also, I love that Jesus custom-tailors his approach to each of them, based precisely on what they need from him.

- He is intellectually provocative with Nicodemus the thinker (John 3).
- He draws out the woman at the well, catching her off guard by asking her to do a favor for him even though he is there to do something for her (John 4).
- His behavior with the blind man in John 9 is downright strange, as it involves him spitting on the guy but then ultimately giving him the same gift my daughter Lenya was able to give to two blind people through the miracle of organ donation—the gift of sight.
- Nowhere is Jesus' unwillingness to pull his approach off the rack so evident as in John 11. His friend Lazarus had died, and Jesus

gave his "I Am" declaration of his identity as the "resurrection and the life"—right before illustrating it by raising Lazarus from the dead (v. 25).

- It has struck me many times how Jesus is so different when comforting Mary than he is while talking to Martha. These two sisters were as different as different could be. Mary was definitely an Enneagram 4 and Martha a 1 or a 3. And Jesus ministered to them exactly where they were, in the middle of their grief. It was a masterclass demonstration of how to win friends and influence people.

Read John 11:1–44 in its entirety. (Yes, I know, it's super long, three whole pages, to be precise. But it is *amazing*. Please don't skip over or blaze through it. Even if you are crazy familiar with the Bible, try to pretend you have never read it. Soak. It. In.)

> Now a man named Lazarus was sick. He was from Bethany, the village of Mary and her sister Martha. (This Mary, whose brother Lazarus now lay sick, was the same one who poured perfume on the Lord and wiped his feet with her hair.) So the sisters sent word to Jesus, "Lord, the one you love is sick."
>
> When he heard this, Jesus said, "This sickness will not end in death. No, it is for God's glory so that God's Son may be glorified through it." Now Jesus loved Martha and her sister and Lazarus. So when he heard that Lazarus was sick, he stayed where he was two more days, and then he said to his disciples, "Let us go back to Judea."
>
> "But Rabbi," they said, "a short while ago the Jews there tried to stone you, and yet you are going back?"
>
> Jesus answered, "Are there not twelve hours of daylight? Anyone who walks in the daytime will not stumble, for they see

by this world's light. It is when a person walks at night that they stumble, for they have no light."

After he had said this, he went on to tell them, "Our friend Lazarus has fallen asleep; but I am going there to wake him up."

His disciples replied, "Lord, if he sleeps, he will get better." Jesus had been speaking of his death, but his disciples thought he meant natural sleep.

So then he told them plainly, "Lazarus is dead, and for your sake I am glad I was not there, so that you may believe. But let us go to him."

Then Thomas (also known as Didymus) said to the rest of the disciples, "Let us also go, that we may die with him."

On his arrival, Jesus found that Lazarus had already been in the tomb for four days. Now Bethany was less than two miles from Jerusalem, and many Jews had come to Martha and Mary to comfort them in the loss of their brother. When Martha heard that Jesus was coming, she went out to meet him, but Mary stayed at home.

"Lord," Martha said to Jesus, "if you had been here, my brother would not have died. But I know that even now God will give you whatever you ask."

Jesus said to her, "Your brother will rise again."

Martha answered, "I know he will rise again in the resurrection at the last day."

Jesus said to her, "I am the resurrection and the life. The one who believes in me will live, even though they die; and whoever lives by believing in me will never die. Do you believe this?"

"Yes, Lord," she replied, "I believe that you are the Messiah, the Son of God, who is to come into the world."

After she had said this, she went back and called her sister Mary aside. "The Teacher is here," she said, "and is asking for you." When Mary heard this, she got up quickly and went to him. Now

Jesus had not yet entered the village, but was still at the place where Martha had met him. When the Jews who had been with Mary in the house, comforting her, noticed how quickly she got up and went out, they followed her, supposing she was going to the tomb to mourn there.

When Mary reached the place where Jesus was and saw him, she fell at his feet and said, "Lord, if you had been here, my brother would not have died."

When Jesus saw her weeping, and the Jews who had come along with her also weeping, he was deeply moved in spirit and troubled. "Where have you laid him?" he asked.

"Come and see, Lord," they replied.

Jesus wept.

Then the Jews said, "See how he loved him!"

But some of them said, "Could not he who opened the eyes of the blind man have kept this man from dying?"

Jesus, once more deeply moved, came to the tomb. It was a cave with a stone laid across the entrance. "Take away the stone," he said.

"But, Lord," said Martha, the sister of the dead man, "by this time there is a bad odor, for he has been there four days."

Then Jesus said, "Did I not tell you that if you believe, you will see the glory of God?"

So they took away the stone. Then Jesus looked up and said, "Father, I thank you that you have heard me. I knew that you always hear me, but I said this for the benefit of the people standing here, that they may believe that you sent me."

When he had said this, Jesus called in a loud voice, "Lazarus, come out!" The dead man came out, his hands and feet wrapped with strips of linen, and a cloth around his face.

Jesus said to them, "Take off the grave clothes and let him go." (NIV)

"I AM" #5

"I am the resurrection and the life."

John 11:25

Before I comment on this fifth "I Am" declaration, I wonder if you would be willing to stop and appreciate with me how mysterious and majestic Jesus is. Like Aslan in The Chronicles of Narnia, "He's wild . . . not like a tame lion." Nothing about any of this was expected. His disciples were bewildered; Mary was disappointed; Martha was disturbed and unglued. And poor Lazarus. He was just getting comfy on his fourth day in Abraham's bosom (I hope you didn't skip over chapter 7, because if you did, that probably makes no sense). Then all of a sudden he was ripped out and shoved back into his body and made to come shuffling out like Imhotep in *The Mummy*. (If you have never seen that Brendan Fraser classic, I apologize . . . for nothing! I must also insist if you *really* want to go down a Brendan Fraser rabbit hole to start with *Encino Man* and then go to *Blast from the Past*. Which will probably lead you to a Christopher Walken binge. Good luck.)

To each of these individuals Jesus tailored his approach. He was for them exactly what they needed.

One of Jesus' names is the Becoming One. He becomes exactly what he knows our hearts long for. Jesus doesn't have what you need; he *is* what you need. And he is not far but close.

Earlier I quoted Buzz, Neil, and CapCom Charlie going back and forth about the final touchdown on the moon. Down below on earth, there was a massive room at Mission Control full of people doing various jobs—but no one gets to talk to the astronauts except the CapCom. The capsule communicator listens to the flight director, flight surgeon, guidance director, and everyone else at Mission Control, but only the CapCom can talk directly to the astronauts. It is that person's job to translate all the craziness of the room and the back room(s), and everything else that's going on, to the astronauts in the air.

JESUS DOESN'T HAVE WHAT YOU NEED; HE *IS* WHAT YOU NEED. AND HE IS NOT FAR BUT CLOSE.

The person selected to be the CapCom is always a fellow astronaut because they know what those in space are facing. They've been through it themselves. For Apollo 11 it was Charlie. He was one of them, so they trusted he knew what they were feeling and that he'd be on their side.

This is why the incarnation is so important. Jesus became a fellow astronaut. He is your friend, and the tears that Jesus wept in John 11:35 are proof. He knew what he was going to do but cried anyway. We do not have a High Priest who cannot relate. He was tempted in every way and was without sin (Heb. 4:15).

Martha thought that Lazarus's resurrection was an event that was going to happen at the end of the age. Jesus wanted her to know it was a person. Jesus is the resurrection. To know him is to know life.

It is so easy to read the Bible to develop a religion based on facts and pieces and parts but miss the person. Jesus is Christianity. It's not about rules, or commandments, or golden streets, or giving tithes. It's all about Jesus. His church is screwed up at times, and his people will let you down. He has not and *never* will fail. The closer you get to him, the more you will be able to make sense of all that is to come.

Jesus can be that one voice in your headset. We don't have to worry about politics, or money, or careers, or what is going on in the world now or in the future. It is his job to listen to everything that is going on. It is our job to stay close to him. To listen to his voice and draw near to him.

Jesus put skin on, breathed our air, died our death, and rose to life. He has proven once and for all he is for us and victorious against anything that would stand against us.

Jesus Christ is Lord!

No matter what you are facing or what is weighing you down today or breaking your heart, Jesus cares. Yes, he will work it all for good and weave gold out of garbage. For sure. But he doesn't just see what he is going to do; he knows the toll it will take on you.

And he loves you deeply. Perfectly. Permanently.

Of course, that doesn't mean it's not hard, scary, and big. The tears that trickled down his face speak louder than words. "Jesus wept" (John 11:35) is the shortest verse in the English Bible, but perhaps it says more than any other. He wasn't crying only for Lazarus, he was crying for himself—Lazarus was his friend; for Mary and Martha—they had just lost their brother; and he was crying for you—he knew the pain of grief and loss that you would someday experience. You have never cried without him knowing and caring.

You matter to him.

And you matter to me.

I pray that, even though things in your life are scary or uncertain or up in the air right now, you know you can rest. You can rest because you are not alone. You can rest because of who Jesus is and what he has done. You can rest because nothing you do can make him love you more or less, and because no matter how hard life gets, nothing that happens to you can destroy you. You are free because he is the resurrection and the life. Not someday. Today. Right now.

CONQUER YOUR INNER SPACE

There is power in knowing Jesus, the one who is the resurrection and life. And if death can't destroy us, what can we possibly be afraid of?

It's no mistake that the promised land, which is itself a picture of life in God's kingdom, is characterized as a land of milk and honey

(Ex. 3:17). Neil deGrasse Tyson noted that those are two of the few things you can eat that don't involve death. Vegetable and animal life has to die for us to eat a hamburger or hummus. Not so with milk and honey. Life in Jesus' kingdom is life and more life. There is peace in that promise.

You and I can now live for real.

No more pretending.

No more posturing.

Real life. Life to the full. Milk and honey.

And because of his giant leap, we can believe God will meet us in our small steps and use our efforts to bless all of mankind.

Good luck and Godspeed.

CHAPTER 16

EIGHT COMES BEFORE ELEVEN

IT TAKES A VERY BIG ROCKET TO GET TO OUTER SPACE. Leaving earth is the easy part. Escaping earth's orbit is quite another thing. We probably never could have done it without Adolf Hitler. Shocking, yes. But let me explain.

NASA began as a small research outfit launching monkeys on fifteen-minute ballistic rides to the edges of space aboard converted missiles built by a former Nazi. It became one of the greatest technological and material enterprises in history, harnessing the work of half-a-million people to send people hurtling from the earth, going from 0 to 17,500 miles per hour in under nine minutes. That is ten times faster than a bullet. NASA plays with fire on every mission.

As I mentioned earlier in the book, it started out as a military entity called the National Advisory Committee for Aeronautics

(NACA), begun at the outset of World War I when we realized we were outmanned by other nations when it came to aircraft. Space was nowhere on the radar at that time. Flash forward to October 1, 1958, when the space race was heating up. NACA became operational as a new civilian organization separate from the military and was dubbed the National Aeronautics and Space Administration—NASA.

That tiny tweak, from NACA to NASA, is huge. I doubt Urban Outfitters would be selling NASA shirts hand over fist if it weren't for dropping the *C*, which stood for "committee." As they say, what's a camel? A racehorse designed by a committee.

You'll never conquer inner space letting every voice in your life have equal volume.

YOU'LL NEVER CONQUER INNER SPACE LETTING EVERY VOICE IN YOUR LIFE HAVE EQUAL VOLUME.

So, who was the former employee of the Führer who developed rockets for America's space program? Wernher von Braun. The man behind the Saturn V rocket that helped us to achieve translunar injection (TLI)—a propulsive maneuver that allowed the rocket to escape the pull of earth's gravity and come under the spell of the moon's gravitational pull. During the early days of the Third Reich, his job was to design rockets that would rain from the sky on Hitler's enemies. The V-2, the world's first guided missile, which caused so much trouble during the Battle of Britain, was his team's creation. But his true passion was rocketry in the name of science.

Yearning to escape the clutches of all dictators who only wanted to unleash evil through his rockets, Wernher sought a way out. He decided the Americans were the best option for both survival and a future. He was also in great danger. SS squads were killing any Germans they deemed disloyal to Hitler. In May 1945, Wernher managed to stage a daring escape for nearly five hundred members of his rocket team and their families, and some of their papers, documents, and drawings. They fled to the Harz Mountains in the south of Germany where they could surrender to the Americans rather than Russians. This was huge.

The newly homeless community of rocket scientists eventually found themselves brought into the US as a part of a program called Operation Paperclip. It was created to bolster the fledgling American space program with the much-needed wealth of wisdom found in former SS officer Wernher and his compatriots. What these Germans had amassed in the name of destruction was now going to be put to a noble purpose. The military's official story was that they hadn't accepted any "ardent Nazis," but there seemed to be a lot of latitude in what was deemed "ardent."

Wernher was moved to Fort Bliss, Texas, just outside of El Paso, where the US military was trying to repurpose the V-2 missiles they had claimed from Germany as spoils of the war. When they test-fired V-2s at the White Sands Proving Ground (now the White Sands Missile Range) in New Mexico, he served as adviser. His popularity in America soared like the rockets he helped launch. He even found opportunities at Disney as a technical adviser for films about space exploration, and a movie was made about his life called *I Aim at the Stars*. A popular comedian suggested that the subtitle could be *But Sometimes I Hit London*. Despite his newfound fame his former life had not been completely forgotten.

Wernher found himself leading the charge at the fledgling space agency. He worked on the Redstone and Jupiter rockets used during the Mercury missions, but his magnum opus was the mighty Saturn V.

The moon landing of Apollo 11 never could have happened without Wernher. And Apollo 11 never could have happened without Apollo 8. Yes, we landed on the moon for the first time on Apollo 11, but we *went* to the moon for the first time on Apollo 8. One of my favorite, easily forgotten moments of the US space program was when

Frank Borman, Bill Anders, and Jim Lovell traveled to the moon on that mission. The case can be made that it was the gutsiest of all the moments in the Apollo program—even more so than the actual moon landing.

Why? They were the first to fly past the 250-mile mark, leaving low earth orbit (all previous Mercury, Gemini, and Apollo missions had gone no further than 250 miles above earth), and the first to go around the moon, see its dark side, and come back to tell the tale. They set a speed record, too, achieving a top speed of 24,226 miles per hour.

There was no certainty and massive risk to this mission. We had never gone so far, never been so exposed. An unprecedented billion people watched or listened. If you look at a photo tracing the mission's trail through space, it is a figure eight. After lifting off from the earth, they took a lap around our planet before they were told, "Apollo 8 you are go for TLI" and given permission for translunar injection, a five-minute-long burn which caused them to reach escape velocity and be propelled toward the moon. They were the first people in history to fully break the bonds of earth's gravity. They slowed down the farther they got away from the earth until they entered the moon's gravitational pull and began to be pulled on by another heavenly body for the first time in history, and then they began to speed up. A quick firing of the engines put them into orbit around the moon—all without either smashing into it or being hurtled out into deep space. Then they did the exact same thing again in reverse, completing the figure eight. The mission plan for Apollo 8 was an eight. This also explains the mission patch with the figure eight incorporated into the design. It is symbolic on many levels.

In August 1968 they hatched the plan to use the Apollo 8 mission, slated for December, as a moon-bound guinea pig. The lunar lander was meant to be tested on 8, but it wasn't going to be ready. Someone came up with the crazy idea of taking the CSM to the moon without a lander, so they would instead leave earth, enter lunar orbit, and come back. It was a crazy idea because they had only four months to prep for the mission when they usually had six.

Frank, Jim, and Bill were also the first to see the earth for the first time from the moon. The photo they took, called "Earthrise," is stunning. It is one of the most famous photos ever taken from space. No human had ever seen our planet like that before.

Not only did the astronauts get to see it in real time but, as they did live broadcasts, America got to see it too.

The flight took place over Christmas, as I mentioned earlier. Apollo 8 entered lunar orbit on Christmas Eve. The astronauts captivated the watching world as they communicated from the moon what the experience and the intense symbolism of the date meant to them. "Before the flight, the NASA public affairs officer told Borman that they expected about a billion people—a quarter of the world's population—to be tuning into their Christmas Eve TV broadcast from lunar orbit," said Teasel Muir-Harmony, the Apollo curator at the National Air and Space Museum. "More humans would be hearing their broadcast than any other human voice in history and he was just told to say something appropriate."

He did just that. Choosing to pull from the Bible for the occasion, the astronauts spoke to the world. Here is a transcript from the end of that broadcast:

WILLIAM ANDERS: For all the people back on Earth, the crew of Apollo 8 has a message that we would like to send to you:
"In the beginning God created the heaven and the earth.
"And the earth was without form, and void; and darkness was upon the face of the deep. And the Spirit of God moved upon the face of the waters.
"And God said, Let there be light: and there was light.

"And God saw the light, that it was good: and God divided the light from the darkness" [Gen. 1:1–4 KJV].

JIM LOVELL: "And God called the light Day, and the darkness he called Night. And the evening and the morning were the first day.

"And God said, Let there be a firmament in the midst of the waters, and let it divide the waters from the waters.

"And God made the firmament, and divided the waters which were under the firmament from the waters which were above the firmament: and it was so.

"And God called the firmament Heaven. And the evening and the morning were the second day" [vv. 5–8 KJV].

FRANK BORMAN: "And God said, Let the waters under the heavens be gathered together unto one place, and let the dry land appear: and it was so.

"And God called the dry land Earth; and the gathering together of the waters called he Seas: and God saw that it was good" [vv. 9–10 KJV].

Borman then added, "And from the crew of Apollo 8, we close with good night, good luck, a Merry Christmas and God bless all of you—all of you on the good Earth." The Creator and his creation were both recognized in this incredible, poignant speech.

It's worth it to not just read it but to listen to it with the words in front of you. Or close your eyes and let the power of it hit you. Check it out on YouTube—NASA's recording of "Apollo 8's Christmas Eve 1968 Message."

Everyone was spellbound. None of the one billion people watching or listening knew it was coming, not even those at Mission Control in Houston. One person in the MOCR reported that it caused the hair on his neck to stand up. Grown men began to weep freely. By the end of the message there wasn't a dry eye in the room.

Not everyone liked it. Famed atheist Madalyn Murray O'Hair filed a lawsuit against NASA objecting to religious texts being read from a government spaceship on the grounds that it violated the separation of church and state and inferred that Christianity was the state religion of America. She lost the suit, but because of the controversy they didn't advertise or televise Buzz Aldrin's decision to take the Last Supper on the moon on Apollo 11. Deke Slayton, who was in charge of crew assignments, discouraged Buzz from making a big deal of it, though he did give him approval for it.

Not to be overlooked is the fact that Apollo 8 was the first time the Saturn V rocket had ever left earth with humans inside. Remember, on an Apollo 1 procedures test three astronauts had died, and for the next almost two years there were no missions. NASA was grounded. I already told you, but Apollo 2, 3, 4, 5, and 6 were either canceled or unmanned, and Apollo 7 used a Saturn 1B rocket, which was dwarfed by its enormously gigantic cousin.

Here is what you need to know about the Saturn V.

Again, this rocket is thirty-six stories tall and generates 180 million horsepower. That is eighty-five Hoover Dams, or enough power to run New York City for over an hour. It is the tallest and heaviest rocket ever built. It takes off with the subtlety of a skyscraper doing the high jump, and it unleashes the energy of a small nuclear bomb in the process. As we've already noted, it goes from 0 to 17,500 miles per hour in under nine minutes, but here's another cool fact. It uses enough fuel to send a car around the world eight hundred times. The launch is so violent the earth heaves, shaking for miles around. It goes without saying that it is really, really loud. Approximately 1 percent of its energy is converted to sound, so if the wind is wrong or cloud cover is too low at launch, windows in the vicinity can be shattered and plaster can "rain from ceilings."

The closest distance people are allowed to stand during a launch is three and a half miles away, so they see the liftoff six seconds before they can hear it. But then they very much feel it as the sound waves hit them, assaulting all their senses and rattling the fillings in their teeth.

The rocket has three stages. The first burns for two and a half minutes and gets it off the ground to about forty miles high, then falls away and burns up in the earth's atmosphere. The second stage of the rocket burns for six minutes and gets it to a height of 115 miles above the earth before falling away. The third and final stage is what burns to get the Apollo spacecraft out of earth's orbit and on its way to the moon at 24,500 miles per hour.

The three brave men of Apollo 8 were buckling into a bolt of lightning for the ride of a lifetime.

There were far more ways for things to go wrong than for things to go right. It was believed there was a one-third chance of success, one-third chance of failure requiring them to return home, and one-third chance of death.

When the first stage burned out, the gravitational force went from 6 Gs to none in a moment, and all three of them thought they were going to shoot through the instrument panel. Instinctively they put their hands in front of themselves to brace themselves. At that moment the second stage ignited. As they suddenly sped up, it was like a high-speed roller coaster taking off, and their outstretched arms were thrown back toward their faces so quickly that the metal rings on their wrists, where their gloves screwed on, cut gouges in their plexiglass face visors. Bill said when he realized what he had done he felt foolish but later found out both of his fellow astronauts had done the same thing. That was when it hit him: when you're riding the Saturn V, everybody's a rookie. Three days before the flight, NASA's safety chief, Jerome Lederer, said the mission would "involve risks of great magnitude and probably risks that have not been foreseen." He explained, "Apollo 8 has 5,600,000 parts and one-and-a-half million systems, subsystems, and assemblies. Even if all functioned with 99.9 percent reliability, we could expect fifty-six hundred defects."

For all these reasons Apollo 8 has been widely considered the most audacious of all of NASA's endeavors. It is the giant leap that made Neil Armstrong's small step in Apollo 11 possible.

THIRSTY FOR YOU

I would now like to bring you to Jesus' fifth statement from the cross.

By this point in the book, I'm sure you are itching to hear about the resurrection. Yes, we've made it through two-hundred-plus pages together and spent much of it on Jesus' agony. But the power of Easter couldn't have happened without the horror of the cross. Hang in there with me. Eight comes before eleven. John 19:28 says, "Later, knowing that everything had now been finished, and so that the Scripture would be fulfilled, Jesus said, 'I am thirsty'" (NIV).

One year I made the mistake of working out before I went to my annual doctor's appointment. I had been told to fast for twelve hours before the appointment, so I had an empty stomach on top of lots of exercise. I remember looking at my blood fill the little tube coming out of my arm when everything started to go fuzzy and pins and needles covered my entire body. The nurse asked me if I was okay and then called out for someone from another room just as I went limp. When I came to, there were ice packs being pressed to my neck and a juice box being pressed to my lips. "Drink this," they said. "It will help."

LAST WORDS #5

"I thirst!"

John 19:28

The medical explanation for Jesus' thirst is the severe depletion of fluids in his system. That would not only have been normal but expected considering what he had been through. Consider the loss of blood after the scourging alone, not to mention the beatings and the crucifixion. Inflammation would have set in all over his body, requiring fluid that he didn't have to spare. He had been kept up all night and very likely been given nothing to drink. I get lightheaded if I stand up too fast.

I can only imagine how Jesus felt. And there was no one there to lovingly press an ice pack to his neck. This explains why, on his mile-long journey to the cross, he kept falling down until Simon of Cyrene did what Jesus invites all of us to do—he picked up the cross and followed him (Mark 15:21). The raging thirst that would have been building in his body must have been extraordinary.

There is prophetic significance to the statement found in Psalm 22:15: "My mouth is dried up like a potsherd, and my tongue sticks to the roof of my mouth; you lay me in the dust of death" (NIV).

If you read Psalm 22, you will find it is full to the brim of specific prophetic references to the way Jesus was dying. Arms spread out. Lifted up. Surrounded by enemies. His limbs hyperextended and bones feeling like they were all coming out of joint.

As I mentioned in chapter 13 of this book, what's most marvelous to me about Psalm 22 is that David wrote it years and years (hundreds) before crucifixion was invented. This is to me one of the many proofs of the resurrection. First, the evidence for the death and resurrection of Jesus is ironclad because of the resurrection appearances, the change in his disciples afterward, and the many, many historical documents. But then, it is bolstered with the prophecies Jesus fulfilled in his coming.

And not just these in Psalm 22, but others. There are more than three hundred in all.

Like the city he would be born in and the city he would die in (Mic. 5:2; Zech. 9:9). Or that he would be buried with the rich but die with the wicked (Isa. 53:9). Every fulfilled prophecy is another eight ball in the corner pocket validating that Jesus is in fact who he says he is, the Messiah and Savior of the world.

I'll leave the calculation of the probability of this sort of stuff to mathematicians, but let me tell you—it is so intellectually enriching to my faith. You don't have to check your brain at the door to be a Christian. Some of the smartest people to have ever lived, and others still alive today, are Jesus people.

Jesus' thirst is not only an example of prophecy fulfilled; it is love portrayed. He voluntarily allowed himself to be thirsty for you. Remember, he is the one who spoke and created all water (Gen. 1:2). Every river, waterfall, lake, and rainstorm exists within his power. And still he sat in thirst. He became thirsty for you.

There is a raging thirst in the heart of every person ever since Genesis 3. That is when we lost our connection to the presence of God, which hydrates and satisfies and nourishes. No matter what success or possession, status or accomplishment you can attain on this earth, nothing can fill the void inside your soul. John 7:37–38 gives an answer to this.

> On the last and greatest day of the festival, Jesus stood and said in a loud voice, "Let anyone who is thirsty come to me and drink. Whoever believes in me, as Scripture has said, rivers of living water will flow from within them." (NIV)

Jesus hung on the cross with his tongue thick, hot, and parched to break down the dam and give us all access to that water. Because he was thirsty, you never have to be again. Once you have that thirst quenched, your new assignment is to help him in bringing that water to a dehydrated world. The water that comes to you is meant to flow through you.

THE CALL OF THE WILD

In another sense, Jesus didn't just come to relieve thirst but to ignite it. Thirst is often used negatively in Scripture, but not always. Some thirsts are caused by sin, but some thirsts come from someplace else. Matthew 5:6 says, "Blessed are those who hunger and thirst for righteousness, for they will be filled" (NIV). So many times, I'm oblivious to these righteous thirsts, and I miss out. I think one of the most important righteous thirsts is the call of the wild.

My son, Lennox, is almost three. His favorite story is *The Call of the Wild*. We watched the movie starring Harrison Ford, and it fascinated him. For the next several weeks at bedtime, he asked me to tell him the story of Buck, the dog in the movie. From memory I would walk through the plot of the movie and tell him the major moments. After doing that for fourteen nights in a row, I went online and bought the book the movie is based on and began reading it to him. Usually he's asleep before a chapter ends, but it always makes his eyes shine when I tell him about the adventures of Buck.

If you have never read the book or seen the movie, it is about an enormous St. Bernard–Scotch Collie mix. He is a domesticated dog that grew up living in the home of a judge in California, but he eventually discovers his true wild nature that has been buried deep down under his bougie, plush life. He has been raised in the city, eaten food out of a dog bowl, and slept indoors. But when the demand for sled dogs skyrockets during the Gold Rush, he is kidnapped by a gardener who sells him to pay off a gambling debt and is thrust into a completely foreign life in the Yukon.

Buck had never seen snow, much less run on it. He had never really been uncomfortable or experienced pain or hunger. But as he is forced to pull a sled, navigate the hierarchy of the other sled dogs, and go from one master to another, something begins to change. Not just physically, as his feet develop thick calluses, his muscles grow powerful, and he sheds the extra fat that accumulated during his days of comfort. Something primitive wakes up inside him—something he didn't even know was there.

The call of the wild.

In the 2020 movie, they personify the instincts embedded into his biology by making Buck see a wolf with glowing eyes whenever he needs to take an action or make a choice contrary to his programmed normal. Lennox loves the wolf with the glowing eyes and always insists that I not forget to mention Buck saw him on top of the cave. By the end of the movie, Buck becomes the true version of himself, unmuted by safety and comfort. He gives sacrificial love and leadership and experiences freedom. He's established his place within a wolf pack, learned to raise his voice, and found someone to love.

I love everything about that story and believe it describes what was on Jesus' mind for you as he endured the thirst of the cross. Just as the thirst for adventure had been activated inside Buck, and he realized the way he was raised was not what he was built for, Jesus calls you. You are invited to shake off the sedentary lethargy that comes from complacency and tap into the wild existence you were built for. Comfort zones don't keep your life safe; they keep your life small.

God is not offended by your wanderlust or zest for life. Your ambition, creativity, or wild heart doesn't surprise or frighten God. He put it there and wants it to be enhanced, focused, refined, and fulfilled to a level that would blow your mind.

There is a man, with eyes that glow, standing by to wake you from the slumber of working for the weekend or living for pleasure, or a life defined by something as small as money or status. He wants you to be free and strong—to run, dance, and laugh. To explore and take risks and reach out and serve. He doesn't want your life to play out within the boundaries of what has been done before, or what you can control or predict. You have been made for adventure. It is your birthright as a child of the King. It's time to go further up and further in.

COMFORT ZONES DON'T KEEP YOUR LIFE SAFE; THEY KEEP YOUR LIFE SMALL.

In Isaiah 55:2–3 we read how the prophet Isaiah envisioned it:

> Why spend money on what is not bread,
> and your labor on what does not satisfy?
> Listen, listen to me, and eat what is good,
> and you will delight in the richest of fare.
> Give ear and come to me;
> listen, that you may live.
> I will make an everlasting covenant with you,
> my faithful love promised to David. (NIV)

Jim Lovell saw his mission to the moon on Apollo 8 as a thrilling, Lewis and Clark–type expedition. He resonated with the language President Kennedy cast, where space exploration was the inevitable new ocean of opportunity that needed to be ventured into. I must admit that kind of language and imagery makes my heart go pitter-patter. I am enamored with Meriwether Lewis and William Clark and their westward journey to find out if there was an all-water route to the Pacific Ocean. And I see it as a picture of the spiritual adventure God wants you to embark on as the Holy Spirit unleashes rivers of wonder in your life. He wants you to imagine and dream and explore. In your home and neighborhood and city and in this world, but also in the vast galaxies of his presence, where there is no end to what you will discover as you walk with him.

You and I should both be afraid that at the end of our lives we might realize that we were satisfied with so much less than what God intended for us to experience, accomplish, and enjoy. We have been given all blessings, potentially, but we must rise up in faith to appropriate them. Wherever you put your foot, God will bless you. How tragic it would be to remain in the pee-filled baby pool of life when your great God bankrupted heaven to open up an ocean of opportunities and adventures Jesus wants you to thirst for. Matthew Henry said, "You have a journey to go on, a race to run, a warfare to accomplish, and a great work to do." All of heaven is willing and able to support you as you give your all to those things.

So why is it so easy to miss the exploration and live with the shades drawn?

Simple. The world gets in the way.

While our feet are planted on this planet, we feel a hypnotizing pull toward the safe, predictable, controllable, and comfortable. In getting up above it, looking at it through the vantage point of heaven, we are able to see things clearly.

Isaiah describes this well in Isaiah 40:12–22:

> Who has measured the waters in the hollow of his hand,
> or with the breadth of his hand marked off the heavens?
> Who has held the dust of the earth in a basket,
> or weighed the mountains on the scales
> and the hills in a balance?
> Who can fathom the Spirit of the Lord,
> or instruct the Lord as his counselor?

Whom did the LORD consult to enlighten him,
 and who taught him the right way?
Who was it that taught him knowledge,
 or showed him the path of understanding?

Surely the nations are like a drop in a bucket;
 they are regarded as dust on the scales;
 he weighs the islands as though they were fine dust.
Lebanon is not sufficient for altar fires,
 nor its animals enough for burnt offerings.
Before him all the nations are as nothing;
 they are regarded by him as worthless
 and less than nothing.

With whom, then, will you compare God?
 To what image will you liken him?
As for an idol, a metalworker casts it,
 and a goldsmith overlays it with gold
 and fashions silver chains for it.
A person too poor to present such an offering
 selects wood that will not rot;
they look for a skilled worker
 to set up an idol that will not topple.

Do you not know?
 Have you not heard?
Has it not been told you from the beginning?
 Have you not understood since the earth was founded?
He sits enthroned above the circle of the earth,
 and its people are like grasshoppers.
He stretches out the heavens like a canopy,
 and spreads them out like a tent to live in. (NIV)

Translation: His glorious power makes the Saturn V look like a bottle rocket.

Jim discovered this on Apollo 8 when he saw the earth for the first time from a quarter million miles away. He remarked how incredible it was that he could block out the entire planet when he looked at it while giving a thumbs-up gesture. "Everything I ever knew was behind my thumb." He referred to earth as "a grand oasis in the big vastness of space."

They went all the way to the moon to discover how little and how precious the earth was.

Michael Collins would later express the same. "T. S. Eliot understood it better than I: 'We shall not cease from exploration / And the end of all our exploring / Will be to arrive where we started / And know the place for the first time.'"

Perspective is everything.

This earth is both small and precious. Your God, who was thirsty for you, is magnificent—and he is love. Rise high today in your heart. Drink deeply. Quench your thirst. Cultivate your thirst. Let the river of God that flows from Emmanuel's side come to you and flow through you.

CONQUER YOUR INNER SPACE

We all thirst at some point in life, but the question is, for what?

If like Wernher your skills and abilities are being used for destruction, change what you are pointing at. When his rockets were aimed at targets on earth, they hurt people and took lives. When they were pointed at heaven, they opened doors and expanded possibility.

The year Apollo 8 went to the moon, 1968, was also the year the first heart transplant surgery was successfully done. It is incredible to think that the year three men left the pull of earth's gravity and came under the power of another heavenly body, another man was opening a human chest for a donated heart to be installed.

God is able to put new life inside your heart so you can feel God's heartbeat when you touch your pulse. Your shadow side and darkest secrets don't scare him. You don't need a different life to build something beautiful.

If you trust him with your heart, what you have landed on, you can launch from.

Before you turn the page, I want to give you a moment to say a prayer. These are the words that guided Wernher into the final twenty seconds of the countdown before the launch of Apollo 11 on his Saturn V. As the seconds ticked away, he put his binoculars down, bowed his head, and prayed words from Matthew 6:9–13 that will give you strength:

> Our Father in heaven,
> Hallowed be Your name.
> Your kingdom come.
> Your will be done
> On earth as it is in heaven.
> Give us this day our daily bread.
> And forgive us our debts,
> As we forgive our debtors.
> And do not lead us into temptation,
> But deliver us from the evil one.
> For Yours is the kingdom and the power and the glory
> forever. Amen.

CHAPTER 17

ROCK[S] OF AGES

I TOUCHED THE MOON ONCE. IT WAS A PART OF A HANDS-ON exhibit at Kennedy Space Center (KSC) in Florida in 2014. Near the end of the tour, just before you exited through the gift shop, there was a display where you could reach your hand into the little hole in the plexiglass and touch a lunar sample brought back from Apollo 17. The germaphobe in me told me not to, but the part of me that loves history said yes.

To have my hand on a piece of another world? Amazing. I would normally say I'll never wash my hands again, but I couldn't wash my hand fast enough. Touring Johnson Space Center in Houston, the Smithsonian National Air and Space Museum in Washington, DC, and Kennedy Space Center Visitor Complex at Cape Canaveral are all ten-out-of-ten recommendations from me. While touching the moon was fun, the most powerful thing in Florida and Texas was standing below a Saturn V that never flew.

The original plan was to make seventeen Saturn Vs, but that was shrunk to fifteen. Of those, two were unmanned, ten were manned, and one was converted into Skylab, a precursor to the International Space Station. When the Apollo program ended, two were left over and went into mothballs before eventually finding homes at Kennedy Space Center and Johnson Space Center for the American people to appreciate.

In the Smithsonian National Air and Space Museum collection, you can see the Apollo 11 command module and Neil's pressure suit.

Quarantine is a way of life for NASA. In all the years that astronauts have been going on missions, they have endured isolation in the weeks prior to launch. No one wants to be sick in space. The only way to keep that from happening is to endure total isolation in the days prior to the launch—also known as the day of "lighting the candle."

For most of us, prior to 2020, quarantine was something from the history books. People were quarantined during the dark days of the plague, or when traveling on ships where scarlet fever broke out. Now it is something we all have memories of. Memories that are a little too fresh. After sheltering in place, some of us discovered firsthand why solitary confinement is a punishment in prison.

Apollo 11 was the first mission where the astronauts not only had to quarantine before but also after the mission. The former because of concern over earthbound infections like the 1968 flu pandemic. The latter because of fear that some sort of moon-borne alien plague would smuggle its way back with the astronauts in the dust on their space suits, or their ship, and unleash an apocalypse on the earth through a lunar virus that medicine had no answer for. If you saw the movie *Venom*, you know this would not be ideal.

The measures in place for recovery when Columbia splashed down were stunning. When navy frogmen jumped into the ocean from a helicopter, they secured a rubber inflatable ring—a giant version of the kind a child would play with in the pool—around the ship so it couldn't sink, then inflated two rafts. One designated frogman carefully opened the door and threw in suits that the three astronauts put on to cover all potential germs, and then they got into the raft with him. He then sprayed them down from head to foot. He also made his way around the charred vessel as it bobbed like a cork in the ocean and sprayed it down as well. Then and only then were they ferried into the aircraft carrier where they were hoisted aboard. On the ship was an Airstream travel trailer. They entered the camper and the door was sealed.

Soon President Nixon (who had been flustered when his request to have dinner with the Apollo 11 astronauts the night before they took off was rejected by the flight surgeon as it would violate their quarantine) was standing in front of them as they spoke to the world from behind a plexiglass window in the trailer. After the interview, navy chaplain of the USS *Hornet* John A. Piirto was brought out to lead them all in prayer:

> Lord God, our Heavenly Father,
>
> Our minds are staggered and our spirits exultant with the magnitude and precision of this entire Apollo 11 mission. We have spent the past week in communal anxiety and hope as our astronauts sped through the glories and dangers of the heavens. As we tried to understand and analyze the scope of this achievement for human life, our reason is overwhelmed with bounding gratitude and joy, even as we realize the increasing challenges of the future. This magnificent event illustrates anew what man can accomplish when purpose is firm and intent corporate. A man on the moon was promised in this decade, and though some were unconvinced, the reality is with us this morning in the persons of the astronauts: Armstrong, Aldrin, and Collins. We applaud their splendid exploits, and we pour out our thanksgiving for their safe return to us, to their families, to all mankind.

The three astronauts, still inside the trailer, were then transported to Hawaii, where the Airstream was loaded onto a plane for the flight to Houston. From there their isolation chamber was transferred to a secure, sealed facility that had been carefully built for them. For three weeks following the moment they lifted off from the moon, they—and the forty-seven-and-a-half pounds of moon rocks they brought home with them—were studied by a team of surgeons who had to make sure that no diseases had stowed away with them.

When the sand in their quarantine hourglass ran out after twenty-one days, they were able to come out. They had a clean bill of health—no alien bacteria. It was then that the moon rocks had their moment to shine. No longer feared as Trojan horses of the end of the world, they became our nation's crown jewels and were treated with all the pomp and circumstance befitting such symbols of conquest. President Nixon sent out a piece of moon rock to the governors of all fifty states and leaders of one hundred thirty-five countries.

People stood in lines wrapped around buildings just to get a glimpse of these rocks. They were symbolic of a twelve-year national fascination with winning the space race. They were literal, tangible spoils from the cosmic war, and everyone wanted to see them with their own eyes.

Many of the moon rocks have gone missing over the years—some lost, others stolen. Imagine—stealing the moon! It's not just the plot of *Despicable Me*; it has really happened. In 2002, a six-hundred-pound safe containing lunar and martian geological material was heisted from the Johnson Space Center in Houston. Officials declared its black-market value to be in the millions of dollars. It was recovered before it could be sold. A piece of the moon that Nixon gave Honduras hit the market with a whopping five-million-dollar price tag. Fortunately, it was also recovered due to an elaborate sting operation code-named (wait for it) Operation Lunar Eclipse.

NASA lore has it that over the years, some rocks were given to girlfriends of astronauts as trophies. They almost certainly weren't taken by Neil. Before takeoff someone asked him whether he would consider bringing a piece of the moon home as a personal souvenir. Neil stiffly replied, "At this time, no plans have been made."

"Would you have the desire?" the person pressed.

"No," Neil went on, "that's not a prerogative we have available to us."

Of course, he could have said, "We can't do it," but when cornered he always spoke in computerese. In 1969 an acquaintance observed, "If you tell Neil that black is white, he may agree with you just to avoid argument." Another added, "No, most likely he won't say anything at all. . . . He'll smile at you and you'll think he is agreeing. Later on you'll remember that he didn't say a word."

His first wife, Jan, also said that silence "is a Neil Armstrong answer. The word 'no' is an argument."

Many, many counterfeit moon rocks have been sold over the years—especially right after the initial moonshot when people were queuing up just to get a glimpse of the pebbles Buzz and Neil brought home with them. Hucksters were trying to pass off earth rocks as lunar samples. In 2009, while doing some testing, the Rijksmuseum, the national museum of the Netherlands, discovered that a moon rock—one of its most prized possessions—wasn't from the moon and wasn't even a rock. It was petrified wood, likely from Arizona. For forty years that piece of Southwestern wood sat next to a plaque boasting it was brought back by Apollo 11. It might as well have been green cheese. The rock had been insured for approximately half-a-million dollars, but its real value was around seventy. *Woh, woh, woh.*

THE CHURCH IN SARDIS

This is all very similar to what Jesus had to say to the church in Sardis.

Revelation 3:1–6 contains his letter to them.

> "To the angel of the church in Sardis write,
>
> 'These things says He who has the seven Spirits of God and the seven stars: "I know your works, that you have a name that you are alive, but you are dead. Be watchful, and strengthen the things which remain, that are ready to die, for I have not found your works perfect before God. Remember therefore how you have received and heard; hold fast and repent. Therefore if you will not watch, I will come upon you as a thief, and you will not know what hour I will come upon you. You have a few names even in Sardis who have not defiled their garments; and they shall walk with Me in white, for they are worthy. He who overcomes shall be clothed in white garments, and I will not blot out his name from the Book of Life; but I will confess his name before My Father and before His angels.
>
> "He who has an ear, let him hear what the Spirit says to the churches."'"

Founded in 1400 BC, Sardis had been at one point one of the great cities of antiquity. Gold and silver coins were first minted there, but it was also a center for the garment industry. They were famous for the way they dyed wool that was used to make jackets, tunics, and carpets. They were very proud, claiming to have discovered how to dye wool which was white as snow prior to their coloring it.

Sardis was a desirable place to live because it had a citadel built on a massive, 1,500-foot plateau that made it virtually impregnable to invasion. It was unscalable on three sides, and the ramp up was very tricky. They boasted that they didn't even post guards at night because of their walls. Such was the fame of Sardis's fortress that the phrase "capturing Sardis" was a widely used metaphor for achieving the impossible. Even so, the Persians overthrew the city about 547 BC when one soldier led the Persian army up a path in the rock face. The Persians' patience and creative approach proved very effective in light of Sardis's overconfidence and set-in-their-ways laziness.

One feature that set Sardis apart was its very desirable cemetery. Their necropolis was a massive graveyard on a large hill that you could supposedly see from seven miles away. It was a very distinct landmark. People loved burying loved ones there because you could always see them and feel near to them.

It was a city people were *literally* dying to move into.

Jesus said the Sardis church was no different. In name they were alive, but in reality, they were dead. No one lives in Sardis today. It is a heap of ruins and its graveyard is its most conspicuous legacy.

THE WALKING DEAD

This morning I met a friend to play tennis. On my way out the door I grabbed a can of tennis balls from a box I bought at Costco. If you have never opened a can of tennis balls, know that it is highly satisfying, even close to addicting. There is a sound, like opening a Coke or a can of Pringles, and then a release of pressure. I love the smell too. Tennis balls are pressurized and sold under pressure so they bounce well. The intensity on the inside starts to dissipate the moment you open them. Depending on how hard you hit and how long you play, you can get one or two days of competition out of a fresh can. After that they go into the hopper to use as practice balls.

It's pretty interesting to think that the solution to a pressure problem is more pressure. It's like fighting fire with fire—which is a thing too. It's called back-burning and is one of the most effective firefighting techniques out there.

That day on the court I removed the plastic cap from the can of tennis balls and found the aluminum lid beneath was missing. The top had already been popped. Someone took three practice balls, put them in a can, and then placed the can in the box of brand-new balls. I am not going to mention any names, *cough* Lennox *cough*, but I had been deceived. Wronged. Sabotaged. Even worse, they were really flat balls. About that time my friend showed up. We decided to suffer through the game with them and play anyway, as they were our only option.

At the risk of sounding melodramatic, it was horrible. We had to run all over the court because the ball would pretty much stop where it landed. By the end of the first set (which I lost), I was not a happy camper.

The balls appeared to be full of life, but they were dead.

It's clear from Jesus' language that he is not a big fan of hypocrisy. He tells the church in Sardis he is ready to open a can because what they are broadcasting is different from what's inside.

LETTER #5 (SARDIS)

"Remember therefore how you have received and heard; hold fast and repent."

Revelation 3:3

The little plaque said "moon rock," but all they had was overpriced petrified wood.

We know things are not always as they appear. What *can* take us by surprise is how circumstances can change so that things are also not what they had previously been.

John Phillips illustrates this idea in his commentary on the book of Revelation: "Astronomers tell us that the light from the polar star takes thirty-three years to reach the earth. That star could have been plunged into darkness thirty years ago, and its light would still be pouring down to earth. It would be shining in the sky tonight as brightly as if nothing had happened. It could be a dead star, shining solely by the light of a brilliant past. The church at Sardis was like that. It had a name, but it was dead. It was shining solely by the light of a brilliant past."

The problem was not that Sardis was weak, but rather that they were boasting that they were strong. Things had changed and they hadn't gotten the memo.

Jesus never dealt cruelly with someone who was struggling in their journey. Isaiah 42:3 says, "A bruised reed He will not break, and smoking flax He will not quench." If you are backsliding, or frontsliding, or in a complete breakdown today, he will heal and restore you if you let him. What he does not tolerate is false advertising—to declare you have it all together when on the inside you are not walking the talk.

You and I join the ranks of the church in Sardis whenever we respond to the brokenness in our hearts with pious talk and a Pharisee's resolve. Strangely, we can end up extremely judgmental when we have strayed from a place of health.

The sins we struggle with look uglier on other people.

THE SINS WE STRUGGLE WITH LOOK UGLIER ON OTHER PEOPLE.

That is what happened to David. When he sinned with Bathsheba, and then had her husband murdered to cover it up, he thought he got away with it. By the time he repented, the child had been born (2 Sam. 12). During that time David went about his normal life as though everything was fine. That would have brought him to the temple where he sang, read the Scriptures, and did his religious duty. His name tag said he was alive, but on the inside, he was carrying around death. He had stolen the moon and gotten away with it. Except he didn't.

Jesus said to Sardis that if they didn't repent, "I will come upon you as a thief" (Rev. 3:3). The truth is, their lack of repentance would prevent them from having experiences he wanted for them. I invite a thief in anytime I choose death over life, or sin over sincerity. The thief steals joy and plunders peace. I end up guilty and anxious, and it not only affects me but everyone around me.

The brokenness inside David perversely made him not more gracious but more legalistic.

When Nathan confronted him using a made-up story about a man who took and killed his next-door neighbor's pet, David demanded the man be put to death. Death for killing a lamb? Indignation and legalism covering up wickedness (2 Sam. 12).

Some of the most awful things are done by some of the most religious people. And ironically, Christians tend to be the harshest on the things we hide in our own lives. For whatever reason having a two-by-four sticking out of your eye often makes you outraged by the speck of dust in the eyes of those you come in contact with (Matt. 7:5). It's not easy to be at peace when your conscience is screaming bloody murder. Whatever you don't process in a healthy way will cause you to project onto other people.

Nathan calmly told David that he was the man—the one who took from his neighbor what was precious. It was then that David finally looked in the mirror and faced what he had been hiding from: himself. What he saw made him sick to his stomach. And after seeing what he had become, he was finally able to move toward healing.

The solution is to wake up, remember all the things you've received and heard, then turn back to God and obey him. Quit faking and pretending you have it all together, and watch as his strength floods into your heart and mind.

David then penned one of the most important songs he ever wrote. A psalm of repentance. When I have gone through the carousel of sin and sorrow, these same words have brought healing to me countless times. Let them wash over you and call you out of Sardis into the experience of an over-the-top good life.

Psalm 51 begins this way: "To the Chief Musician. A Psalm of David when Nathan the prophet went to him, after he had gone in to Bathsheba."

> Have mercy upon me, O God,
> According to Your lovingkindness;
> According to the multitude of Your tender mercies,
> Blot out my transgressions.
> Wash me thoroughly from my iniquity,
> And cleanse me from my sin.
>
> For I acknowledge my transgressions,
> And my sin is always before me.
> Against You, You only, have I sinned,
> And done this evil in Your sight—
> That You may be found just when You speak,
> And blameless when You judge.
>
> Behold, I was brought forth in iniquity,
> And in sin my mother conceived me.
> Behold, You desire truth in the inward parts,
> And in the hidden part You will make me to know wisdom.
>
> Purge me with hyssop, and I shall be clean;
> Wash me, and I shall be whiter than snow.
> Make me hear joy and gladness,
> That the bones You have broken may rejoice.
> Hide Your face from my sins,
> And blot out all my iniquities.

Create in me a clean heart, O God,
And renew a steadfast spirit within me.
Do not cast me away from Your presence,
And do not take Your Holy Spirit from me.

Restore to me the joy of Your salvation,
And uphold me by Your generous Spirit.
Then I will teach transgressors Your ways,
And sinners shall be converted to You.

Deliver me from the guilt of bloodshed, O God,
The God of my salvation,
And my tongue shall sing aloud of Your righteousness.
O Lord, open my lips,
And my mouth shall show forth Your praise.
For You do not desire sacrifice, or else I would give it;
You do not delight in burnt offering.
The sacrifices of God are a broken spirit,
A broken and a contrite heart—
These, O God, You will not despise.

Do good in Your good pleasure to Zion;
Build the walls of Jerusalem.
Then You shall be pleased with the sacrifices of righteousness,
With burnt offering and whole burnt offering;
Then they shall offer bulls on Your altar.

This kind of repentance isn't just cathartic; it's catalyzing. It is not just about feeling bad; it's about getting free. Don't mishear what I am saying. This did not change anything David had done; it changed David.

You can change too. You don't have to stay stuck in Sardis. The doors are locked only on the inside. If you will hear what the Spirit is saying, you can overcome.

IF YOU WILL HEAR WHAT THE SPIRIT IS SAYING, YOU CAN OVERCOME.

CONQUER YOUR INNER SPACE

All you need to do is what I did that day in Florida when I touched the moon: reach out. Only you must reach out and touch a far greater and far older rock—the Rock of Ages.

> Rock of Ages, cleft for me,
> let me hide myself in thee;
> let the water and the blood,
> from thy wounded side which flowed,
> be of sin the double cure;
> save from wrath and make me pure.

This old hymn says it all for me, especially when the third verse reminds me that I have nothing at all in my own hands that I can bring; it is simply to *his* cross I cling.

Before you move on, quiet yourself and read these lyrics through one more time and sit still with them for a minute or two. You will go into your day ready for the challenges, stress, and blessings waiting for you, dressed to the nines where it matters most—on the inside.

CHAPTER 18

ON SHUTTLES AND ROADS

ABRAHAM LINCOLN DIDN'T LIVE LONG ENOUGH TO BECOME AN astronaut, but he did have a hand in the space program in an unusual way.

NASA has always relied on dispersed manufacturing. No one facility builds all the pieces for a rocket or vehicle. During the Apollo program, when the spacecraft was being built in the early sixties, its pieces were manufactured in different congressional districts in the country. The space shuttle was the same way later on. Its twin solid fuel engines were built in Utah. The plan was to transport them to Florida by train, so they could be no wider than what could be transported on the four feet, eight-and-a-half-inch width of the train tracks that ran between Utah and Florida and had to fit through the tunnels along the way. It's interesting to think that a major design feature on a spacecraft that was literally going to be key in building the space station in actual outer space was limited by something as archaic as a train.

After the Civil War in the 1860s, when the groundwork was being laid for the transcontinental railroad, Lincoln made the final decision on how wide the train tracks were going to be. He chose the now-ubiquitous four feet, eight-and-a-half inches to match the gauge of the most important Eastern railroads, setting the standard for North American tracks. This width was originally pulled from the width of the tram lines in England, which had been retrofitted from preexisting paths made for horses and carts. These paths were all precisely the same four feet, eight-and-a-half inches in width because that was how wide the roads built by the Romans were, and that was what they based theirs on.

If you are keeping track (pun intended), this all means that a critical engineering aspect of one of the most sophisticated human inventions ever built was decided based on a Roman path designed more than two thousand years ago.

Adam Morgan and Mark Barden use this idea in their book, *A Beautiful Constraint*, to point out that some aspects of modern life are the way they are not because they are best, but because they are rooted in the past. They have simply become such an integrated part of the way we all work now that they are too difficult, or too expensive, to change, even though better solutions might be available.

I agree wholeheartedly. It's always worthwhile to ask if something is being done simply because it has always been done that way or because it is the right thing to do.

Square, Airbnb, Uber, *Seinfeld*, Facebook—these were all successful because they broke free of what was normal, expected, and accepted. They allowed themselves to be a different size than the path that was handed to them.

Beware: As you lead and invent and are willing to see things differently, you will likely be ridiculed, and maybe even written off. If you break free from what is conventional and celebrated you will initially face shock, and maybe even criticism, before you ever see adulation and then imitation.

It's worth it. God is doing a new thing, and I want you to know you have every right to see that the fact that something has never been done before might just mean you are meant to be the first to do it.

Let's look at the idea of the train track from Utah to Florida—not as a negative, but as a positive. Viewed in this light, you might feel profound gratitude and appreciation for how the events of the past prepared and paved the way for the future.

When I first read about the connection between Roman roads and the space shuttle, I felt like I had been hit by lightning—which as you might remember is what happened to Apollo 12 right after launch. This was not only because quirky historical facts like that are what I live for but also because of the spiritual import of a two-thousand-year-old road shaping the engines of a vessel that would soar through the heavens. I shook my head. As I sat there in stunned silence, it dawned on me that Jesus is that road. He said so himself.

"I AM" #6

Jesus said to him, "I am the way, the truth, and the life. No one comes to the Father except through Me."

John 14:6

As I thought about the implications of these words, I immediately thought about the word *way*. I grabbed a Bible dictionary and my suspicions were confirmed. The word for *way* in Greek is *hodos*. It appears 101 times in the Bible. And though it can mean *way*, it can also be translated *means of access*, *approach*, *entrance*, or *road*.

Eugene Peterson captured this perfectly when he rendered John 14:6 in *The Message*: "Jesus said, 'I am the Road, also the Truth, also the Life. No one gets to the Father apart from me.'"

The only means by which you can soar through the heavens is by traveling down a two-thousand-year-old road. This is the exploration that will lead to true transformation.

HONEST DOUBTS

The context for this stunning, all-important announcement from Jesus is even more powerful when you understand that it came in response to something we often look down on. Doubt. You probably aren't proud of your doubts, and you aren't likely to think highly of yourself when you find them creeping in. But doubting is not only normal as a human, it is to be expected. And if it hadn't been for doubt, we never would have been given this vital piece of information where Jesus clarified, in no uncertain terms, that he is not just a way to get to heaven but the only one. Information that is incredibly important to know.

We have Thomas to thank. Yes, Thomas, the disciple who has gone down in history as the doubter. I'll have you know that Thomas's superpower wasn't that he had doubts (like I said, that's all of us) but that he was honest about them. All the disciples were doubters; they just did a good job of pretending they weren't.

Thomas was up front and open about what he didn't know. As a result, he was given answers he wouldn't have otherwise heard. That is what happened in John 14. Jesus was giving a speech about how after his death on the cross and resurrection he was going to leave them and go to heaven. They began to freak out, wearing it all over their faces. Jesus responded with a pep talk:

> "Let not your heart be troubled; you believe in God, believe also in Me. In My Father's house are many mansions; if it were not so, I would have told you. I go to prepare a place for you. And if I go and prepare a place for you, I will come again and receive you to Myself; that where I am, there you may be also." (John 14:1–4)

If you are a Christian, this world is not your home. You live here and are meant to view yourself as in the world but not of it. You are a resident alien. Max Lucado said, "The greatest calamity is not to feel far from home when you are, but to feel right at home when you are not."

Jesus was assuring them it was going to be okay. Then he followed up with a little more information that assumed they *were* tracking with him because he also wanted to be sure they were being honest with themselves: "And where I go you know, and the way you know" (v. 4).

As he said this you can almost imagine eleven of his disciples nodding their heads as if to say, "Yes, yes we do. We do know exactly where you are going." Like the penguins of *Madagascar*.

Thomas wouldn't have it. He couldn't believe the disciples were just going along with all this like the universe wasn't about to come to an end. He butted in and let Jesus have it straight: "Thomas said to Him, 'Lord, we do not know where You are going, and how can we know the way?'" (v. 5).

I picture Jesus feigning surprise and turning to the other eleven as if to say, *Is he right?*

In my mind they act surprised by the question, and for a split second seem like they are either going to pretend like it's not true or feign outrage at Thomas for outing them in their phony know-it-all attitudes. In the end it seems they shrugged and owned up to the fact that he was right and accepted the truth of their ignorance.

Without Thomas's honest, doubt-fueled interruption, the world might never have received Jesus' to-the-point, blunt, and helpful clarification. He used plain words: *You do know the way. It is me. I am the way. I am the road. And I am the only one.* Translation: *There is no getting to God except through me.*

In our day such stark positioning is thought to be intolerant. It causes our "everyone gets a trophy" cultural hairs to stand up on the backs of our necks if someone suggests that a sincere person could be sincerely wrong. We want everyone to get to be right. Except that Jesus is saying that the only way to get right with God is to go through him.

There is just one way to heaven because heaven only sent one Son. That one is Jesus, and two thousand years ago he walked the dusty roads of the Roman Empire so that today you could launch into the freedom of being fully alive through his Spirit. And when you die and leave this world, you have his footsteps to walk in, many mansions prepared for you to dwell in, and ultimately a newly created heaven and earth for you to live in.

TWO THOUSAND YEARS AGO HE WALKED THE DUSTY ROADS OF THE ROMAN EMPIRE SO THAT TODAY YOU COULD LAUNCH INTO THE FREEDOM OF BEING FULLY ALIVE THROUGH HIS SPIRIT.

In that final, awesome day, there will be many questions to ask each other—when we arrived, how we died, what happened to us during our time on earth. But you won't need to ask anyone how they came to be in heaven. Every single one of us will have this in common: we will all have gotten there through Jesus—the only road.

BEFORE TIME [ZONES]

Though he would not live long enough to see the railroad completed in 1869, Lincoln didn't just inadvertently influence the Space Shuttle Program when he signed the legislation into law establishing the transcontinental railroad. He also influenced how we tell time today. When the final golden spike was laid of the seven-million-plus spikes it took to put down rails over the 1,800 miles from San Francisco to Omaha, the North American continent was, for the first time, accessible by train from coast to coast. This changed the nation in unfathomable ways. Humans went from traveling on land three miles per hour (walking speed), or a little faster if on an animal or wagon, to sixty miles per hour. Communication also became instant as telegraph lines were installed simultaneously with the rail lines. Cross-country travel went from a dangerous, daunting, months-long affair to something that could be done reliably in days. It also led to the creation of time zones from coast to coast.

Before the train began crisscrossing the country, time-telling was localized and could be different from place to place. But to keep everyone on the same schedule—for instance, in boarding a train—they needed a unified system. Every time you glance at your Apple Watch to check the time, the time you see is a reflection of what people long ago created based on the newly built railroad that Abraham Lincoln set into motion.

Jesus doesn't just want to change your life after you die. He wants to change the way you spend your minutes and your hours here on earth. How you walk, forgive, love. How you respond to getting dumped. We don't just come to the Father through Jesus when we leave this world but while we walk in it. No one is more ready to live than someone who is ready to die.

Does any of this seem impossible?

There is a way. His name is Jesus.

CONQUER YOUR INNER SPACE

Take a moment to reflect on the ancient Roman roads that shaped an aspect of the space shuttle and ask God to shape your life through the dusty, two-thousand-year-old—but still vibrant and relevant—words of Christ. As you think about what lies in front of you today, what is one simple thing you can do to take up your cross and follow him?

CHAPTER 19

STEAK AND EGGS

MY BODY KNOWS WHAT DAY IT IS WHEN I EAT BREAKFAST. Well, at least it knows what day it *isn't.* If I am eating eggs and beans and peppers with spinach, or Magic Spoon protein cereal, I know for certain—it's not Saturday.

I got into this diet called the slow-carb diet about six months ago. I read about it in a book by Tim Ferriss called *The Four-Hour Body.* I have actually started using a lot of things recommended by Tim. Cold baths, for one thing. Alaska Bear sleep masks from Amazon for $12.90. The Roost portable laptop stand. He lost me on psychedelic mushrooms, though.

The slow-carb diet for me has been the easiest approach to maintaining my weight because it is so simple. Don't eat anything white or that can be white—so no rice, no potatoes, no pasta, no bread, no quinoa, no grains, no sugar. Simple enough. Also no drinking calories, with the exception of really dry wine. Just proteins, fats, legumes, black coffee, and club soda.

It's a less-fussy keto.

The program encourages you to not overthink it and be willing to eat the same meal every day. I'm okay with boring. It's one less thing to think about. So most every morning I scramble eggs with either a meat or fake meat made out of plants and throw in some onions and peppers and then some spinach. If I'm in a rush I'll just heat up a can of Amy's low-sodium vegetarian chili. And *voilà*. High protein, slow carbs, good flavor.

For lunch it's almost always a burrito bowl with no tortilla or rice but some sort of meat and all the veggies. And dinner is usually cauliflower rice and chicken breast or some other meat.

My favorite rule about the diet is that you have to do a cheat day once a week. In Tim's words, "Everyone binges eventually on a diet, and it's better to schedule it ahead of time to limit the damage." His advice is to absolutely go nuts.

Most weeks I pick Saturday as my cheat day.

Tim said that many of the girls he has dated have been absolutely horrified at what he does to himself on his cheat days. For instance, he'll eat bear claws and éclairs and chocolate croissants until he just about makes himself sick.

I do the same throughout the day, though I am more about fresh-baked cookies, chips and salsa, Taco Bell, Ben & Jerry's . . . but the day always starts with waffles.

Daisy, Clover, and I are the waffle-making committee. We have it down to a science. We fry bacon and cook a huge stack of waffles, usually to some New Orleans brass band's jazz playlist going off the Sonos system. And we dance and laugh and pull waffle after waffle out of the iron and pop them in a two-hundred-degree oven, where they will wait for the feast to commence.

In the summer we eat this glorious meal outside in the backyard at a table under string lights. It is my best re-creation of the outdoor environment from the show *Parenthood*. In the winter we eat on the table supported by the cross that we talked about way back in chapter 5.

This meal is my favorite of the week. By far. And by the time we sit down to eat, it's not as much breakfast as it is brunch. It is always so wonderful to breathe deep and know that it is Sabbath, a day to rest, relax, delight our souls, and rejoice at the week behind us.

Part of the joy and the glory is knowing that the day before us is full of foosball, bike rides, and the hot tub. Maybe if we feel really energetic, we'll go for some tennis or get the stand-up paddleboards wet and we will cannonball into a body of water.

Every Sabbath is a picture of the delight and glories of heaven where the mountains will run with wine (Joel 3:18) and the trees will clap their hands (Isa. 55:12), and I will eat Cap'n Crunch for breakfast on a Wednesday and not gain a pound. Between now and then I have to curb Saturday's freewheeling eating by 7:00 p.m. or I will need to take a TUMS before bed to avoid heartburn.

BREAKING THE CODE

Buzz, Neil, and Michael had steak and eggs in their stomachs when they left the earth for the moon. Alan Shepard had eaten bacon-wrapped steak and eggs the morning before becoming the first American in space in 1961. (He flew to the moon on Apollo 14 after his ear surgeon released him and allowed him to return to the astronaut pool. Shepard also has the distinction of being the first man to play golf on the moon; he brought a golf ball and a makeshift 6-iron club and drove three balls that sailed, in his words, "miles and miles.") His preflight steak and eggs became a NASA tradition for astronauts bound for the heavens. It was chosen for the most important meal of the day from that moment forward.

It is likely that Alan chose this particular meal for two reasons. The high-protein, low-residue makeup of the meal kept him full while making an in-flight number-two potty emergency less likely. And Alan was a navy man. The navy has a long-standing preference for steak and eggs as the breakfast of champions on big days. In fact, steak and eggs was on the breakfast menu at one of the most critical moments in American history: the Battle of Midway.

Norman Jack "Dusty" Kleiss, a veteran of World War II, said in the Netflix documentary *Greatest Events of WWII in Colour*, "The day of battle had come. I dressed and went to the officers' mess. The delicious smell of steak and eggs wafted through the halls, a telltale sign the cooks expected us to have a bloody morning."

The breakfast sent them a message.

"Every time they served us steak and eggs on the aircraft carrier you knew that was going to be a rough, rough day."

Pearl Harbor was just six months behind them. It was a day that, as President Franklin Delano Roosevelt put it, would "live in infamy." Nearly the entire Pacific Fleet had been wiped out in a surprise Sunday morning attack in Hawaii. However, the aircraft carriers had been out to sea and were spared.

The Japanese intended to finish the job at Midway. Their plan was to lure the American naval fleet back to the island by attacking it with a small force, assuming that the response would be to steam out with the remaining ships. Only then would their four concealed aircraft carriers, lurking a few hundred miles off the island, come out, crush the US fleet, and win the Pacific War once and for all.

But a code breaker in Hawaii, Thomas Dyer, had intercepted the transmissions and knew what was going to happen. He tried desperately to get his superiors to listen as he sounded the alarm.

No one would listen.

He persisted and got through to Chester Nimitz, the commander of the Pacific Fleet under FDR. They ran a test and found out that indeed Midway was going to be a trap, so Chester and our remaining forces showed up early, in force and fury, and caught the Japanese off guard. In a period of thirty minutes, they devastated the Japanese strike force. The battle had more twists, turns, and dramatic moments than I have time to do justice to, but the bottom line that day was, the Imperial Japanese Army—in that one battle—lost four aircraft carriers, nearly three hundred planes, and as many as three thousand men, including Japan's most experienced pilots.

It was the decisive turning point in the Pacific theater of World War II and gave our nation a much-needed sense of momentum and victory after the tragedy at Pearl Harbor.

This also revealed how vital the right equipment was to victory. It gave us the incentive to scale up naval production in incredible ways, and by 1943 we reached a point of churning out a state-of-the-art aircraft carrier every single month.

At 10:25 a.m. the day of the battle, the Japanese were winning the war. Five minutes later, the pendulum had swung almost completely to the other side.

Almost never do you find the tipping of history so swift and so decisive in such a tiny moment. But none of it could have happened if the code hadn't been broken.

That is why God has had you tiptoeing in hushed admiration through the agonizing moments of the cross in the pages of this book. It is our Midway.

Jesus broke the code against us by nailing it to his cross. Colossians 2:14 says that "Christ has utterly wiped out the damning evidence of broken laws and commandments which always hung over our heads, and has completely annulled it by nailing it over his own head on the cross" (Phillips).

JESUS BROKE THE CODE AGAINST US BY NAILING IT TO HIS CROSS.

LAST WORDS #6

"It is finished!"

John 19:30

The sixth statement Jesus spoke from the cross was perhaps the most powerful. Certainly it was the most triumphant. It's three words in English, "It is finished," but just one in Greek: *tetelestai*. This has been called "the greatest single word ever uttered." It was not the cry of a victim but the shout of a victor. And as it thundered from his lips it reverberated through the corridors of heaven and hell alike, causing angels to rejoice and demons to tremble.

The code had been broken.

Satan's chokehold on humanity had come to an end. No more would man be separated from God. No more would sin reign unchecked. No more would death be able to terrorize.

It.

Is.

Finished.

Mark's account tells us that hearing this word, *tetelestai*, and the way Jesus said it was instrumental in the conversion of the centurion (the Roman soldier in charge of his execution). Mark 15:39 says, "So when the centurion, who stood opposite Him, saw that He cried out like this and breathed His last, he said, 'Truly this Man was the Son of God!'"

In James Stalker's book *The Trial and Death of Jesus Christ* he says, "[This word] may be said to comprehend in itself the salvation of the world; and thousands of human souls . . . have laid hold of it as the drowning sailor grasps the life-buoy."

But what does it mean?

Tetelestai, a very common, everyday word in that day, was used in more than one context. It was a word used by servants when they finished a job their master gave them. "*Tetelestai*. I did the job you wanted me to do." It was a word used in art. "*Tetelestai*. Nothing more is needed. It's perfect—*voilà!* Quit now." It was a word used by priests. The law required that sheep brought to be sacrificed be without spot or blemish, as you were supposed to give God your best (Ex. 12:5). When the priests inspected the animal, if it was suitable, they would say, "*Tetelestai*. This sacrifice is without flaw. It is perfect."

What's wild about this is that Jesus died on Passover—the holiday connected to the full moon, the same day that the nation was all bringing their Passover lambs to be slaughtered. Josephus says that between the ninth and the eleventh hour (three to five in the afternoon) lambs were brought to the temple to be sacrificed.

Jesus died right around 3:00 p.m. That means that thousands of people were in line with lambs in their arms. And just as the first priests were saying, "*Tetelestai*," and slitting the throats of the lambs, right outside the city Jesus said the exact same thing: "It is finished."

He was, as John the Baptist said, "the Lamb of God who takes away the sins of the world" (John 1:29).

PAID IN FULL

One final use of the word was in banking. When you finally paid off something you had financed, they would stamp "*Tetelestai*" on the note. Paid in full.

Archaeologists have found many ancient business documents on papyrus marked with the word *tetelestai*.

When we first moved to Montana, I wanted to buy a new bicycle. It was a black-and-turquoise Bianchi steel and carbon fiber road bike with Ultegra components. It was sleek and old-fashioned in all the right ways but with new technology in the beneficial places. It nodded to a different time of Italian cycling tradition while tipping toward modern invention.

I was smitten.

But the price was more than I could afford. The store offered a no-interest layaway plan, so I put a down payment on it. Every couple weeks throughout the winter and early spring I would bring by fifty or one hundred dollars and chip away at the balance. It felt like it would take forever, but I am a big believer in the one-bite-at-a-time principle. As unappetizing as it may seem, if you keep chewing, you can eventually eat an elephant.

Months later I was still hundreds and hundreds of dollars away from the finish line on the transaction. My plan was to have it paid off by summer. And then in March I broke my femur.

I told the whole story in my second book, *Swipe Right*, where of all things I compared the snowmobile accident that nearly killed me to having sex. I told you I was weird.

We were just over one year into the new ministry at Fresh Life, and this accident pushed us to our limits in every way. I preached with the help of pain pills on Easter Sunday, one week after the accident. They brought out a little lounge chair so I could speak with my battered, full-of-titanium leg propped up.

I have watched the video footage from that message. It is not pretty.

At one point, while slurring my words and trying to focus on what I was saying, I looked really put out when somewhere in the room a baby began crying. I looked up in confusion, and then frustration, and then sarcastically quipped, "Someone didn't get the memo."

My show-must-go-on impulses don't always serve me well. Rather than play hurt, it would have been far better to let someone else speak.

Lesson learned.

There were actually a lot of lessons to learn in that season. Ambulances are expensive. So are surgeries. And physical therapy. And prescriptions. And the lost income from missed outside consulting work and speaking engagements, which was a necessary subsidy in the early days of our church plant.

It all ate at the extra money in my budget. And it seemed all but impossible to make any headway toward my dream ride.

I tried not to think of it, but I was discouraged. The summer would come and go, and I would not be riding the bike of my dreams. I would be limping down the sidewalk forever. To add insult to injury, one day someone stole my cane.

I was in San Diego speaking at a surf camp. It was the first time I had been back on a plane or booked to speak. After clearing it with my physical therapist, I decided to try to go surfing. It would be good to strengthen my weakened left leg that had atrophied to half the size of my right one.

I've never been an extraordinary surfer, but I do have good balance and have always been able to pop up on the board with no problem after catching a wave.

It was humiliating.

I caught wave after wave but when my brain would tell my body to spring up, and I tried to lunge up with catlike reflexes, my left leg refused to cooperate. It was clumsy. I was clumsy. And I never got off my knees.

When I came back to shore, the cane, which I had left lying on my towel, was gone.

The only thing worse than having to use a cane is not having a cane to use when you need one.

I arrived home from that trip demoralized.

The next day was my birthday.

Exactly twelve years before I began this book, we had a few friends over to our condo to celebrate that birthday. I was still using a cane and was bruised inside and out by the accident. To my surprise, at the end of the party my friend Kevin went to his car and returned with a bicycle that had a ribbon on the handlebars. Not just any bicycle. It was the black-and-turquoise Bianchi steel and carbon fiber forked road bike with Ultegra components.

My bicycle.

"But how can it . . . I haven't . . . how did you . . . what is . . . ?" I sputtered like a speedboat running on the wrong kind of gasoline. Kevin had the biggest grin on his face as he moved toward me with the bicycle. He handed me a piece of paper and said, "Ron and Jan wanted you to have this." Ron and Jan were the owners of the store where I had been making payments on the bicycle. Mom-and-pop owner-operators if there ever were some.

"Happy birthday!"

On the paper was the name of the bicycle shop, Glacier Cyclery & Nordic in Whitefish. Below that was the model number and make of the bicycle. Then my name and the original down payment I had made—a couple hundred dollars. Below that was every amount of money I had paid, fifty dollars here, one hundred there. Below each was the new balance I owed, adjusted each time. Like an old-timey accounting book from a different era. The final figure I owed was still a thousand dollars. But that number was scratched out. Written over all the numbers, in a bold pen, were three words:

Paid in full.

Every single time I rode that bike, for years and several hundred-mile century rides, I flew down the road because of a price someone else paid.

When Jesus said "*tetelestai*," he was saying that your attempt to pay your own bill was now and forevermore unnecessary.

Your bill has been paid.

WHEN JESUS SAID *"TETELESTAI,"* HE WAS SAYING THAT YOUR ATTEMPT TO PAY YOUR OWN BILL WAS NOW AND FOREVERMORE UNNECESSARY.

How? His blood became money. Ephesians 1:7 tells us, "In Him we have redemption through His blood, the forgiveness of sins, according to the riches of His grace."

You might not have thrown this book to the ground in your hurry to run around the room celebrating, but please tell me this makes you want to! You don't have to earn God's approval, and you can't lose it. It doesn't rise and fall with your parenting ability, financial success, social media clout, or lack thereof. It is as unchanging and constant as the moon's nightly rise.

Salvation can't be added to or subtracted from. It's once and for all. A perfect work of art. A masterpiece. The cross is Jesus' last word on your standing before God. It can't be improved upon or altered in any way. No aspect of conquering inner space is more challenging than truly believing that and seeing yourself the way God sees you.

I say this because nothing is so deadening to the soul as a performance mentality—in other words, attaching your identity to your activity. It goes up and down with your good days or bad days. But you can't ever be loved any more than you are right now.

The devil loves to condemn us, to try to rub our noses in what has already been forgiven. It's his name—Satan, Accuser of the Brethren (Rev. 12:10).

Know this: God takes no joy in you wearing shame that Jesus already wore. It doesn't bless God to see us try to pay penance for something he already paid. It hurts his heart.

Hold your head high, wear your ring, and go to the feast (Luke 15:21–24). Celebrate what Jesus did for you!

At Mission Control in Houston after Apollo 11 flew, a sentence from JFK's May 25, 1961, message to Congress was flashed on the large screen: "I believe that this nation should commit itself to achieving the goal, before this decade is out, of landing a man on the moon and returning him safely to the Earth." An Apollo 11 logo appeared on the NASA screen, alongside Kennedy's statement, with the words, "Task Accomplished—July 1969." They'd honored their late president in the best way they possibly could—by completing the mission he had sent them on.

Jesus spent much of the last portion of his three-year ministry preparing his followers for his death. Matthew 16:21 says, "From that time Jesus began to show to His disciples that He must go to Jerusalem and suffer many things from the elders and chief priests and scribes, and be killed, and be raised the third day." As he hung on the cross and spoke the words, "It is finished," there were no steak and eggs in his stomach—just the bread and wine that was his final meal before being arrested. His task was accomplished.

Guilt? Finished.

Condemnation? Finished.

Loneliness? Finished.

Trying to measure up? Finished.

Insecurity? Finished.

Separation? Finished.

After his resurrection, Jesus went to great lengths to convey to Peter just how real the shame-destroying power of forgiveness was. Peter had failed and failed big, denying Jesus three times after vowing he would die before betraying him.

When he first realized what he'd done, "he went out and wept bitterly" (Matt. 26:75).

You can probably relate on some level or other to failing God and feeling unworthy. Peter said to his friends, "I'm going fishing," which wasn't so much a way to kill time as it was a way of saying, "Party's over." This was his job before Christ. Picking up the fishing game was a way of saying, "There is no way Jesus would want me still on his team."

Jesus sought him.

John's gospel says he cooked a meal over a bed of coals for Peter (John 21:9). This is significant. Peter denied Jesus over a fire (Luke 22:54–57), and so Jesus forgave him over one. He was saying to Peter what our iPhones regularly announce to us: "You have a new memory." He no longer had to be triggered by his self-inflicted trauma whenever smoke wafted into his nostrils.

There is a connection between smell and memory. I have no doubt that for the rest of Peter's life, the smell of smoke would cause him to be triggered in the way any perfume or other aroma can take you back to a painful memory.

How tender of Jesus to redeem a fragrance. To overwrite an association.

Breakfast can send a message. And what Jesus was saying to Peter in this meal, he says to you: *When the stone was rolled away in the garden on that Easter Sunday, with it I rolled away your shame.*

Good night, failure. Good morning, grace.

CONQUER YOUR INNER SPACE

Perhaps today is a good day to have breakfast for dinner. Plan on it. It can be as simple as Froot Loops (steer clear of Cap'n Crunch if you value the roof of your mouth) or as elaborate as waffles from scratch. Obviously if you can swing it, the NASA-approved meal of steak and eggs would be more than appropriate.

Put some worship music on as you cook, or if you choose to go to

IHOP, as you wait for your order and before you feast, say a prayer. Thank God for the power of *tetelestai* and that it is finished and your sins are paid in full!

And if you don't mind, post a photo with this book and your meal and tag me (@levilusko) so I can share the joy!

CHAPTER 20

TATTOOED SOUL

THE LAST MAN TO STAND ON THE MOON DID SO IN 1972 AS A part of Apollo 17: Gene Cernan. We haven't been back since. In my lifetime, the moon has once again sat out of reach, but at the time of this writing, it's coming back. I am so excited for the return! As we discussed earlier, the new mission is called Artemis; it will include the first woman and the first person of color on the moon.

I've been so eager for my daughters to experience this incredible science of space exploration—especially Clover, who has long revered Amelia Earhart.

Her reaction was priceless when I told her that in December 2020 China landed an unmanned spaceship on the moon and planted a flag there *using a robot*.

"That doesn't count," she said, outraged that humans weren't present. "If I ever went there, I would take it down."

Apollo 17 was the sixth flag placed by a human being and was also unique in that it was the first time an astronaut who hadn't been a pilot was included. Harrison Schmitt was a geologist, and the only scientist from NASA's Group 4 who got to go to the moon. Gene and Harrison hopped in a lunar vehicle they brought along. It had been packed like a jigsaw puzzle under their lunar module, and they had to assemble it on the moon's surface in their space suits. They drove about twenty-one miles, exploring more of the moon than ever before. Compared to Buzz and Neil's twenty-one hours, they spent three days on the moon, did three different excursions or extravehicular activities (EVAs), and got almost no attention from the people of planet earth. That was basically five decades ago.

When we do return to the moon it will not be as a destination but as a layover on the eventual path to going to Mars. Though it is much farther away, we know a hundred times more about Mars now than we knew about the moon when we sent humans to the lunar surface. And the moon will be a critical base of operations and fuel station for vehicles coming and going to the Red Planet. It will be a stepping-stone.

THE CHURCH IN PHILADELPHIA

With that in mind, read the words Jesus wrote to the church in Philadelphia in Revelation 3:7–13.

> "These things says He who is holy, He who is true, 'He who has the key of David, He who opens and no one shuts, and shuts and no one opens': 'I know your works. See, I have set before you an open door, and no one can shut it; for you have a little strength, have kept My word, and have not denied My name. Indeed I will make those of the synagogue of Satan, who say they are Jews and are not, but lie—indeed I will make them come and worship before your feet, and to know that I have loved you. Because you have kept My command

to persevere, I also will keep you from the hour of trial which shall come upon the whole world, to test those who dwell on the earth. Behold, I am coming quickly! Hold fast what you have, that no one may take your crown. He who overcomes, I will make him a pillar in the temple of My God, and he shall go out no more. I will write on him the name of My God and the name of the city of My God, the New Jerusalem, which comes down out of heaven from My God. And I will write on him My new name.

'He who has an ear, let him hear what the Spirit says to the churches.'"

Many years ago I had the chance to travel to the East Coast and spend some time in Philadelphia. Though I had always been a fan of their cream cheese and Rocky Balboa, I had never been there before. Philadelphia is also known as the City of Brotherly Love, so I was thinking that I'd maybe get hugs everywhere I went. I saw the Liberty Bell (it was smaller than I thought it would be), Independence Hall, and Benjamin Franklin's grave. I even had a great cheesesteak. But also while there I was told that the city has one of highest murder rates in the country—an honor that has also been given to our nation's capital in Washington, DC. For this reason Philly has earned the nickname "Killadelphia." One local pastor I met said it's also jokingly referred to as "the City of Brotherly Shove."

Much like the violence in Pennsylvania, the church in ancient Philadelphia was dealing with violent opposition, primarily from the Jewish community. But let's back up.

Philadelphia was a small city on a hillside about thirty miles southeast of Sardis. It was the youngest of the seven cities, founded in 189 BC by a king who loved his bro—thus the name City of Brotherly Love—and was nicknamed "Little Athens" because it had so many temples in the city.

It was a center of Greek culture built on an important stretch of the imperial road, so it could hellenize (spread the language and Greek way of life to) the whole area of Lydia. They were located on a junction of several major trade routes, which earned it another nickname—"Gateway to the East." This connects with Jesus' comments about a great door of opportunity being open.

They were known for their vineyards and their wine production, and the patron deity of the city was Bacchus (Dionysus), the god of wine.

We have no idea how the church there began; it's possible that it was an offshoot of the ministry going on in Ephesus. Eusebius recorded that the church was led in the first half of the second century by Ammia, a prophetess, and the work there thrived under her gifted leadership.

These people didn't act one way sometimes and then another way other times. They were reliable and consistent. What you saw was what you got. And, as we will see, that means more to Jesus than a church with more works but that had lost love (like Ephesus), or had lost doctrinal purity (like Pergamum), or had lost sexual purity (like Thyatira), or had died (like Sardis).

Whatever could be said about Philadelphia, they weren't hypocrites. They were the same people on Sunday as they were on Friday night.

Jesus' first remark to this faithful church was that they were dependable. This might not seem like that great of a compliment, but as a leader I can tell you—there is perhaps no greater virtue to be found than in those who serve consistently, quietly, and faithfully. And really, don't we all want relationships defined by this?

It was important to Jesus, who in verse seven identifies himself as both holy and true. *Holy* is a word we are familiar with. The Greek word is *hagios*, which means "set apart," or other than. It's fine china.

It is also one of the fundamental attributes of God and the foremost thing we are called to be as Christians. And a counterintuitive way to true happiness.

God said, "Be holy; for I am holy" (Lev. 11:44).

Sanctified is another word for holy. It is God's will for your life (1 Thess. 4:3).

God also calls himself "true"—an interesting word that can be defined as sterling, real, or not fake. Jesus is saying, *I am holy, and genuinely that way. And so are you.*

Philadelphia was dependable, not fool's gold like some of the seven churches were.

NAME-BRAND FAITH

I once got ripped off on eBay buying Diesel "designer jeans" from Italy—except they had been made in someone's basement in Cleveland.

Jesus is looking for the real thing—the genuine article—not pretend holiness. He does not want his name on a counterfeit.

I'll never forget my first tattoo.

Someone told me that if you want a tattoo, you should design it, put it in a drawer, and wait for a year. Only get it done if you still like it after the year is over. I did this with a design based on my four daughters. But when my daughter Lenya suddenly and unexpectedly left this world and went home to heaven, I ended up getting a tattoo of an anchor instead, since it symbolized the hope we have of being reunited in heaven. I did not wait a year or deliberate a long time about that. As I wrote in my first book, *Through the Eyes of a Lion*, we had the anchor symbol painted on her casket and

JESUS IS LOOKING FOR THE REAL THING—THE GENUINE ARTICLE—NOT PRETEND HOLINESS.

emblazoned on her tombstone. And both Jennie and I wanted it on our bodies, as well, as a reminder of our connection to her through the Holy Spirit.

The one who now holds her lives in me.

I eventually did get the original tattoo I had planned—an olive branch for Alivia, a daisy flower for Daisy, the lion for Lenya, and a wild clover flower for Clover—each individually inked in the four quadrants of an X over my left bicep. After Lennox was born I was thinking I would need to add a tattoo for him until Alivia pointed out that the X that surrounded his sisters was already representing the son we didn't know we were going to have. How cool is that? I picked the X as a design feature having no clue that it would be more than appropriate for a boy with an X in his name.

LETTER #6 (PHILADELPHIA)

"I have set before you an open door, and no one can shut it."

Revelation 3:8

Jesus found nothing to chastise in the church in Philadelphia and everything to praise. They, along with the church in Smyrna, get a glowing report card. And their reward? Two things. First, Jesus promised to write his name on them—to tattoo it on them, if you will.

Companies fiercely protect their signature, their logos, making sure if their brand is on something that it is per their standards. If they find out someone is using their name without authorization, a cease-and-desist letter from an attorney quickly follows. No doubt Jesus, looking at how some "followers" of his sometimes live (including me), would prefer if we didn't tell anyone that we are his people. *Not Philadelphia*, he says in effect. *I'll gladly put my name on you. What is going on in your church matches up to my name.*

And secondly, Jesus opened up a door of opportunities to them that no one could shut. In other words, their faithfulness had for them become a stepping-stone, a pit stop on the way to new and awesome opportunities they couldn't experience any other way.

The kingdom of God, as it is journeyed through, releases more and more treasures to those who persevere. There is no end to it. The higher you go, the more there is to know. You can never run out of room to grow. What's true of the love of God is true of the will of God. It has no end.

Paul said to the Ephesians, who, like the Philadelphians, were crushing it:

> May you have the power to understand, as all God's people should, how wide, how long, how high, and how deep his love is. May you experience the love of Christ, though it is too great to understand fully. Then you will be made complete with all the fullness of life and power that comes from God.
>
> Now all glory to God, who is able, through his mighty power at work within us, to accomplish infinitely more than we might ask or think. (Eph. 3:18–20 NLT)

I love how so much of what God has for us looks small until we get into it.

One of my favorite books in the Narnia series is the final one, *The Last Battle*. Listen to this excerpt that takes place after King Tirian is thrown into a small hut and discovers an entire world inside:

> Tirian had thought—or he would have thought if he had time to think at all—that they were inside a little thatched stable, about twelve feet long and six feet wide. In reality they stood on grass, the deep blue sky was overhead, and the air which blew gently on their faces was that of a day in early summer. . . .

"The door?" said Tirian.

"Yes," said Peter. "The door you came in—or came out—by. Have you forgotten?" . . .

Tirian looked and saw the queerest and most ridiculous thing you can imagine. Only a few yards away, clear to be seen in the sunlight, there stood up a rough wooden door and, round it, the framework of the doorway: nothing else, no walls, no roof. He walked toward it, bewildered. . . . He walked round to the other side of the door. But it looked just the same from the other side. . . .

"Fair Sir," said Tirian to the High King, "this is a great marvel." . . .

"It seems, then . . . that the stable seen from within and the stable seen from without are two different places."

"Yes," said the Lord Digory. "Its inside is bigger than its outside."

"Yes," said Queen Lucy. "In our world too, a stable once had something inside it that was bigger than our whole world."

Hello.

There is no running out of things to learn and things to experience and things to see as you walk with King Jesus. Those in the church of Philadelphia must have known that the reward for growth is the realization of how much more growing there is to do. In this sense, it is only when you think you can see that you are blind. When you admit you are still blind in certain ways, your eyes are opened to knowing that there is much that is yet out of sight (John 9:39–41).

Once you get to the moon, the possibility opens up to then go to Mars and beyond.

I love the freedom that comes from knowing that the doors Jesus opens, no one can shut; and that the doors he has shut, no one can open (Rev. 3:7). There is such freedom in knowing that when you walk in Christ you are unstoppable. Through the years I have been serving Jesus, I have often comforted myself with the fact that he will build his church and no one can stop him (Matt. 16:18). It has given me peace leading through storms of every single kind—death threats, legal battles, issues with properties we were trying to buy, outright slander, staffing issues, unfavorable and unfair newspaper articles, and beyond. Through every kind of attack we have faced, God has been faithful to see us through. And no matter what you are up against, he will do the same for you.

WHEN YOU ADMIT YOU ARE STILL BLIND IN CERTAIN WAYS, YOUR EYES ARE OPENED TO KNOWING THAT THERE IS MUCH THAT IS YET OUT OF SIGHT.

At the end of his letter, Jesus adds that he will keep the Philadelphians from the "hour of trial," which is going to come upon the whole world to test those who are still living on the earth.

Is this a reference to the rapture? I hope so. But even if he sees fit to allow us to be on the earth during whatever dark days lead up to Christ's ultimate triumphant return, remember this: no weapon formed against you can prosper (Isa. 54:17). Be one of the shining bright lights and seek the renewal of this world by participating in Jesus' mission. To be in Christ is to be safe no matter what danger you face.

G. K. Chesterton once said, "There is no such thing on earth as an uninteresting subject; the only thing that can exist is an uninterested person." I think curiosity is one of the most important character traits you can have. Bill Gates once said that he counts his willingness to be confused as one of his greatest assets. New subjects always confuse him and make him feel like a fraud. Many adults simply refuse to be confused. He presses on and eventually figures things out. What is your comfort level with confusion? I believe the amount of confusion you can tolerate will determine how much God can do through you. To do great things by his hand is to "walk by faith, not by sight" (2 Cor. 5:7). Confusion is guaranteed.

I hope hearing these words about vistas yet unexplored awakens some sense of adventure in your soul. If you are found faithful, like those in the original City of Brotherly Love, your reward will be greater levels of responsibility and opportunity. There will be new devils for the new levels you face—but that's par for the course of greatness.

Perhaps the greatest thing you must overcome in your quest to walk through open doors or to discover new vistas is not just the opposition around you but the darkness within you. In that great list of struggles I have faced (death threats and all), the most challenging and nonstop source of conflict in my ministry is managing *myself.* It's a full-time job.

THE RIGHT NAME

In his book *The Obstacle Is the Way*, Ryan Holiday pointed out that Apollo astronauts were trained in not panicking more than anything else. He described how this training was critical to their survival:

> At 150 miles above Earth in a spaceship smaller than a VW, [panic] is death. Panic is suicide.
>
> So panic has to be trained out. And it does not go easily.
>
> Before the first launch, NASA re-created the fateful day for the astronauts over and over, step by step, hundreds of times—from what they'd have for breakfast to the ride to the airfield. . . . They did it so many times that it became as natural and familiar as breathing. They'd practice all the way through, holding nothing back but the liftoff itself, making sure to solve for every variable and remove all uncertainty.

> Uncertainty and fear are relieved by authority. Training is authority. It's a release valve. With enough exposure, you can adapt out those perfectly ordinary, even innate, fears that are bred mostly from unfamiliarity. . . .
>
> John Glenn, the first American astronaut to orbit the earth, spent nearly a day in space still keeping his heart rate under a hundred beats per minute. That's . . . a man who had properly cultivated, what Tom Wolfe later called, "the Right Stuff."

Then Holiday quoted Publilius Syrus: "Would you have a great empire? Rule over yourself." The question, of course, is how do we lead ourselves well? Jesus taught and demonstrated that if we want to live right, we need to listen right. "He who has an ear, let him hear what the Spirit says to the churches" (Rev. 3:22).

The Greek word for "breath" is *pneuma*. Through prayer you allow God to give you a second wind that can keep you from panic when the noise of life's engines is roaring all around you. Prayer is often treated as a last resort. "All we can do is pray; it's in God's hands now," we say with resignation. Prayer should be your first resort!

The first time the transit authority in New York City ever used its communications system for anything besides subway operations was February 20, 1962, when it asked the passengers aboard its trains all over the island of Manhattan to pray for John Glenn during his flight. "The call to prayer would be repeated every ten minutes for almost five hours." It's intuitive to pray for big stuff. Your Father wants you to talk to him about that, but not just the big things; we can talk to him about everything. If it matters to you, it matters to him. The more you lean in, the better you will get at listening. Prayer taps you into power that can calm you in the midst of the craziness of life and help you overcome your ability to self-sabotage.

Listen up!

The poet John Greenleaf Whittier said, "Of all sad words of tongue or pen, the saddest are these: 'It might have been!'"

Those are words of regret. We all have our "might have beens."

Today, Christ has set before you doors of opportunity. No one can close them, but you can also choose to not walk through them. You are not left on your own to navigate them by yourself as in the secular Stoicism that Marcus Aurelius pointed to. As you overcome yourself and tune in to heaven's frequency, you have the wind of the Holy Spirit at your back!

In 2021 a giant vessel called the *Ever Given* got stuck in the Suez Canal, blocking traffic in one of the most important shipping lanes on earth. Experts saw no easy way to get it out. Fortunately, God had a plan: the moon. The full moon arrived for Easter and it altered the tides and raised the water level in the canal by eighteen inches, allowing the *Ever Given* to be removed and the logjam to be broken. When the moon was involved, doing what was all but impossible before was done easily. This is what your life can look like. No pushing and pulling and straining. There is a relaxed confidence that comes from God's power.

Speaking of regret, according to surveys, one of the primary reasons people in America regret a tattoo they have gotten is because they were impaired when they got it. (Bad idea.) But high on the list is because of whose name is written on it—e.g., they have an ex's name on them in permanent ink.

Don't die with the wrong name written on your life.

CONQUER YOUR INNER SPACE

We are a brand-oriented culture. Pepsi or Coke? Visa or MasterCard? Nordstrom or Neiman Marcus? A far more important question than

"What's in your wallet?" is, Whose name is written on your heart? It's not whether you have a Samsung or Apple phone that matters. It's whose logo is on your life. Do you belong to Jesus Christ? Have you allowed *him* to write his Word on you?

Why not allow Jesus Christ, the one who died on the cross to pay for your sin, to cleanse you from all the stains of your sin and fill your life with his love?

One word of caution before we go. The very fact that the church in Philadelphia was doing so well put them in heightened danger and in Satan's crosshairs.

If all is going well for you today, rejoice! And beware. But you can also be sure that with faith like this, the moon may be only a stepping-stone.

CHAPTER 21

NASA MEETS NAPA

IT'S IRONIC. ONE OF THE LEAST DANGEROUS THINGS A HUMAN does on earth is actually the most dangerous thing in space.

Astronaut Peggy Whitson is not only the first woman to ever command the International Space Station, but she has also spent more time in space than any other American. In total she has 665 days under her belt—to put that in perspective, that's long enough to make it to Mars and back. One of her trips lasted 289 days. Peggy also holds the Guinness World Record for most space walks by a woman, so I believe her when she says spacewalking is one of the hardest things you can do. Yes, in her opinion, the most dangerous thing isn't the aliens, or asteroids—it's good old-fashioned, one-foot-in-front-of-the-other walking.

Walking in space requires you to leave the safety of the spacecraft or station. And that means transition. You have to put on a space suit—which is technically a one-person spacecraft. It also means pressurization issues when doors are being opened and closed. It's the same in a submarine. Submarines can go down so deep that if they weren't pressurized, they would be crushed, along with everyone inside.

Before you can go out of the spacecraft you must first make sure the suit is sealed up airtight and that all life-support systems are working correctly. The Apollo space suits were all x-rayed before going off to space. If so much as a stray needle were left in one of the suits it could be catastrophic if the breach allowed the one-man spacecraft to leak. And suiting up can't be accomplished solo. You are dependent on someone else to help seal you up. A zipper or clasp not sealed properly on earth, where they train in a pool, could mean drowning; in space, it could mean suffocating or having your tears boil. Each suit was personally custom-tailored to the astronaut's body and was incredibly heavy. With a price tag of $100,000 (more than $740,000 in today's dollars), each one was made of twenty-one layers that provide protection against fire, cold, the vacuum of space, puncture, and abrasion. According to NASA, "Each Apollo mission required fifteen suits to support the mission. For the main, or prime, three-man crew, each member had three suits: one for flight; one for training; and one as a flight backup in case something happened to the flight suit. . . . The backup three-man crew each had two suits."

A vintage space suit on earth, including the backpacks and water-cooled underwear (not including the astronaut himself!), comes in at 280 freaking pounds. Of course, in an environment where there is one-sixth gravity, it's not as heavy, but still, it's bulky. I put on a spacesuit glove once while on a tour of NASA in Houston. The material was so thick I could hardly flex my fingers closed. I can't imagine how hard it would be to perform the delicate work they're required to do on satellites or space stations for hours on end while hanging on to the outside of a space station wondering if a meteor will come zipping by. Especially because they become even more inflexible when pressurized. After spending just over two hours in a suit like this, an exhausted Gene Cernan said they had "all the flexibility of a rusty suit of armor."

Nowadays most space walks take place on the ISS and have the benefit of an air lock. Once the astronaut is loaded up in the air lock on their way out of the craft and into space, the inner door of the chamber is closed and the room is depressurized. Then and only then the outer door is opened. At that moment it gets intense, because every time you open the hatch you are in danger. You can snag, hook, or otherwise hurt your space suit—all of which would be disastrous. If you don't get tethered in properly there is the very real threat of simply floating away.

How scary that would be. *Shudder.*

If your body is exposed to space, you will die instantly. It's a bit like what would happen if you or I in our sinful state were exposed to God's glory. Instant death. To handle space and God's glory you must be hidden inside something that protects you. That's why NASA painstakingly builds suits, and that's why Jesus offers his righteousness for us to step into and be safe inside. He offers life support.

A TIME TO WALK

The very first space walks took place during the Gemini missions. The Gemini program was sort of like the puberty of the American side of the space race. The one-man Mercury space capsule used in the program before Gemini was exciting because it was all new, and during those missions we proved we could go to space—but that was pretty much all we did. The men were "Spam in a can" as the tiny, one-person Mercury capsule gave them very limited piloting ability. They had some buttons to press but were mostly along for the ride. The original plan, in fact, was for it to be completely controlled from the ground.

This chafed the hide of the proud test-pilots-turned-astronauts, and, famously, it was the reason flying legend Chuck Yeager—the first to ever break the sound barrier—scoffed at the notion of working for NASA: because it was a job that a chimp could do. This is unfair and smacks of jealousy to me, but it was simply the birth of the American space program, and newborns don't do much. Unlike Mercury, the two-person-crewed Gemini had thrusters to alter its orbit and move where its pilot desired, short of leaving the earth's gravity (that would require a larger rocket, like what Wernher von Braun was cooking up). The Gemini missions were each designed to test and perfect a series of small movements that would be necessary if Apollo was to be undertaken, like rendezvousing, docking, and extravehicular activities (EVAs).

Russia beat us to the EVA punch, but on June 3, 1965, Ed White became the first American to walk in space when he exited his Gemini 4 capsule connected to a twenty-five-foot-long, gold-tape-wrapped umbilical cord that fed him oxygen. He flew freely, moved around with the use of a zip gun, and took pictures. He was so ecstatic that he had to be told multiple times to come back inside when the scheduled EVA was over. When he finally returned, he said, "This is the saddest moment of my life." He did so well on that mission that as a reward he was assigned to the crew of Apollo 1, where he perished.

Dying is not the only thing that can go wrong. In his book *An Astronaut's Guide to Life on Earth*, astronaut Chris Hadfield talks about getting anti-fog cleaner that had been applied to the inner glass of his space helmet into his eye while he was out on a space walk. Getting soap in your eye is never a picnic, but don't forget that (a) he had no way to rub his eye because his hands were in gloves and his head was in a bubble helmet and (b) tears don't fall in space. So no matter how much he blinked the tears, which should have flushed the invasive soap from his eye, it remained pooled in the center of his cornea.

Chris said it burned so badly that he temporarily lost the vision in his eyes. And if that isn't frightening enough, there's the added psychological factor of remembering he was outside the space station hurtling through low earth orbit at 17,000 miles an hour, held in place only by a safety tether, and even if he wanted to go back inside, he had no way to because he was blind. His ability to calm himself down and remain cool was the only way he got through that walk.

In your relationship with God, it is the opposite. Walking might be the most dangerous thing you can do in space, but in your soul, *not* walking is the most dangerous thing. This is what Jesus references in the final "I Am" declaration, in John 15:1–5:

> I am the true vine, and My Father is the vinedresser. Every branch in Me that does not bear fruit He takes away; and every branch that bears fruit He prunes, that it may bear more fruit. You are already clean because of the word which I have spoken to you. Abide in Me, and I in you. As the branch cannot bear fruit of itself, unless it abides in the vine, neither can you, unless you abide in Me.
>
> I am the vine, you are the branches. He who abides in Me, and I in him, bears much fruit; for without Me you can do nothing.

There are many places Jennie and I love traveling to, but I don't know of any more relaxing place than Napa, California. It is by far the most chill. Something about an agrarian culture is instantly soothing. Being in a place like New York City is stimulating; you feel the energy, power, and intensity the moment you hit Broadway, the stem of the Big Apple. Napa has the exact opposite impact on me. I found my blood pressure would drop the moment we'd land at Charles M. Schulz Sonoma County Airport and walk past the statue of Snoopy and Charlie Brown—which I hope makes you think of outer space and inner space for the rest of your life.

WALKING MIGHT BE THE MOST DANGEROUS THING YOU CAN DO IN SPACE, BUT IN YOUR SOUL, *NOT* WALKING IS THE MOST DANGEROUS THING.

Over the years we have toured a number of vineyards. Mile after mile and row after row of perfectly planted, purposeful grapevines growing their sweet fruit would call to me to relax my shoulders and breathe deeply. On each of our visits Jennie and I have felt ourselves settle into a deep rest—full of the joy in Christ, with each other, and through the beauty of our surroundings.

"I AM" #7

"I am the vine, you are the branches. He who abides in Me, and I in him, bears much fruit; for without Me you can do nothing."
John 15:5

I have often thought of John 15 as we've soaked it all in. The grapes are only growing because the branches are connected to the vines. If you broke off any one of them, the life-giving sap would immediately stop flowing and the branch would stop producing. It also is not lost on me that the vines growing up from the ground and following the trellis form a cross.

It is the same for you and me. We can only produce fruit as we walk with Jesus, allowing his life to flow through us. Paul describes it in Acts 17:28: "In him we live and move and have our being." The goal is to abide. To constantly remain in Jesus and Jesus in us. To allow his strength to flow through us step by step. It is this constant connection of prayer and worship, meditation and conversation that gives us the power to avoid the dangers inherent in this life.

Just as in a space walk, there are hazards that can threaten your walk with Jesus—things that can trip you, snag you, and stop you. Verses in Hebrews 12 put it this way:

> Do you see what this means—all these pioneers who blazed the way, all these veterans cheering us on? It means we'd better get on with it. Strip down, start running—and never quit! No extra spiritual fat, no parasitic sins. Keep your eyes on Jesus, who both began and finished this race we're in. Study how he did it. Because he never lost sight of where he was headed—that exhilarating finish in and with God—he could put up with anything along the way: Cross, shame, whatever. And now he's there, in the place of honor, right alongside God. (vv. 1–2 MSG)

With urgency and seriousness, you and I need to be on guard against anything that could compromise the purity of our connection to Jesus the Vine. Like an astronaut with a life-sustaining tether, so we must never become complacent in our connection to the Holy Spirit. Or, as in space, the result could be disastrous. Hebrews 2:1 reminds us, "We must give the more earnest heed to the things we have heard, lest we drift away." He is the tether; we are the astronaut. If we are disconnected from him, we will drift off into the blackness of space.

A TIME TO REST

The first time Jennie and I visited Napa it was harvest time. The vines were lush and leafy and the grapes were full on the vine, plump and picture-perfect. The workers were scrambling to get them off the branches and into crates because there is a very narrow window of time when they are perfect. In one instance we saw them working through the night with headlamps to get them harvested in time. We stayed at an Airbnb that had a small vineyard on the property growing Cabernet Sauvignon grapes. They were sweet and delicious. I noticed that none of the branches were straining. There was no pressure or trying. They simply remained.

Remain. We need this word for our culture and the frenetic speed at which we run to fulfill the need to accomplish and do and be. There *is* a place for effort and efficiency and honoring God with hard work. But we need to learn the superpower of resting in Christ. It is not doing things *for* God that will lead to fruit, but walking *with* God. "Walk with me and work with me—watch how I do it. Learn the unforced rhythms of grace. I won't lay anything heavy or ill-fitting on you" (Matt. 11:29 MSG).

I am a type-A personality and pragmatic to a fault, so I have the hardest time with resting. I have to remember that what God wants to grow in me, I can't do on my own. I need to allow him to do it. There is no Amazon Prime option.

Growing fruit is a slow process. It can't be hurried. What has helped me the most is intentional resting. Sitting still is hard. I have to work hard to not work hard. Listening to leaves rustling. Listening to the sound of birds. Interacting with my kids without directing the play. Soaking in what my wife is saying without thinking what I am going to say next or glancing at my phone. Giving the Holy Spirit the space and time to speak to me without rushing on to the next box I need to check off so I can feel that *I am* because *I did*. When I incorporate this new rhythm in my life, it allows me to find fruit on my own branches.

And fruit can't be faked. It can only be grown. Cultivated. Waited for.

The second time we visited Napa it was winter, and the vines had all been mercilessly and cruelly hacked back. They looked like skin-and-bones versions of themselves. I felt for them. Their former glory was gone.

Why were they cut? Because the vinedressers loved them and wanted them to be ready for the next growing season.

Being pruned is not punishment; it is a reward. If anyone grows fruit, the Father prunes so they can bear more fruit. Without pruning, all the energy that should be used to grow the grapes would go into unnecessary growth.

BEING PRUNED IS NOT PUNISHMENT; IT IS A REWARD. IF ANYONE GROWS FRUIT, THE FATHER PRUNES SO THEY CAN BEAR MORE FRUIT.

If you listen to God as you walk with him, he will help you know what relationships, distractions, and preoccupations are holding you back from the fruit you are meant to grow. Pruning isn't always about cutting away bad things. Often, he trims off good things that have become enemies of the best things. Pruning never feels fun in the moment, but the result down the road is effectiveness that wouldn't be possible without it.

I have found that nothing is as good for my soul as taking a walk with God. Literally. And without my phone. When my legs start walking, my thoughts start moving. I bring a piece of paper and a pen with me and take off. It saved my life when my daughter went to heaven, and still, to this day, it's one of the best ways I can keep my soul healthy. I have shed more tears, prayed more prayers, had more great ideas, and found more clarity in frustration while walking than at any other time. I highly recommend it.

If you need something new in your walk with God, maybe just add a literal walk. In Genesis Adam and Eve would walk with God in the cool of the day (3:8). God still wants to walk with you today.

Buzz Aldrin acknowledged this on his historic walk. There is a reason he chose these twenty-seven words contained in John 15:5 out of all the hundreds of thousands of words in the Bible to bring with him for his Communion service on the moon. "I am the vine, you are the branches. Whoever remains in me, and I in him, will bear much fruit; for you can do nothing without me" (GNT). With all he was up against in this first-ever moonwalk, he smartly chose the comfort and the encouragement that came from knowing he wasn't alone. He had help, and so do you.

CONQUER YOUR INNER SPACE

Perhaps you feel like you are stuck in Gemini going through a form of puberty. It might be actual puberty, or a form of metamorphosis between caterpillar and butterfly at work, in school, or in your development as a human. These seasons get old. But God wants you to see the beauty in your chrysalis.

Because Gemini wasn't new like Mercury, or triumphant like Apollo, the American public was uninterested. Who wants to watch puberty? After Ed White's first EVA, viewers began calling TV networks to complain that coverage of the missions was interrupting their football games and television shows. They got comfortable. "Americans no longer half-expected the whole thing to blow up," said one *LIFE* magazine article, accounting for the drop in interest. The smoother the mission went, the more people tuned out.

NASA has a word to describe something going according to plan—*nominal.* We often use this word in a negative way. But the best thing that can be said of a rocket launch is that it is nominal: nothing is outside of the expected ranges.

I hope you see the power of being content while going through transition. I hope you see the necessity of Gemini moments in life. Don't be afraid of "nominal" as you walk with Jesus—there's nothing more exciting than things going according to the plan. What gets people's attention on the outside is usually not good for those on the inside.

Below the verse from John 15, on the little handwritten index card

Buzz brought to the moon to read during his Communion service, he also wrote out the words to Psalm 8:3–4:

> When I consider thy heavens, the work of thy fingers, the moon and the stars, which thou hast ordained;
>
> What is man, that thou art mindful of him? And the son of man, that thou visitest him? (KJV)

I can't imagine what it must have been like to read that verse *on* the moon. But I do know that, no matter what you are going through, it will be good for your soul if you meditate on it, maybe even as you look at the moon on a dark night.

And what if you made your own little note card to take with you on today's adventures? Write a verse or two on it, and fold it in half like Buzz did before you put it in your pocket or purse. When you bump into it, remember God has given you a passport to wonder.

CHAPTER 22

RADIO SILENCE

BEING ALONE CAN BE SCARY.

Perhaps every time your kids are with your ex you feel paralyzingly by yourself—each of those days and especially the nights. Or it could be the fact that you have to travel for work and end up on the other side of the country from your family. Maybe someone you love has died and you feel helplessly, hopelessly by yourself. You might even experience loneliness in the middle of a crowded room. Though fear can accompany aloneness, we can also find extraordinary preciousness in these lonely moments, especially when they open the door to experiencing the presence of God.

Michael Collins was the only astronaut of his era to have dark brown eyes; most astronauts he worked alongside were blue-eyed. This wasn't specified in any criteria that was laid out, though there were height and weight requirements. Initially there was also a rule that you had to be a test pilot. This was decided by President Dwight D. Eisenhower.

When the space program first began, they didn't know what to look for in an astronaut. It was suggested that bullfighters might be good astronauts because they are brave. Some said scuba divers because of how they handled pressure, race car drivers because of hairpin turns at breakneck speeds, or mountain climbers because of the thin air.

This is what they were looking for as they searched:

> Intelligence without genius, knowledge without inflexibility, a high degree of skill without overtraining, fear but not cowardice, bravery without foolhardiness, self-confidence without egotism, physical fitness without being muscle-bound, a preference for participatory over spectator sports, frankness without blabbermouthing, enjoyment of life without excess, humor without disproportion, fast reflexes without panic in a crisis.

This reminds me of the lists many people use when looking for a potential spouse. Who could possibly stack up against such high expectations? It certainly makes me grateful to know that when God looks for someone whose life he can use, he is not looking for ability but availability. He actually gets a kick out of using people who are not necessarily the bravest or tallest or the most connected but instead the most surrendered. If, like me, you have never been the first one picked at kickball, take heart—what man rejects, God selects!

He sees things the way my daughter Clover does.

"Why aren't you taking any?" she asked, looking at my empty bag when we were collecting seashells.

I said, "These are all damaged, and I can't find any that are intact."

She looked up at me confused, gingerly holding her own bulging bag full of what I had overlooked, and then said, "Dad, broken shells are beautiful too."

God has a soft spot for broken things.

GOD HAS A SOFT SPOT FOR BROKEN THINGS.

In 1958, Eisenhower and his advisers eventually landed on the option of recruiting test pilots. So that was where they began their search. Combing through the records of 508 test pilots, they found that 110 met the requirements, including the height and age limitations of no taller than five foot eleven and between twenty-five and forty years of age. Strenuous, and in many cases unnecessary, testing of every sort conceivable (rectal probes were involved) whittled down the 110 to 69 and then 32 before finally 7 astronauts were named—the famed Mercury Seven. There would be another six waves of additional astronauts after them, bringing the total number of astronauts to seventy-three by the last Apollo mission, though only forty-three of them would get the chance to fly. In the second group, "the New Nine," was Neil Armstrong. Group three, "the Final Fourteen," included Edwin "Buzz" Aldrin Jr. and Michael, who had worked as a test pilot at Edwards Air Force Base before he became an astronaut—though by the time he was brought in, that was no longer a mandatory criterion.

And up in space, Michael was now all alone.

As they say, in space, no one can hear you scream. While he orbited the moon, no one could hear him at all. As Buzz and Neil descended to the moon in the lunar module Eagle to establish Tranquility Base, he was left in Columbia. For more than twenty hours he was all by himself in a way no one in history had ever been. It must have been eerie. When your travels take you to the moon, and your companions leave you all alone, the days are long.

It is incomprehensible to think of the solitude he experienced, being a quarter million miles from home to start with, but then loading his two crewmates into another vessel and watching them separate. They all three privately viewed this mission to land on the moon as a fifty-fifty gamble on whether they would successfully return. There were just too many things that could go wrong. It all sounded good on paper, but moon gravity had never been experienced in real life, and they would have critically low fuel, even if everything on the descent went right—which, as we know, it did not. There were program alarms and the speed issue that brought them in too fast and directed them toward a boulder field. Nobody knew what was going to happen.

President Nixon prepared remarks to be read in the event of the death of Buzz and Neil. The letter came to light for the first time in 1999. Fortunately, he never had to read it.

> Fate has ordained that the men who went to the moon to explore in peace will stay on the moon to rest in peace.
>
> These brave men, Neil Armstrong and Edwin Aldrin, know that there is no hope for their recovery. But they also know that there is hope for mankind in their sacrifice.
>
> These two men are laying down their lives in mankind's most noble goal: the search for truth and understanding.
>
> They will be mourned by their families and friends; they will be mourned by their nation; they will be mourned by the people of the world; they will be mourned by a Mother Earth that dared send two of her sons into the unknown.
>
> In their exploration, they stirred the people of the world to feel as one; in their sacrifice, they bind more tightly the brotherhood of man.
>
> In ancient days, men looked at stars and saw their heroes in the constellations. In modern times, we do much the same, but our heroes are epic men of flesh and blood.
>
> Others will follow, and surely find their way home. Man's search will not be denied. But these men were the first, and they will remain the foremost in our hearts.

> For every human being who looks up at the moon in the nights to come will know that there is some corner of another world that is forever mankind.

While Buzz and Neil were down on the surface of the moon, Michael orbited above for twenty-one and a half hours, hoping and waiting and thinking and praying they would come back so they could return to earth together. Think about the most alone you have ever been. Then put that feeling on steroids. It wasn't just the quarter-million-mile distance to those he loved on earth, or the 2,100 miles of lunar rock separating him from his crew, but the deafening silence. During that period his ship went around the moon thirty times. Each of those thirty times he would be out of reach of radio transmission for forty-eight minutes. This is "going dark" on a completely different level. Passing around the dark side of the moon Michael Collins was blocked off from all communication. Total radio silence, for forty-eight minutes, thirty times. There was not a human being he could talk to. He was profoundly, decidedly, and undeniably alone. In his extremely well-written book *Carrying the Fire*, he put it this way:

> Far from feeling lonely or abandoned, I feel very much a part of what is taking place on the lunar surface. I know that I would be a liar or a fool if I said that I have the best of the three Apollo 11 seats, but I can say with truth and equanimity that I am perfectly satisfied with the one I have. This venture has been structured for three men, and I consider my third to be as necessary as either of the other two.
>
> I don't mean to deny a feeling of solitude. It is there, reinforced by the fact that radio contact with the Earth abruptly cuts off at the instant I disappear behind the moon. I am alone now, truly alone, and absolutely isolated from any known life. I am it. If a count were taken, the score would be three billion plus two over on the other

side of the moon, and one plus God only knows what on this side. I feel this powerfully—not as fear or loneliness—but as awareness, anticipation, satisfaction, confidence, almost exultation. I like the feeling. Outside my window I can see stars—and that is all. Where I know the moon to be, there is simply a black void; the moon's presence is defined solely by the absence of stars. . . .

Although I may be nearly a quarter of a million miles away, I am cut off from human voices for only forty-eight minutes out of each two hours, while the man in the skiff—grazing the very surface of the planet—is not so privileged, or burdened. Of the two quantities, time and distance, time tends to be a much more personal one, so that I feel simultaneously closer to, and farther away from, Houston than I would if I were on some remote spot on earth which would deny me conversation with other humans for months on end. . . .

I am sixty nautical miles above Tranquility Base, traveling at about thirty-seven hundred miles per hour. If Neil or Buzz were able to see me, they would find that I would come up over their eastern horizon, pass almost directly overhead, and disappear below their western horizon. The entire pass takes thirteen minutes.

Michael's greatest fear in those moments was not for his own life but the lives of his friends. In the days leading up to the launch, the pressure on him was so great that he developed tics in both eyes. As Michael orbited, one author said:

[He] waited for a call from fellow astronauts Neil Armstrong and Buzz Aldrin to say their lander craft had successfully blasted off from the Moon.

The message would banish Collins's deepest fear: that he would be the only survivor of an Apollo 11 disaster and that he was destined to return on his own to the United States as "a marked man." . . .

> Should the engine fail to ignite, Armstrong and Aldrin would be stranded on the Moon—where they would die when their oxygen ran out. Or if it failed to burn for at least seven minutes, then the two astronauts would either crash back on to the Moon or be stranded in low orbit around it, beyond the reach of Collins in his mothership, Columbia.

Collins later wrote: "Of such possibilities are nightmares bred." Clipped to his space suit was a notebook containing all eighteen emergency procedures. On the day of the actual mission, Collins said he was "sweating like a nervous bride."

Michael died on April 28, 2021, at the age of ninety, while I was in the midst of editing this book. I grew to respect him greatly while researching this project and felt like I knew him through his reflections and those of his fellow astronauts. He has been called the most eloquent person to ever ride a Saturn V, and his book *Carrying the Fire* is extraordinary. That night, when I looked up, the moon was a waning gibbous, one of the three fullest phases.

Poetry.

Good night, moon.

FINAL WORDS

When I think about the radio silence Michael Collins endured, I can't help but think about Jesus' dark hours on the cross and how there are answers in them for your lonely moments. The gospel writers specifically point out the astronomic abnormalities that accompanied the crucifixion. When it should have been its brightest, the sky went black. Noon felt like midnight. It was almost as though the sun was embarrassed to illuminate such a miscarriage of justice. Homicide is the murder of a man. This was that but more; it was deicide, the murder of God. I wonder if in those final moments, the almost-full moon became visible in the darkened noon sky as Jesus suffered silently. As I said when we first started this journey, the moon and the cross cannot be separated. And like Michael's mind, Jesus' mind was not on himself as he endured the pain and agony. You were on Jesus' mind.

The first three times he spoke, it had nothing to do with himself.

- "Father, forgive them." (Asking for his murderers to be forgiven.)
- "Today you will be with me in paradise." (Taking care of the thief on the cross.)
- "Woman, here is your son." (Taking care of his mom.)

It wasn't until he spoke for the fourth time that he even acknowledged the excruciating pain he was feeling. As the radio silence came to an end and he passed back into communication, he said three things that give us access to his agony.

- "My God, my God, why have you forsaken me?" (How could anyone read these words without feeling his distress?)
- "I thirst!" (This is heartbreaking as well.)
- "It is finished!" (A stirring cry of victory.)

Last of all came his final words in Luke 23:46:

> When Jesus had cried out with a loud voice, He said, "Father, 'into Your hands I commit My spirit.'" Having said this, He breathed His last.

When Jesus said these words, he disappeared. Radio silence. He was gone.

LAST WORDS #7

"Father, 'into Your hands I commit My spirit.'"

Luke 23:46

No one saw this coming. Jesus died after just six hours. People who were crucified usually didn't die that quickly. It was plenty shocking to all those at the foot of the cross, as is clear from scanning through the four different camera angles we get of this ultimate moment in the Gospels. Think of a film set where they have a GoPro-type camera on the dashboard of a car, another camera mounted outside the open window, a giant boom camera, and another on a drone. You see so many different things when you pay attention to the four different angles.

The angle from Matthew 27:50–53:

> Jesus cried out again with a loud voice, and yielded up His spirit.
>
> Then, behold, the veil of the temple was torn in two from top to bottom; and the earth quaked, and the rocks were split, and the graves were opened; and many bodies of the saints who had fallen asleep were raised; and coming out of the graves after His resurrection, they went into the holy city and appeared to many.

The angle from Mark 15:38–39:

> Then the veil of the temple was torn in two from top to bottom. So when the centurion, who stood opposite Him, saw that He cried out like this and breathed His last, he said, "Truly this Man was the Son of God!"

The angle from Luke 23:47–49:

> When the centurion saw what had happened, he glorified God, saying, "Certainly this was a righteous Man!"

> And the whole crowd who came together to that sight, seeing what had been done, beat their breasts and returned. But all His acquaintances, and the women who followed Him from Galilee, stood at a distance, watching these things.

The angle from John 19:30:

> When Jesus had received the sour wine, He said, "It is finished!" And bowing His head, He gave up His spirit.

The spectacle was over for those who had come merely to be entertained. Besides just wanting something to do, they had more or less been motivated to show up by the religious leaders. As Jesus hung on the cross, he had done the exact opposite of everything they expected. Where they were sure he would be spiteful, he had been magnanimous. When they taunted, he had prayed—for them. As they tried to rile him up, he took care of his mom, who was standing beside them. Talk about taking the air out of their tires. And now, before it ever really got going, it was over. They couldn't wait to put this day behind them. What had they said in front of Pilate's hall? "His blood be on us and on our children" (Matt. 27:25)? One thing is for sure: this day had gone nothing like they thought it would, and they were happy to get out of there. And as soon as the ground stopped shaking, they did! But man, it must have been weird walking home with all the newly resurrected people walking around the city like zombies!

For the centurion who put Jesus to death, this moment marked the exact second when everything changed. He bowed his knee that day, declaring Jesus righteous—declaring him the Son of God. He had a salvation moment similar to that of the penitent thief on the cross. Think of it! The moment on Jesus' death certificate was the exact same time written on this man's birth certificate—he was born again.

He worked for Rome, the city where one day Michael Collins would be born, but on this day he became a citizen of a different empire. According to tradition this centurion would go on to live for Christ all his days. He was a changed man. He had put thousands to death but had seen nothing like this. But first he had one more necessary, but unpleasant, responsibility to carry out.

FINISH IT

Unknown to any of them there on top of Skull Hill, Jesus' enemies had been lobbying hard in the halls of power to get this show on the road. Pilate had acquiesced to speeding up the execution, and the orders were given. The two thieves crucified beside Jesus had their legs broken and they sank into their fractures and began to suffocate. The centurion looked up at the now-silent figure on the central cross and made a judgment call: *Jesus' legs would not be broken*. Still, Rome required verification of death, so "one of the soldiers pierced His side with a spear, and immediately blood and water came out" (John 19:34).

The word *immediately* is the key. Had he been alive, water and *then* blood would have run out. Since water and blood both ran out right away, it indicated that blood was already in the pericardium.

The centurion had no idea that he was fulfilling prophecy that day. The prophets foretold that in all the savagery and brutality the Son of God would endure throughout his crucifixion, not one bone would be broken. There might be something out of joint, but his skeletal structure would be whole (Ex. 12:46; Num. 9:12; Ps. 34:20; John 19:36).

Jesus' crowded hour went according to plan. Not just down to the manner, but the timing as well.

This split-screen aspect projects both the sinfulness of those involved and the sovereignty of God. Peter recognized this on the day of Pentecost when he preached, in Acts 2:23: "Him, being delivered by the determined purpose and foreknowledge of God, you have taken by lawless hands, have crucified, and put to death." There were several things going on that day: humans were doing their worst, but God was doing his best.

But it all was different for Jesus' friends. This small group included the beloved apostle John, Nicodemus, Joseph of Arimathea, and Mary Magdalene. And of course, Jesus' mother, Mary. They watched in a mix of shock and grief as he died. But Jesus' mother probably caught something no one else saw that day. The final statement Jesus spoke from the cross, like the first and the middle one, was a prayer. Several of the Gospels mention that he yelled, "It is finished," but then bowed his head as he gave up his spirit. It seems with this final exhalation he spoke once more and then perished. Many, maybe even most, missed the final thing he had to say.

But not Mary. She would have been able to read those words on his lips the same as if they came out of her lips. And as she did, I am sure she had flashbacks going back thirty years. The words Jesus spoke were not only a prayer; they were a prayer pulled straight from Scripture. Psalm 31:5:

> Into Your hand I commit my spirit;
> You have redeemed me, O Lord God of truth.

Praying Scripture is powerful. There is incredible freedom in praying God's Word back to him. Think of it like a train. A train bound to a track can travel swiftly across the country, but a train on the ground can go nowhere no matter how fast it spins its wheels. When you pray God's Word, it is like putting track under yourself. You gain confidence and speed from praying according to what you know to be his will. That is not to say you can only pray God's words; you can pray anything! But when your prayer life is guided by the track he has given in his Word, you aren't selfishly motivated or just spinning your wheels on something that is a fabrication of your imagination. There is safety in it.

THE PRAYERS GO UP, THE BLESSINGS COME DOWN

It is a special privilege to put a little one to sleep. I wrote in my first book, *Through the Eyes of a Lion*, about how crazy bedtime can be with teeth brushing, pajama putting on, and all the wrangling that happens as you try to move toward bed. But once the nighttime sips of water have been sipped, and in general the RPMs have gone down and energy begins to subside, something magical happens. The never-ceasing-to-move human slows down and transforms into something like an angel before your eyes as they give in to sleep. In these moments, if you are present and paying attention, some unmistakable gems can be unearthed. Like when my daughter Lenya mused about flying horses getting you to heaven.

Since the girls are older and stay up later and are now capable of putting themselves to bed, usually Jennie and I rotate the responsibility of putting Lennox down. But when Jennie broke her ankle on Mother's Day, I put him to bed every single night for thirty days in a row.

It was beautiful in an unexpected way. Some nights I thought he would never fall asleep. I would be sitting on the beanbag next to his bed for forever, with my boy's squirming body parts randomly shooting out, taking unexpected elbow shots to my nose. But more than anything I felt God's love and the beauty of childhood, and I was many times overcome by the wonder of life and breath. Every night we'd read some books, ending with a Bible story, then I'd hold his hand and pray for him.

My very favorite moment in the almost-asleep-but-not-totally-out phase is where the dreamworld has already closed his eyes but his ears are still hearing what's going on around him. Playing in the background through this routine is "Lennox's Bedtime Playlist" from the Alexa-powered Echo Dot speaker in his room. It contains his favorite songs in this order: the *Onward* title song, "Echo" by Elevation Worship, "Good Grace" by Hillsong UNITED, "Home on the Range" by Gene Autry, "The Blessing" by Elevation Worship featuring Cody Carnes and Kari Jobe, and "May the Lord Bless You" by Maranatha! Praise Band. On a night when he is especially tired, he goes out during "Good Grace." Occasionally he falls asleep during "Home on the Range." Once I have seen him fall asleep during "Echo." That was like a unicorn sighting. Afterward I'd tell Jennie what song he'd gone out to. When I told her he was out by "Echo," it was with the smug satisfaction of a pro.

Jennie shrugged. "He must have been exhausted."

"Or I am just that good," I replied.

On the rare occasion he makes it through the entire playlist, I resort to the sound of rain. Those are nights when my sanctification is tested. Most often he falls asleep in between the verse and chorus of "The Blessing."

We have been singing "May the Lord Bless You" by Maranatha! to the girls since they were young. The new Elevation Worship version, "The Blessing," came out during the pandemic and swept through the world and our home in the most incredible way. One is fifty years old and the other hot off the press, but both songs pull from the same Aaronic blessing in Numbers 6:24–26.

Lennox saw things differently when we first tried to mix it up and play the new one instead of the one by Maranatha! Praise Band. "This is not the real blessing," he said.

"No, it's good, buddy."

"I don't like Cody Carnes."

"Yikes, buddy. He's a friend of mine and amazing."

He finally came around. The hypnotic power of the song gets him nearly every time. I usually sing it over him as I pray for him before I tiptoe out of the room. But my favorite moments are when, half asleep and half awake, he sings along. His lips barely move, but I can see them: *Amen. Amen.*

If I see him fading, I always say, "I love you, buddy." And it puts shivers in my spine when from dreamland he murmurs back, "I love you, Daddy." I usually put my right hand on his head at some point while we are singing as a way of blessing him.

I picture Mary beside Jesus' crib. He is drifting off to sleep. A thousand times in a row, she prays over him, *Father, into your hands I commit my spirit.* His little lips eventually pick up on it and murmur back before drifting off. *Amen.*

Maybe she had no idea at the time, but now, as she stands at the foot of the cross watching him pass out of radio signal to those on this earth, she must have realized she had been given the privilege of being the one who would prepare him to die.

But Jesus wasn't gone.

He had just gone home.

He was out of communication with his friends temporarily but not permanently.

Just as Buzz and Neil finished their mission, planting a flag near Tranquility Base and returning in the Eagle to Columbia, so Jesus would capture the Enemy's flag and, in a moment of power and triumph, return to his body, the temple that had been torn down (Mark 13), and ascend to heaven, from whence he will return to judge the living and the dead.

You don't have to fear death, the devil, or anything else hell can throw at you. He has secured power for you greater than the 7.5 million pounds of thrust that powered the Saturn V.

Today, say the words Jesus died with on his tongue, and you will live with the strength that comes from your life being in his hands.

TODAY, SAY THE WORDS JESUS DIED WITH ON HIS TONGUE, AND YOU WILL LIVE WITH THE STRENGTH THAT COMES FROM YOUR LIFE BEING IN HIS HANDS.

It's not just a prayer for dying; it's also for living.

Father, into your hands I commit my spirit.

If you are a parent, I challenge you to get into the rhythm of praying with and blessing your sons and daughters. It is your right and responsibility. You have no idea how much they are absorbing. The other night I went to pray for Lennox and I reached out to hold his hand.

"No!" he objected. "That's how you pray for me. But I want a blessing."

Stunned, I realized he knew exactly how a blessing was supposed to go. I asked him, "How would you bless me?" He put his right hand on my head and said, "May God's face shine upon you."

Let it be.

ALL SMILES AND GIGGLES

Michael Collins's hours of loneliness ended when the Eagle rose from the moon and rendezvoused with the Columbia in lunar orbit. It looked like a spider whose legs had been plucked off, because it now lacked its descent stage—which remains on the moon to this day. The prodigal LM had come home. As they approached Columbia he snapped a photo of their arrival. In the photo is the moon, the Eagle, and, in the background a half-illuminated earth (like the one on the spine of this book). What is stunning about the photo is that every single living person at the time—both on earth and the moon—*except* Michael Collins was in the photo.

Once Columbia and Eagle had successfully docked, Michael Collins's time of silence and solitude came to an end. There was a joyous reunion. Here is how he described the moment:

> The first one through is Buzz, with a big smile on his face. I grab his head, a hand on each temple, and am about to give him a smooch on the forehead, as a parent might greet an errant child; but then, embarrassed, I think better of it and grab his hand, and then Neil's.
>
> We cavort about a little bit, all smiles and giggles over our success, and then it's back to work as usual.

I get emotional thinking about this scene and thinking about Jesus' reunion with his Father in heaven at the end of his mission. He told his disciples he was going to the Father when he ascended. What a moment that must have been! But that's not all. All those who have committed their souls into his hands have the promise of life during life and life after death. This is our living hope! When we shuffle off this mortal coil, we have everything to look forward to. What a reunion awaits us. And I know I won't have Michael's restraint; I won't be shaking my Lenya lion's hand—I'm gonna take ahold of her face and kiss her right on her head! It will be all smiles and giggles. Until that day . . . back to work as usual.

JUST DO IT

Once upon a time a condemned murderer was about to be put to death by a firing squad. The man's name was Gary Gilmore. He was asked by his executioners if he had any final words. He said, "Let's do it."

Many years later a design team called Wieden+Kennedy was urgently trying to come up with an ad slogan for a shoe company's first big TV campaign. Burning the midnight oil was not just tolerated at Wieden+Kennedy, it was celebrated. They believed that some of the best ideas were unlocked by the urgency and crisis of approaching deadlines. With only hours before their idea was due, Gary's famous concise speech came up in conversation. They tweaked the sentence only slightly and it became "Just Do It." The next morning they submitted the idea to Nike, and, as they say, the rest is history. Well, almost. Nike cofounder Phil Knight hated it at first. In a 2015 interview for *Dezeen* magazine, Dan Wieden recalled that Phil said, "We don't need that s—." Wieden continued, "I said, 'Just trust me on this one.' So they trusted me and it went big pretty quickly."

Those three words, inspired by last words, have reverberated and ricocheted around the world and defined a sports mentality for a generation. And they don't seem to be letting up.

And Knight warmed to them with time. In a recent interview, author Simon Sinek recalled a speech in which Phil Knight explained what "Just Do It" means to him. He said, "When you're out there when it's cold and wet and you're the only one out on the street, we're the ones standing under the lamppost cheering you on. And that's what 'Just Do It' means."

Boom. The last words have taken on a life far larger than what they initially meant in the moment.

This is the conclusion of our examination of Jesus' seven final statements from the cross with his last words before dying, "Father, 'into Your hands I commit My spirit'" (Luke 23:46). But they are not an end; they are a beginning. From the launchpad of the cross, they propelled all humankind into a new era of forgiveness, hope, and grace.

As you find life in his words, and power in his person, you have more than someone standing under the lamppost cheering you on—you have the Spirit of God inside you propelling you forward, illuminating your every step, and supplying you with divine energy to do all you are meant to do. If you ever doubt that, consider the lengths he went to in order to save you. When he died on the cross, it wasn't for himself. It was for you.

Though President Kennedy didn't live to see his vision come true in 1969 when Neil took that first step, what the president foretold rang true: "It will not be one man going to the moon—if we make this judgment affirmatively, it will be an entire nation."

What JFK predicted of Neil's small step is infinitely true of Christ. In a very real sense, he hung on that cross and endured the darkness and radio silence as a representative for all of us.

In your most lonely and pain-filled moments he is rejoicing over you like I sing over Lennox as he falls asleep (Zeph. 3:17). He will never leave or forsake you, and he loves you to the moon and back! His Spirit is urging you on to the great adventure before you. If you pay attention, you'll hear him whispering in your ear: *Let's do it!*

CONQUER YOUR INNER SPACE

If you would allow me to pull a Sarah Young, I would like to speak to you as I believe Jesus is trying to today. He is calling you into the wild:

Dear heart,

I know where you are right now, and I am here with you. You might feel alone, but that is not the truth. I am here, and I have always been waiting for you to meet me here. Waiting for you to trust me with your whole heart. But even if you feel like you can't muster up any strength to bring yourself to me, bring whatever you can. I can work with a little. I just want you.

So if you feel like you're broken up in a million pieces, and you don't have anything to give, come to me just as you are. Trust me. I don't expect you to do anything, or be anything; just run to me. Let me hold you, let me speak to you, let me love you. Love you back to life. I am able. I am strong. I am enough for you. Walk with me. Enjoy me. I'm funnier than people make me out to be. Don't overthink it. Let's go. Let's walk together. Let's live life together. You are precious to me. I love you to the moon and back.

xoxo Jesus [*I told you I was fun*]

CHAPTER 23

MASKS AND THERMOMETERS

THERE WAS A PANDEMIC HAPPENING. IT ORIGINATED IN CHINA in July 1968 and lasted through 1969 into 1970. There were two waves, with a mutation in between, and the second was much more serious than the first.

Its technical name was the H3N2 virus, but in its day, it was simply called the Hong Kong flu. This virus killed one hundred thousand people in the United States and between one and four million people worldwide. It is largely forgotten now but was majorly disruptive in its day.

On July 5, 1969, one and a half weeks before launch, the Apollo 11 crew held their final press conference before liftoff, wearing "bright blue metal face masks" to shield them from germs. They also sat "enclosed in a three-sided plastic tent 50 feet from the nearest newsman. The tent included a series of blowers which pushed air from the back of the astronauts onto the assembled newsmen." And as I mentioned previously, their flight surgeon refused to allow the three members of the crew to attend a preflight dinner in the White House. President Nixon was not impressed.

They weren't taking any chances after what happened on Apollo 8.

It's not every day you get a disease from a president, but it's possible President Lyndon Johnson spread the H3N2 virus to at least one of the Apollo 8 astronauts when, twelve days before their mission, they attended a send-off dinner at the White House in their honor.

NASA's self-important flight surgeon Dr. Charles Berry didn't feel good about the men being in close quarters with so many people so close to their launch, but what the president wanted, the president got. Days after the dinner the president was hospitalized. He spent his final weeks in the White House in bed with the flu—the worst, he said, he had felt in his life.

During the flight Frank Borman became sick, experiencing flu-like symptoms, including diarrhea and vomiting at the same time. Not good under any circumstances, but especially bad in space. All the above-mentioned particles flew through the air until they exploded against equipment panels and his two fellow astronauts, Bill Anders and Jim Lovell. There the globules became brown and green stains everywhere, including their white flight suits. The stench was so bad that Bill put on a gas mask.

"You're not supposed to use those," Jim objected.

"To hell with that, I'm using it," Bill shot back, cranking the oxygen up full blast.

The two of them opened towelettes and hunted down each and every unpleasant floating projectile while caring for their commander.

Frank rallied and recovered, and the mission didn't skip a beat. The three of them didn't tell NASA until he got better, out of fear that they would abort the mission.

That is nothing when compared to what happened on Apollo 7 when all three astronauts got head colds. Because of the lack of gravity, noses don't run, creating challenging congestion and pressure. Due to the misery, they became adversarial with each other, quarrelsome with Mission Control, and even rebellious—refusing to wear their helmets on reentry as protocol dictated so they could instead relieve the pressure in their heads.

Though the hardware functioned perfectly, the men's attitudes completely broke down—and, what's more, the world saw it happen in real time. Despite the fact that the mission itself was a flawless victory, NASA's Christopher Columbus Kraft determined that none of the three would ever be allowed to fly for NASA again. They were permanently grounded.

What a contrast between the two missions, and what a difference attitude makes.

It is also worth noting that the crew of the Apollo 8 mission has the distinction, in the entire Apollo program, of being the only crew to not suffer any divorce. All three marriages remained intact. I find it highly interesting that on a crew where spirits were strong even in the midst of adversity, and where they demonstrated moments of servanthood, you can find a correlation to success in marriage.

THE CHURCH IN LAODICEA

I walked through the doors of a medical complex and was instantly scolded. "Outside! You have to wait outside until I take your temperature."

It was late spring of 2020, and this protocol was new to me. I was just picking something up that had been left at the front desk for me. I had no idea. The last time I saw my doctor, it was through a FaceTime call. Coming to the office in person had changed drastically.

Eventually the receptionist approached me where I had been banished outside and pushed the thermometer to my forehead.

"You're hot," she said.

It was eighty degrees outside. I gestured to my Jeep, which had the top removed, and explained that I had driven in with the sun beating down on me.

"Oh, I see," she said. "Yes, it is hot out. I guess that makes sense." Then she somewhat reluctantly let me in.

This new postpandemic world would take some getting used to. Perhaps as you read this there are no longer signs indicating that masks are required, but as of this writing moment, I am staying at a hotel in Texas where anywhere indoors a mask must be on your face, and on the way in an automated machine took my temperature as it dispensed hand sanitizer.

Interestingly enough, both heat and masks come up in Jesus' remarks to the church in Laodicea. It's in Revelation 3:14–22:

> "Write to Laodicea, to the Angel of the church. God's Yes, the Faithful and Accurate Witness, the First of God's creation, says:
>
> 'I know you inside and out, and find little to my liking. You're not cold, you're not hot—far better to be either cold or hot! You're stale. You're stagnant. You make me want to vomit. You brag, "I'm rich, I've got it made, I need nothing from anyone," oblivious that in fact you're a pitiful, blind beggar, threadbare and homeless.
>
> 'Here's what I want you to do: Buy your gold from me, gold that's been through the refiner's fire. Then you'll be rich. Buy your clothes from me, clothes designed in Heaven. You've gone around half-naked long enough. And buy medicine for your eyes from me so you can see, *really* see.
>
> 'The people I love, I call to account—prod and correct and

> guide so that they'll live at their best. Up on your feet, then! About face! Run after God!
>
> 'Look at me. I stand at the door. I knock. If you hear me call and open the door, I'll come right in and sit down to supper with you. Conquerors will sit alongside me at the head table, just as I, having conquered, took the place of honor at the side of my Father. That's my gift to the conquerors!
>
> 'Are your ears awake? Listen. Listen to the Wind Words, the Spirit blowing through the churches.'" (MSG)

As I mentioned earlier, in space it's 250 degrees in the sun and -250 in the shade. It's a lot of hot, and a lot of cold, but not much in between. (For this reason the Apollo spacecrafts have to spin like a rotisserie chicken on the way to the moon to alternate hot and cold. This procedure was nicknamed "Barbecue Mode.") But what comes directly in between these two extremes, Jesus finds offensive about the church in Laodicea.

Nothing great has ever been accomplished with half of someone's heart. Perhaps the sin that keeps you from greatness is indifference, apathy, and half-heartedness. Going through the motions is what Jesus sees in this church.

Laodicea was a wealthy city forty miles from Philadelphia. It was the southeasternmost of the seven churches—in the Lycus Valley with sister cities Hierapolis and Colossae. It benefited from its location on a trade route at the intersection of two important roads. And because of all the money coming in and going out, Laodicea was widely known for its banking establishments, in addition to glossy wool cloth and a special eye salve developed there. It was built on a plateau that made it nearly impregnable. One thing that put the area on the map was its water. In the area there were icy cold mountain brooks and naturally occurring hot springs. The former were refreshing; the latter were healing. Jesus worked all of this into his comments. I love that, by the way, because it shows clearly that God wants to speak to you through your environment. He talked to them about cloth, eye ointment, and money, and, most famously, he told them that you jump in pools to feel refreshing cold water and jump in hot tubs to soak in hot water, but nobody wants to experience water that is tepid and lukewarm.

LETTER #7 (LAODICEA)

"You are neither cold nor hot. I could wish you were cold or hot."

Revelation 3:15

For the people in the church at Laodicea, success had made them apathetic and unwilling to be energetic. They were lukewarm. Present and accounted for. Taking up a seat but unwilling to move a muscle.

It is interesting to note that Jesus' "I stand at the door and knock" message, which has been used by preachers all over the world, was not spoken to the lost but to the found (Rev. 3:20). He had been barred from their gatherings and they hadn't even noticed.

How easy it is to become a Laodicean. To slowly creep toward indifference. To become soft and spiritually fat and distracted. To not even notice because of how hypnotic the passing of time can be, how far you can drift from where you started. It's called backsliding. And it might be happening to you right now.

John Stott said, "Perhaps none of the seven letters is more appropriate to the church of the twenty-first century than this. It describes vividly the respectable, nominal, rather sentimental, skin-deep religiosity which is so widespread among us today. Our Christianity is flabby and anaemic. Like the Laodiceans, we appear to have taken a lukewarm bath of religion."

USE IT OR LOSE IT

In March 2019, NASA and the European Space Agency began a unique experiment studying how to keep astronauts healthy in space. Scientists have known for years that spaceflight can negatively impact human health. Without the stabilizing effects of gravity, muscles waste away and bodily fluids drift upward. Bone and muscle atrophy reduce cardiovascular function and force bodily fluids to move to the upper part of the body. It can also lead to weakness, dizziness, stuffy heads, puffy faces, motion sickness, inner-ear disturbances, compromised immune systems, and back pain. NASA has learned to do a lot of work to compensate for this. Astronauts on the International Space Station currently work out for two-plus hours a day on carefully created exercise equipment built to operate in microgravity.

They wanted to test out different treatments for these conditions without sending people to space, so they paid participants in Cologne, Germany, to lie in bed for sixty days straight. Their beds were tilted slightly downward to simulate a low-gravity environment. The scientists say their findings could someday be used to protect astronauts on long-haul spaceflights.

The twelve male and twelve female participants, who received about $18,500 each, had to eat, sleep, exercise, and bathe while in a head-down tilted position. One half would visit a centrifuge in a laboratory from time to time. The centrifuge acted as an artificial gravity chamber as its spinning rig simulated gravity, pushing blood toward the participants' lower extremities. This helped scientists see whether the simulator was in any way helpful in minimizing the effects of lying down in one position for a prolonged period of time.

If you're not intentionally moving forward, you're automatically moving backward. It's the nature of life. There is no such thing as standing still. It requires constant reinvigoration of spiritual muscles to avoid atrophy and deterioration.

This is why you need to fight to focus and be intentional about remaining on your toes. It's easy to be lulled to sleep through prosperity and affluence and forget that the blessings that have come to you aren't just for you. If you don't reinvest your miracles, you'll reach the end of them.

It's also why it's important to regularly make yourself uncomfortable. Spend time with people who aren't like you. Do things for people who can't do anything to benefit you. Sacrifice in ways that get your attention and shock you out of your lethargy. Remember—comfort zones don't keep your life safe; they keep it small!

IF YOU DON'T REINVEST YOUR MIRACLES, YOU'LL REACH THE END OF THEM.

Notice Revelation 3:17, which says, "You say, 'I am rich, have become wealthy, and have need of nothing'—and do not know that you are wretched, miserable, poor, blind, and naked." What a contrast to what Jesus said to Smyrna, the persecuted church: "I know your works, tribulation, and poverty (but you are rich)" (Rev. 2:9).

To Smyrna it was, "I know you are poor but you are actually rich," but to Laodicea it was, "You are rich but actually poor."

As we mentioned, Laodicea was a very wealthy city, and no doubt most in the church were well-off. But we must learn the difference between physical wealth and spiritual wealth.

Because they had money in the bank and nice clothes on their backs, lived in nice homes, and had an extravagant church building they sat in once a week, they figured they were fine. What they didn't realize was that it's possible to be wealthy in this world and to arrive bankrupt in the next one.

That's not to say that being wealthy is a sin. Your financial status has nothing to do with how you will live in eternity. First Timothy 6:17 says, "Command those who are rich in this present age not to be haughty, nor to trust in uncertain riches but in the living God, who gives us richly all things to enjoy."

Those in this church had trusted in their uncertain riches—to their detriment. If you would have asked them to describe themselves, they wouldn't have said they were wretched, miserable, poor, blind, or naked.

But that was how Jesus saw them.

I feel for and am challenged by those in the Laodicean church, who from an outward perspective were doing really well. There was no sexual immorality going on; none of the compromise or outright scandals the other churches dealt with. They weren't dead like Sardis, perverse like Thyatira, or unloving like Ephesus. The noteworthy thing about Laodicea was that there wasn't really anything noteworthy. Their spirituality lacked intensity, neither hot nor cold. Forgettable and ineffectual but continuing on because it's habitual.

I grew up thinking I was fine. I was the son of a pastor, grew up in church, and knew the Bible. Then one day my eyes were opened to my true state. I realized my hypocrisy, my sin, and I was crushed. I turned to Christ and was saved. I have experienced a lot of trauma in my life and have done a lot of work to see how that shaped me, especially at key moments. This has allowed me to not be subconsciously directed by hidden wounds in my life. The decision to take honest inventory of where I really am and who I want to be has given me the chance to grow and develop. It's helped me become aware and in control of the dark side of my moon—the "shadow sides" of my personality—so they don't control me without me realizing it.

Your spiritual existence is not meant to just be about going through the motions. Praying before meals. Looking both ways before crossing the road. God wants our souls to be touched by fire. For our hearts to burn within us. To come to the end of ourselves. To push into new horizons.

He wants you to be curious. To stay hungry. To ask questions. For you to be small in your eyes so you can see clearly how large he is. To get out and smell the brokenness of this world and realize you are meant to be a part of the solution. This is what outer space has to do with the nitty-gritty of your daily life.

Perhaps the problem is that we love a safe, small God because he is not threatening and doesn't call us to change very much. He is there for us to pray to in the event that we face a crisis we can't handle with our money or fix with our eye salve.

Religion that doesn't change us or move us makes God sick.

What is the answer? It's repentance. "As many as I love, I rebuke and chasten. Therefore be zealous and repent" (Rev. 3:19). Turn around; come to Christ. Repent.

Martyn Lloyd-Jones put it this way:

> Repentance means that you realize that you are a guilty, vile sinner in the presence of God, that you deserve the wrath and punishment of God, that you are hell-bound. It means that you begin to realize that this thing called sin is in you, that you long to get rid of it, and that you turn your back on it in every shape and form. You renounce the world whatever the cost, the world in its mind and outlook as well as its practice, and you deny yourself, and take up the cross and go after Christ. That is repentance.

If you can continue going through the motions of a religion that doesn't even require God's attendance for it to be considered successful, you are well on your way to living in a Laodicean cottage by the sea instead of taking your rightful seat at the conqueror's table.

What you need is a fresh glimpse of who God is.

I love the scene in *The Avengers* where Loki tells Hulk he should back off because Loki is a god. Hulk smashes Loki against the ground like a rag doll about five times and then mutters disdainfully, "Puny god."

When your spirituality is mechanical, stale, and lukewarm, it is probably because you are worshiping a puny god. How easy we go from *OMG* to *meh*, from the untold millions who watched Apollo 11 to the Apollo 13 that wasn't even televised—until there was something wrong. When it was smooth sailing, it was forgettable.

The further you are from God, the better you feel about how you are doing. I know from experience. In those moments, I offer my safe, stale, prepackaged prayers like a good little boy and think to myself how thankful God must be. I feel entitled to whatever selfishness or indulgence comes naturally because I haven't missed my devotions in more than a week. With such stellar attendance, how could God not be impressed with me?

What's needed is a fresh vision of his glory.

There are 10^{24} stars in our galaxy. There are more bacteria microbes in your stomach than there are people who have ever lived.

From the infinitesimal to the infinite, there is so much of God's glory he wants you to discover. Life is far too big of a deal to be defined by anything so trivial as a paycheck or career or standing on social media. And yet how easy it is to be caught up in the blessings of God and, in the process, completely forget about God himself.

FROM THE INFINITESIMAL TO THE INFINITE, THERE IS SO MUCH OF GOD'S GLORY HE WANTS YOU TO DISCOVER.

The one whose heart is open, let them listen carefully to what the Spirit is saying now to the churches.

And no matter what happens—head colds, pandemics, trials, pleasures, whatever—remember this: choosing your spirit is your responsibility. A failure to steward it properly will keep you on the ground, and you are meant to fly.

CONQUER YOUR INNER SPACE

We only have a little bit to go before the end of the book. But don't worry, we don't have to say goodbye yet because it is *so* awkward to say goodbye to someone you bump into again before you get in your car to go home, and I don't want to make it weird.

I am so proud of you for making it this far! What a journey. Can you believe there was ever a time when you thought you didn't like space?

I thought it might be cool if you would jot down a few of your favorite things you have learned in the fantastic quest to conquer inner space. It could be a big thought, one of the sevens, or something God has indirectly spoken to you during our nearly five hundred pages together.

I'll tell you one of mine.

This exploration has taught me to not hide the dark side of my moon. I'm learning that my asteroid-streaked surface is not something to be embarrassed about but embraced so I can see God use it for his fame!

CHAPTER 24

BLOOD, SWEAT, AND TEARS

I ALMOST LOST MY EYE ON THE FOURTH OF JULY.

The entire ride to the hospital I kept my left hand clamped over my left eye, thinking, *Maybe if I never take my hand away everything will be okay.*

Thoughts raced through my head like a windstorm picking up dust in the desert. *How will I be able to read my Bible while preaching? Will I have to get a glass eye? Do they make patches that are cool? Will anyone want to listen to a pirate preacher?*

"I think I am blind in this eye, Jennie," I said grimly through clenched teeth. "I have a bad feeling about this." She was driving my Jeep, and I felt her accelerate, as though if we arrived at the hospital sooner my eye would have better chances of not being permanently damaged. Sitting in the passenger seat I had to turn my head more than ninety degrees to see her out of my right eye. When I did, I saw her face was distraught, but I could see her straining to remain calm as she yet again was forced to face my going into a medical ordeal.

The most bizarre thing about this situation was the fact that she had driven me to the emergency room with a hand clamped over my eye on one other occasion. Only it was my right eye that time, and it wasn't a firework that attacked me; it was a piece of seaweed.

Ironically, I write this today from Monterey, California, the city where Jennie was born. The accident happened at Asilomar State Beach, only a few miles from where I am sitting. I thought I could make Alivia, Lenya, and Daisy laugh by swinging a giant piece of seaweed around while we were watching the sunset.

I have a video of that day in 2011 of us singing on the rocks on the beach. The next photo in my phone, I am in the ER and there is a long, ugly cut from my forehead right down to my cheek, across my right eye. I had jumped off a rock and swung this seven- to eight-foot-long section of seaweed, trying to make it snap like a whip. (In my mind I was Indiana Jones, and I have been since I was seven or eight.)

It worked. The seaweed snapped, all right, directly across my eye—much like how young Indy in *Indiana Jones and the Last Crusade* cut his chin the first time he snapped a whip. Fortunately, my eye must have closed just before the seaweed hit it because there was no damage to the eye, just a nasty welt running for four inches across it. Like Scar from *The Lion King*. It stayed like that—swollen and an ugly purple—for a week. I had to preach at a church in San Jose the next day. I can only imagine what the people thought this visiting pastor was up to in his spare time. Just a seaweed incident gone wrong. Nothing to see here.

You had to be there.

The weirdest thing about all this is that at dinner on that Fourth of July in 2020 I had told the story about the seaweed and the eye injury. Just before heading to the ER holding my other eye.

YOU'LL SHOOT YOUR EYE OUT

The firework that got me was called Blood Sweat & Tears. Daisy, Clover, and Liv had gone with me to pick out the explosives. Montana does a lot of things right, and I have always considered the fireworks a part of that list. It was a great shock moving from California, where you would have to be a pyrotechnician to purchase what anyone can buy in Montana.

I was about to set myself up for thirty-seven seconds that would change my life.

After choosing sparklers and fountains and other things from the kids' section, I picked out some big firework cakes. Those are boxes with mortar rounds inside. When you light the fuse, they go off for a minute or two, shooting round after round that go up hundreds of feet into the air, exploding in massive fireballs. I had picked out a few and asked for a recommendation from the man running the stand. With a twinkle in his eye, he grabbed Blood Sweat & Tears and set it down in front of me. "This one is my favorite. It's a real crowd-pleaser."

Excellent, I thought. *I will save it for the end.* Jeremy Camp and his wife Adrienne and their kids were with us, and everyone was having a great time. I hadn't lit very many fireworks yet at that point and was enjoying standing on the deck, watching everyone have fun.

As we neared the end of the fireworks pile, I thought to myself that it was a perfect time for the big guns. "Here's a good one, everybody," I said, grabbing the big black-and-green box and lighting the fuse before sprinting like crazy to get to the deck before it blew.

I looked it up later and on the website the description states that Blood Sweat & Tears is a thirty-seven-second-long show. True to the man from the stand's promise, it was an impressive display. Like Disneyland epic. Until it went horribly wrong. Halfway through the firework discharge, it malfunctioned and a mortar round shot out of the box sideways and came straight toward where we were all standing on my friend's back deck. I was next to Daisy with my arms crossed when it hit me—directly in the bull's-eye of my left eye socket. I had no idea what happened but found myself on the ground. It had hit me hard enough to knock me down so aggressively that I bloodied my left knee, which hit first.

The worst part was that there were still about twenty seconds to go in the show, so it was some time before anyone noticed me on the ground groaning. I made my way to the side of the house, braced myself against the glass, and cried out. Eventually Jeremy heard me and said, "Dude, are you okay?"

"Something hit me in the eye," I managed.

Everyone gathered around me. I heard lots of different suggestions ranging from "Get him some Tylenol and ice" to "We need milk and water."

They got me inside and my friend Tim, who had experience as a volunteer EMT, asked me to take my hand away so he could look at my eye. My hand shook as I pulled it back. I could see nothing out of it. Just black. Through my right eye I saw his eyes scanning. After studying it for a moment he turned to Jennie and said, "You need to take him to the hospital."

Today is Thanksgiving. Let me pause this story long enough to tell you that as I look back across the last four months to this most gruesome Fourth of July, it is with thanksgiving that I remember it. I am not so far away that I don't still shudder a little bit. I am profoundly grateful as I remember how scary and terrible this all was.

Back to the bumpy ride in the Jeep on the way to the emergency room.

Deep down I felt like I would never again see out of my left eye. That is what occupied my mind for much of the drive. And the closer we got to the hospital, the more it began to dawn on me where we were going. *That* hospital. The one where my daughter Lenya was both born and declared dead. An entrance and an exit.

We had been back to the hospital on the day my first book, *Through the Eyes of a Lion*, released. We brought balloons and copies of the book for employees and patients and made an occasion of it. This was a way for us to "run toward the roar," facing a fear so it couldn't hold any power over us anymore. But I had never been into the emergency room itself. That would mean being in or near the room where we were told there was nothing more that humans could do. "Declared dead" is such a mean phrase. I believe with all my heart that the only declaration over my daughter that stands is resurrection and life. The Lion of the tribe of Judah has prevailed, and death does not get the last word. Nonetheless, what happened happened, and somehow my damaged eye and I were going to have to deal with it.

THE LION OF THE TRIBE OF JUDAH HAS PREVAILED, AND DEATH DOES NOT GET THE LAST WORD.

Jennie got me out of the car and steered me toward the entrance of the ER. I felt myself begin to panic. Between the anxiety I was experiencing over my traumatic memories and the agony I was in over my vision, I thought my heart might just crunch like a cherry underfoot. And the mask I was forced to wear because of the ongoing pandemic compounded the feeling of being smothered. Hands sweaty. Heart like a war drum. Vision began to swirl. Hot everywhere.

As we arrived at the front desk, Jennie explained that I had been injured by a firework. "Name, please," the woman said.

"Levi Lusko," Jennie said.

"Oh! You're my pastor!" she gushed.

"So good to see you," I managed. "Please help my eye."

We were shown to a triage room where the wait felt interminable.

When the doctor finally arrived to assess me, he introduced himself as Dr. Dallas and said, "So I hear you were injured by a sparkler?"

"Not exactly," Jennie said. "A mortar round hit him in the eye."

The doctor's face revealed his surprise. I couldn't help thinking that Dr. Dallas seemed like a fake name, something a male stripper would use on the stage. He was young, handsome, and extremely kind.

Dr. Dallas became much more serious when he found out this was not sustained in the line of sparkler duty. He asked me to move my hand so he could get a look at my eye. He turned on a bright overhead light, and when I pulled my hand away and opened my eye, it wasn't pitch black like before; it was bright white. I could make out his silhouette, but it was like I was looking through milk. "There is some debris we will need to remove," he said.

Lovely.

Simultaneously they scrubbed gravel from my knee and removed what looked like crumbs of burnt toast, like what you would find when cleaning out a toaster, from my eye.

"I think there is a scratch in your cornea," he told me. "You will need to get a CAT scan and consult with an ophthalmologist to know what happens next."

When he left, the floodwaters spilled over, and I began to cry. Lenya's corneas had been donated two rooms over, and now it seemed as though I was going to need a new one. I gave myself over to this emotional storm, and instead of fighting it, I let it do its worst. Soon it began to pass.

Jennie held my hand, and we rode the wave together. Almost immediately I began to do an emotional one-eighty and find the humor in the absurdity of it all. Jesus did say to pluck out your eye, after all. How can you get the speck out of your neighbor's eye if you don't first take the firework out of your own? And on and on I went.

Eventually they came and got me for the CAT scan. It was a huge relief to discover that there was no tear in the cornea, but the best news Dr. Dallas had to give us was that the eyeball itself had not ruptured. There was no telling what damage had been done to my vision or what my sight would be like long-term, but, according to the ophthalmologist, the biggest concern was the pressure inside my eye. They gave me drops for that and ordered me to go to his office the next morning. A Sunday. With that, Dr. Dallas sent us out into the night with a patch and some pain medication.

It was a fitful night's sleep with dreams of ruptured eyeballs and blackened toast crumbs.

In the ophthalmologist's office the next morning, I was told the pressure in my eye, which should normally be ten to twenty was fifty. And that pressure could do permanent damage. So more drops were given to dilate, some to relax the muscle and others to specifically deal with the pressure. I felt sick. Dizzy. Off. Then he went on to explain what the Bible teaches—when the eye is not well, the whole body is not well (Matt. 6:22). Thanks, Doc.

Over the next week I went back to the ophthalmologist every single day for pressure checks, scans of the back of my eye, and other tests I didn't fully understand. As the pressure slowly started to subside, the milkiness of my vision began to subside as well. It went from like looking through heavy cream, to whole milk, to 2 percent, skim, and then watered-down milk. Finally, it was just blurry, like I was looking through water.

When I went back for my one-month checkup, he finally admitted to me that even though the pressure was under control and there appeared to be no specific damage to the components of my eye, there was no guarantee my vision would ever fully return. It was going to be a waiting game. Theoretically my eye was fine, but it wasn't focusing. He said that in cases of trauma, sometimes the muscle just doesn't contract and expand. It's as though it is stunned.

Basically my eye was inappropriately dilated all the time. The really weird thing was that I was eventually given permission to wear contacts again, and when I put the left one in, it made my vision worse, not better. He explained that the shape of my eye might have changed, affecting the prescription, and that we would wait one more month to assess if it was still shifting.

Over the next thirty days I began to sense my eye starting to focus on objects, but I had to try—it took effort. For instance, if I wanted to pull something from the background into the foreground, I had to think about it. Eventually I could do it, but it was like a slow camera. I asked what would happen if I never rebooted and the eye stayed dilated. Then I wished I hadn't. The doctor told me that the best option would be an artificial pupil that would split the difference between dilated and not. Yikes.

Fortunately it didn't come down to me needing a bionic eye; one day the muscle just started to work. It wasn't like a light switch turning on; it just took less and less effort to focus, and then one day I wasn't really paying attention, and I noticed it was doing it on its own. The wildest surprise was when I went in for the final assessment on my prescription and they discovered that the shape of my eye had indeed changed. For the better! The eyesight in that eye had improved so much they had to lower the prescription down to a weaker power.

I can say with all honesty and no trace of exaggeration that I am better for having had a firework shot directly into my eye. Not metaphorically; it actually, literally helped me see better.

Now, let me be clear. I don't recommend it. So don't go canceling your LASIK and buying bottle rockets instead. Results are certainly not typical. Jennie and I have mused many times that in heaven, when we watch the replay, we will see that an angel slowed the firework down so that it softened the blow and prevented my eye from being shot out.

All I know is that the view from this Thanksgiving is one of gratitude—to God for his protection and healing and for those who were praying. *I am eternally grateful* (said in my best "aliens from *Toy Story 2*" voice). And I'll never look at burned toast the same way again.

YOU SAY GOODBYE, I SAY HELLO

It's ironic that many astronauts have said that space smells burnt, like a steak or toast that's a bit too well done. One astronaut said, "To me, it's sort of like brimstone—as if a witch has just been there." Others have said it was like the smell of gunpowder from a bomb . . . or a firework. That connection aside, the real reason I told you this story is that the name of the firework that took me out has everything in common with both NASA's exploration of outer space, Jesus' bloody death, and the ongoing quest to conquer inner space. All three require blood, sweat, and tears. It's the price of victory.

Winston Churchill candidly put it out there in one of his more famous speeches when he said that was all he had to offer the British people.

It's also what was given by the brave men and women at NASA, which had only existed for eleven years at the time of the moonshot. With nothing short of a herculean effort, the men and women poured it all out. They laid everything on the line collectively, and humans were able to land on the moon. It is the greatest thing mankind has ever done.

To fully understand what made it possible, you can't just look to those who were directly involved; you must also consider those who paved the way.

If you visit the Smithsonian National Air and Space Museum, you can stand below the impossibly small, burned Hershey's Kiss command module that Neil, Michael, and Buzz returned home in. The

Smithsonian also boasts the Wright brothers' flyer, complete with a dummy lying on the wing, mimicking how it was originally flown in North Carolina.

Neil and the Wright brothers have something in common (other than the fact that they're all from Ohio).

Neil brought a piece of the Wright brothers' airplane with him on the Apollo 11 mission—a small bit of cloth from the wing of the Kitty Hawk Flyer (the 1903 Wright Flyer), which he carried with him when he stepped out of the lunar module onto the surface of the moon.

Think about that.

What I love about that gesture is the humility of it. It communicates that we are only here because of what they did back there. Orville and Wilbur flew for the first time in December 1903. They flew one hundred twenty feet. Their flight lasted twelve seconds. Sixty-six years later a piece of their wood-and-fabric flyer would make a quarter-million-mile trip to the moon. Less than one lifetime separated the feeble yet significant first flight from the voyage out of this world.

Here's the point. Jesus said that greater things than he did, you will do (John 14:12). His death on the cross was not an end; it was just the beginning. He intended his mission to be a Kitty Hawk that would propel you into a life of service, creativity, and beauty the likes of which the world has never seen. The blood, sweat, and tears he paid as he suffered and died for you and me was meant to be a launch point, only the inauguration.

The Last Supper didn't end anything; it began everything. You have a clean slate, a fresh start, and a new heart. No weapon formed against you can prosper (Isa. 54:17). All heaven is for you, and the gates of hell can't separate you from the love of God (Rom. 8:38–39). Raise the bread and wine, and toast the joy of being a child of the kingdom of God, a citizen of heaven.

FOR ALL MANKIND

Six and a half hours after landing, Neil and Buzz stepped onto the surface of the moon. They were able to do so because of the combined blood, sweat, and tears of the four hundred thousand people at NASA and on behalf of a nation that was willing to pursue a dream. More people watched that single event around the world than any other live event in human history at the time. It was eight years after JFK cast the vision for the moonshot.

The mission patch for Apollo 11 is unique. It was designed by Michael and has a bald eagle clutching an olive branch. Normally the eagle holds arrows when you see it, a symbol of power. The olive branch points back to the Old Testament story of Noah, where it is used as a symbol of peace after the "war" of the flood (Gen. 8:11). Showing a triumphant eagle holding an olive branch was a way of communicating that the United States had come to the moon in peace, not as an act of war. That is pretty cool imagery, but the unique part is not what is on the patch but what is missing.

Google "Apollo mission patches" and see if you can spot what's not there. I'll wait.

Did you see it? Names. There are no names on the patch. Breaking the trend of all NASA patches, Neil's, Michael's, and Buzz's names are nowhere to be found. This was the crew's way of tipping their hats to the fact that while they got to be a part of it physically, it was much bigger than them. It was bigger than the four hundred thousand people who worked on it or the late president who dreamed it up. It was even bigger than just the United States or the Russian competitors who egged us on. It was about the spirit of adventure and exploration that God put in us when he made us in his image and gave us a universe to explore.

WE COME IN PEACE FOR ALL MANKIND

Among other things that took place that day, including the observance of the Last Supper, Buzz and Neil erected an American flag on the Sea of Tranquility. If you were able to fly to the moon you would see something shocking. Old Glory is still on the moon, but due to the UV radiation it has been exposed to over the last fifty-plus years, it is probably completely bleached white.

This is a stunning picture of what Jesus launched when he fulfilled prophecy by dying a bloody death and rising again eternal, immortal, and dwelling in unapproachable light—though your sins were as scarlet, he has washed them white as snow (Isa. 1:18).

And through his substitutionary death and unstoppable life he has given you all you need to live a life of adventure, exploration, forgiveness, love, and peace.

LET THEM LAUGH

On July 17, 1969, the day after Apollo 11 blasted off toward the moon, the *New York Times* ran an apology correcting a 1920 editorial mocking Dr. Robert Goddard. Robert was an early proponent of space flight. Since then he has been considered the father of modern rocketry, but prior to his theories being proved right, he was an easy target and the butt of jokes by "experts."

He dreamed of the day a rocket could be sent to the moon. And he theorized it would be liquid fueled. In 1920, the Smithsonian published his paper, "A Method for Reaching Extreme Altitudes," which outlined his theories of rocket propulsion; it was met by ridicule from the press, including the *New York Times*. Professionals preached that his far-fetched notions would never get off the ground.

It'll never work.

Those three words were spoken over me the first week we started our church by someone who sought to dampen my expectations. I put them in the tank, and they have remained fuel for me ever since. I don't want to do anything with my life that doesn't have a touch of impossible to it, and I hope you don't either.

The morning after Robert's dream came true, the *New York Times* acknowledged its misinformed position and humbly admitted he had been right all along. Sadly Robert died in 1945 and didn't live long enough to be proven right. That's okay. He was ahead of his time, and you can be too.

If it's not until heaven that your life "makes sense," the eternal, final review will matter most. Buzz and Neil had actual gold in their visors to keep the bright lights from the sun out of their eyes. They looked at the moon through gold.

In that coming final assessment, some of our works will be determined to have been built on wood, hay, and straw—others on silver, precious stones, and gold (1 Cor. 3:12). Shoot for gold. Process your life and choices and thoughts through an eternal golden visor.

Let people scratch their heads at the calling God has put inside you. Never let anyone saying "It can't be done" stop you from aiming for the sky. Your unbelieving family won't understand why you are raising your kids the way you are. Snarky, disillusioned people show up in every great endeavor. You are in good company.

Jesus was dismissed as crazy by his own family (Mark 3:21). His followers were similarly written off as lunatics—a word that has its origins in the idea that the moon could make you crazy, like I mentioned earlier. To combat the pressure you are going to face outside, you will need pressure on the inside—and the Holy Spirit is willing to supply that essential air pressure. Today and again tomorrow. It's there when you fold the laundry; and after your toddler messes it up, it's there when you refold it. He will refill you when you have a low moment and after a great victory that intimidates you because you don't know how you'll top it.

If you take up your cross and follow heart and soul after the one who hung the moon, you will be laughed at.

I say, let them laugh.

You can't get to outer space if you haven't conquered inner space. And tonight, when you see the moon, remember this: It isn't the critic who counts. It's the lunatics who end up changing the world, one small step at a time.

As Jesus said,

"I am the vine, you are the branches. Whoever remains in me, and I in him, will bear much fruit; for you can do nothing without me."

Psalms 8:3,4

"When I consider thy heavens, the work of thy fingers, the moon and the stars, which thou hast ordained; What is man, that thou art mindful of him? and the Son of Man, that thou visitest Him?"

Buzz Aldrin wrote Bible verses and prepared comments to guide his remarks for the lunar surface broadcast. His actual words from the moon omitted any references to the Bible or his planned Communion service, as advised by Deke Slayton.

LM OTERO/AP PHOTO

Wernher von Braun, NASA's director of the Marshall Space Flight Center, in his office in Huntsville, Alabama, in 1965 sitting in front of a row of rocket models. Fifth from the left with the red escape tower is a Mercury-Redstone (used by Alan Shepard on the first manned launch by NASA) and the far right is a Saturn V sticking through the ceiling tiles.

ARCHIVE PHOTOS/STRINGER

UNITED STATES

On September 12, 1962, President John F. Kennedy pledged to put a man on the moon and return him safely home by the end of the decade in a speech at Rice University in Houston, Texas, attended by more than forty thousand people.

NASA

Wernher von Braun explains the launch system to his good friend President John F. Kennedy at Cape Canaveral, Florida. The date is November 16, 1963, six days prior to the president's assassination.

NASA

John F. Kennedy Jr., on his third birthday, stands at salute in front of the casket of his father, President John F. Kennedy. His mother, Jacqueline Kennedy (center), and sister Caroline Kennedy are accompanied by the late president's brothers, Senator Edward Kennedy (left) and Attorney General Bobby Kennedy (right).

NEW YORK DAILY NEWS

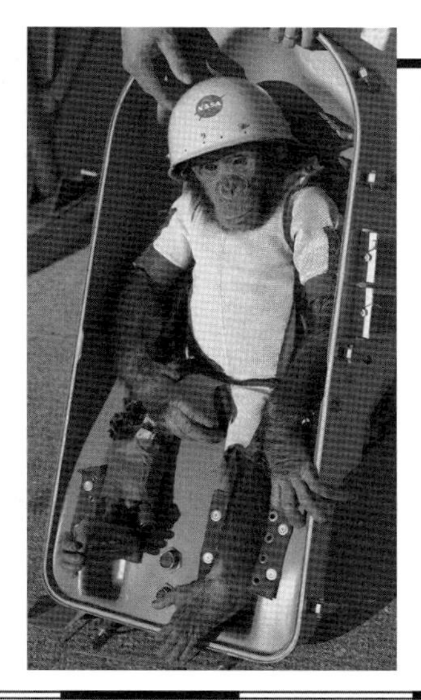

Before there were space men there were space monkeys. Ham, NASA's first "astro chimp," was sent into orbit on January 31, 1961.
NASA

Dr. Robert H. Goddard and a liquid-fueled rocket, which was fired on March 16, 1926, in Auburn, Massachusetts. It flew for only 2.5 seconds, climbed to 41 feet, and landed 184 feet away in a cabbage patch. Dr. Goddard has been posthumously considered the father of modern rocketry and an indispensable part of the moon landing.
NASA

Ed White (left), Virgil "Gus" Grissom (center), and Roger Chaffee (right)—the crew of Apollo 1, who perished in a fire on the launchpad during a test on January 27, 1967.
NASA

Aldrin takes the first "space sel[illegible] extravehicular activities (EVA) on Gemini 12 on November 12, 1966. He set his camera on the edge of the hatch, pointed it in his direction, and took the photo.
NASA

Gleaming white and bathed in light. The Saturn V is the most magnificent and powerful rocket that has ever been launched by man. Hollywood-style spotlights had a stunning effect that only enhanced its mystique. The 363-foot Saturn V was a three-stage vehicle that stood taller than the Statue of Liberty. This picture is from Apollo 8, the first manned flight of the Saturn V.

NASA

The Crawler-Transporter that moved the Saturn V from the Vehicle Assembly Building to the launchpad. You can see a man next to one of the beast's enormous treads. This photo is from Apollo 13.
NASA

With the Apollo 11 Saturn V rocket and mobile launcher on its back, the Crawler-Transporter makes its way out of the Vehicle Assembly Building—the largest single-story building on earth.
NASA

Houston, Texas. The front row was nicknamed "the trench" supposedly because the pneumatic tubes (like what you use at an outside bank deposit) piled up on the ground reminiscent of spent Howitzer casings. The three crews of flight controllers led by a flight director worked in shifts designated by colors; during Apollo 11 the colors were blue, red, and white. Mission Control retires colors the same way professional sports retire numbers.
NASA

Steve Bales at his guidance, navigation, and control station (GNC) at Mission Control Center at Johnson Space Center. Steve was twenty-six years old in a room where the average age was twenty-eight. The decision to land or abort when the program alarm went off during the Apollo 11 mission was left to him.
NASA

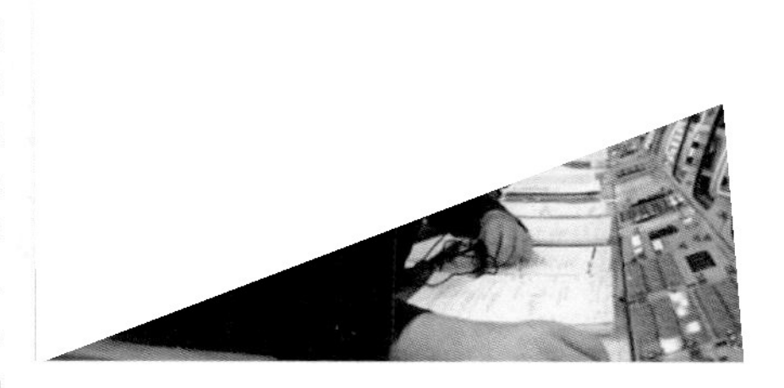

For every ... see there was ... support you couldn't. The flight dynamics support team is pictured here during Apollo 11. Jack Garman, who knew what the 1202 alarm was in moments, is second from the left.
NASA

Earthrise. One of the most iconic photos of all time. This was the first time the earth was seen from the vantage point of the moon on Apollo 8 in December 1968.

NASA

The dark side of the moon, which was seen for the first time by man's eyes on Apollo 8 in December 1968, is far more battered and jagged than the side we see from earth.

NASA

Astronaut Peggy Whitson was the first woman to ever command the International Space Station. She has spent 665 days in space.

NASA

Human computer. NASA mathematician Katherine Johnson at her desk at Langley Research Center in 1966. One of the "hidden figures" of the space program, NASA could never have landed on the moon without the help of many people like Katherine who worked behind the scenes on the orbital mechanics that made spaceflight possible.

NASA

All hands on deck. It took the combined efforts and sacrifice of four hundred thousand people to get from the earth to the moon, including these workers who are assembling the Apollo 1 Module in 1966 in Downey, California.

RALPH MORSE/THE *LIFE* PICTURE COLLECTION/ SHUTTERSTOCK

Mission patch for Apollo 11. Designed by Michael Collins, it has a bald eagle clutching an olive branch with the moon in the foreground and the earth in the background.
NASA

Apollo mission simulator. Astronauts, engineers, and backup crews spent as much as fourteen hours a day, six days a week practicing. They prepared for hundreds of potential problems thrown at them by Dick Koos, simulation supervisor (SimSup). William A. Anders, Michael Collins, and Frank Borman (from top of stairs) are about to enter in the photo at the top; in the bottom photo you can see them inside the simulator.
NASA

The LLRV (Lunar Landing Research Vehicle) is a machine that simulated the powered descent in the lunar module. It nearly killed Neil Armstrong when he ejected from it seconds before it crashed. Here Armstrong trains for the mission at NASA's Lunar Landing Research Facility.
NASA

Heading to the moon. The Apollo 11 crew and support teams in the hallway of the Manned Spacecraft Operations Building as they head to Launch Complex 39A for the first manned lunar landing mission.
NASA

Liftoff of Apollo 11. Generating 7.7 million pounds of thrust, the Saturn V launch vehicle wears the Apollo spacecraft like a witch's hat as it leaves earth heading for space—going from zero to 17,500 miles per hour in under nine minutes.
NASA

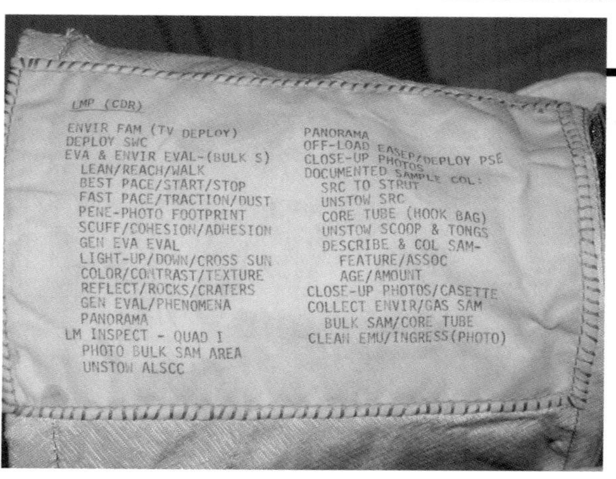

Checklist. Aldrin and Armstrong each had a checklist sewn onto the wrist of their space suit so they never lost sight of all they had to do on the lunar surface. This is Aldrin's checklist.
NASA

A million people flocked to the beaches around Cape Kennedy, Florida, to witness the Apollo 11 launch on July 16, 1969.

RALPH CRANE/THE *LIFE* PICTURE COLLECTION/SHUTTERSTOCK

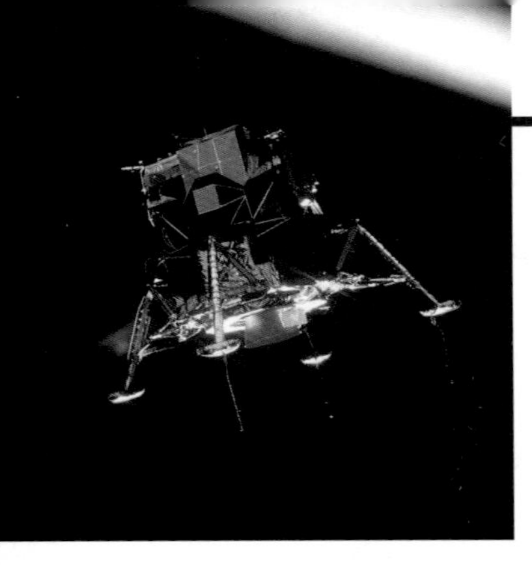

Contact light. The Apollo 11 lunar module, Eagle, photographed by Collins. The long rod-like protrusions under the landing pods are lunar surface sensing probes. Upon contact with the lunar surface, an indicator light told the astronauts to cut the engines. When Jim McDivitt returned from Apollo 9, the lunar module's first manned flight, he sent the manufacturer a note that said: "Many thanks for the funny-looking spacecraft. It sure flies better than it looks."

NASA

Perhaps the most famous photo of the Apollo 11 mission taken as Aldrin posed before the camera, with Armstrong in the reflection of his gold-plated visor, along with the flag and the Eagle.

NASA

Armstrong and Aldrin return to rendezvous and dock with Columbia after a successful moon landing in the Eagle as Collins looks on. The earth is in the background.

NASA

No normal day. Armstrong is all smiles after a historic small step on the moon. Charts are duct-taped to the wall of the LM Eagle indicating various switch and fuse positions for the controls.

NASA

Space food. The food NASA prepared improved dramatically over the years but was never why anyone became an astronaut. These dehydrated meals were from Project Gemini in 1963.

RALPH MORSE/THE *LIFE* PICTURE COLLECTION/ SHUTTERSTOCK

Moon car. The Lunar Roving Vehicle (LRV) on the moon during Apollo 17. Built by the Boeing Company, the LRV was light, battery powered, and had a range of fifty-five miles. On this mission the astronauts managed to break a fender and repair it with duct tape.

NASA

Lunar samples being observed after Apollo 16 by geologists Don Morrison (left), Fred Hörz (right) and William (Bill) Muehlberger (center). The larger rock is the biggest sample ever taken from the moon and is known as "Big Muley."

NASA

This stamp, commemorating the success of Apollo 11, enraged Aldrin's father, who thought it should read "First Men on the Moon."

US POST OFFICE

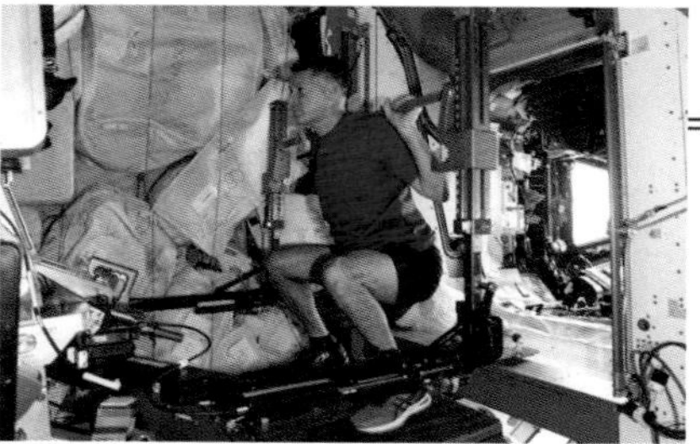

Astronauts on the International Space Station currently work out for two-plus hours a day to offset the negative effects of microgravity on the human body. Here astronaut Shane Kimbrough does some squats in August 2021.

MEGAN MCARTHUR

Task completed. On July 24, 1969, after a successful landing on the moon, the Apollo 11 command module Columbia splashed down in the Pacific Ocean and was retrieved by a specially trained recovery team. The astronauts had traveled half-a-million miles in eight days and made it safely home by the end of the decade, as President Kennedy predicted.

NASA

Armstrong, Aldrin, and Collins were kept in quarantine for twenty-one days after they left the lunar surface. In this photo from Hawaii, they are shown reuniting with their wives, Pat Collins, Jan Armstrong, and Joan Aldrin, from the window of the Airstream trailer that would transport them to the quarantine facility in Houston.

NASA

On August 13, 1969, New York City welcomed the Apollo 11 astronauts with a ticker-tape parade down Broadway and Park Avenue in what was, at that point, the largest parade in the city's history.

NASA

Armstrong, Collins, and Aldrin were swarmed by thousands in Mexico City on one stop of their forty-five-day Giant Leap Tour after the moon landing.

NASA

END HERE

WHILE NEIL AND BUZZ WERE ON THE SEA OF TRANQUILITY, President Nixon called them from the White House, his voice relayed into the radios in their space suits. One thing in particular he told them explodes in my heart every time I read it, because, in a profound and eternal way, it is true of Jesus' accomplishment. With excitement palpable in his voice, he told Neil and Buzz, "Because of what you have done, the heavens have become a part of man's world."

The heavens, dear friend, are a part of *your* world! I hope you never take that for granted—not heaven, which is where your Savior and citizenship are, or the heavens that are the home of the moon, God's eternal night-light reminding you of his goodness and presence in your life. Every time you see it, I hope you'll do what I do—reach out and try to grab it between your thumb and pointer finger and know that the one who hung it is holding you. The moon is your invitation to wonder, a reminder of your inner space left to be conquered, and a symbol of God's love for you.

No one has ever taught us to explore and conquer inner space; we've now discovered it by exploring outer space.

In the most famous photo of all the Apollo 11 pics, you can see Buzz standing, posing before the camera, and Neil in the reflection of his gold-plated visor, along with the flag and the Eagle. It is astounding. If you look at it carefully you will notice that Buzz is sort of holding his left arm weirdly. It almost looks like he is inspecting his left wrist. He is. There is a reason for this. He had a mission objective checklist for their time on the lunar surface sewn onto it. It's like having a grocery list written on your arm. Thirty or so things he needed to accomplish before leaving for earth. And as Neil took the photo, he was checking where he was at on the list and what he needed to do next.

They stuck to the plan, making sure to not get distracted, because their time on the moon was limited.

So is your time on this earth. The clock is ticking.

Seven letters. Seven assurances of who Jesus is. Seven last words. For you. For me. For the kingdom.

I pray that when you close this book you will see your mission all around you.

Your objectives are clear:

- "Love . . . God with all your heart."
- "Love your neighbor as yourself."
- "Go into all the world."
- "Go everywhere and announce the message of God's good news to one and all."

You don't need to "stick to this plan" and accomplish these things so God will love you. That kind of living is finished. The task has been accomplished. You and I get to focus on these things because we have been forgiven and freed and are now fully alive. That is the message of the Last Supper on the moon.

I want to say to you, as we part, the same thing spoken at 4:15 a.m. on July 16, 1969, when Deke Slayton, director of flight crew operations, spoke to the crew of Apollo 11 the day of liftoff. He knocked on their bedroom doors and said, “It’s a beautiful day; you’re *go.*”

It is a beautiful day.

And you are *go.*

ACKNOWLEDGMENTS

AFTER SERVING AS FLIGHT CONTROLLER DURING THE MOON landing, Gene Kranz said, "I pray that my children will someday feel the triumph, the joy, and the shiver I felt the day we painted the Moon with our Star-Spangled Banner." He was so excited on the day of the Apollo 11 launch that he took the stairs instead of the elevator, lest a malfunction in the notoriously finicky elevator kept him from the moon landing.

I'm not Gene Kranz's child, but throughout the writing of this book I definitely felt the triumph, the joy, and the shiver of both the moon shot and the day my Savior died for me on Skull Hill. This book is meant to be my love letter both to the cross and to space.

The shiver began for me years ago when I first read Tom Wolfe's *The Right Stuff*. I was gripped by the magic of the original Mercury 7 and the beginnings of manned space flight in America. I started to nurse my new addiction, taking my family to Kennedy Space Center in Florida and seeing a Saturn V for the first time. It blew me away.

Things got real in May of 2019 when Brant Cryder, Scott Harrison, and I were standing on my back deck and Brant told me that Buzz Aldrin brought Communion with him to the lunar surface. It became like the movie *Inception* for me. The token had been planted in my safe and I couldn't stop thinking about it. I wrote down a phrase that came to me: "the last supper on the moon." I knew it would eventually become a book, but I sat on it for over a year until the volcano erupted.

Also in May of 2019, Jeff Henderson gave me the book *American Moonshot*, which poured gas on an already flickering fire. I waited a year to begin writing, and by the time I did, it seemed the moon was haunting me. I would look up and it would be there—in broad daylight, out a plane window, in the night sky. I couldn't get away from it. Jennie teased me, telling me the moon was saying, "Write a book about me." It was such a lofty idea, and I wasn't sure that I could pull off what I envisioned: a history book with devotional impact. It was a secret idea I carried with me and thought about constantly.

Finally, on my thirty-eighth birthday, I began creating the manuscript for this book. (Starting is the hardest part.) After that, I didn't let off the gas, writing every day except Saturdays until the sixty-thousand-word rough draft was done.

In the midst of writing, Jeff also told me about BBC's *13 Minutes to the Moon* podcast by Kevin Fong, which was extraordinary and really helped me understand the masterful complexity of the powered descent and how close Apollo 11 came to not landing on the moon at all. The podcast mixes in the voices of those involved with the space missions and audio from the actual events and makes it all come to life. I listened to the episodes of the first season again and again and never got tired of its magic.

Louie Giglio is my deep space–nerd friend and perhaps the only person more likely than me to get a NASA tattoo. His "Indescribable" and "How Great Is our God" talks shaped my love of the cosmos. Listening to whales and stars sing every night with him at the helm of the DJ console on tour was unforgettable. Touring Kennedy Space Center and watching a rocket launch with Shelley and Louie will always remain one of the top moments of my life. I am also indebted for all the space things he has given me or pointed me toward. Without his influence I am sure this book would not exist, or it would be a sickly creature too weak to stand.

Steven Furtick was the first person outside my immediate family I told about the idea of this book, and it was on a really low day for me. He immediately spoke life into it and told me, "This is a big book," and, as always, lifted me up with his words and friendship.

I am thankful to my dad, Chip, for always showing interest when something new from history lights me up and being up for a phone call to share in the excitement of that day's discovery. Coffee is on me out back at sunrise or wherever in the world we find ourselves.

I am thankful for my agent, Austin, for being a longtime ally, friend, and adviser. I respect your opinion and appreciate your wisdom, encouragement, and expertise greatly.

Thank you to my executive assistant, McKenzie, for your strength, grit, and grace in and through all the chaos. Your work and insight on this project, and all the other things, is greatly appreciated.

Thank you, Elisha, for another extraordinary cover and for guiding all the art in this book with an eagle eye, a steady hand, and masterful touch. From combing through thousands of NASA images to sending emails at 3:08 a.m., this book would not be what it is without you. Oh, also, the NASA Apollo 11 press kit you found? Massive.

Thank you, Katelyn and Zai, for being fluent in Lusko and for the passion and joy you have shown and continue to show to tell the story.

Thank you, Kevin and Colton, for your help on the curriculum shoot and for freaking out with me over how awesome space can be.

I also am grateful for Damon and Debbie on the W team. Thank you for the unceasing enthusiasm you showed for this project; your belief in this strange new "franken monster" gave me strength throughout the long, winding road to completion. She is indeed a big girl and does sit low in the water, and I still can't believe you gave me the runway and the latitude to attempt to pull it off. You guys are my kind of crazy. Damon, your notes on the first pass were amazing. Debbie, it's been a long, beautiful road with many memories. Congratulations on your retirement. Now let's dream about the next project! LOL.

Thank you, Lysa, Shae, Leah, and Joel for *not* caring about space and for allowing me to see this book through your eyes. Thanks for letting me convince you space is awesome. Your help, as always, is appreciated greatly. Next time in person!

I am grateful for the exquisite schematics and diagrams at the front of the book, which were drawn by Marcin Szpak from Poland.

On the editing side this was a relay race with many involved. Thank you, Meaghan Porter, Whitney Bak, Jennifer McNeil, Stephanie Newton, and Carrie Marrs: all of you played a part and this was a beast to bring to the finish line. Here's to all the spinning beach balls and prayers for recovered files. At the end Stephanie and Carrie went above and beyond and worked, I am sure, to the brink and then some. I appreciated in particular Carrie's effusive cheerfulness and kindness at every turn.

Finally, thank you to Jennie, Alivia, Daisy, Clover, and Lennox for always grabbing the moon with me and for showing excitement every step of the way. Your love, support, encouragement, and cheerleading mean the world to me. Wherever I go and whatever I do, the best part of my day is always coming home to you.

Thank you once again, dear reader, for trusting me enough to buy this book and, most of all, for trusting me with your time. I will never take the honor of writing for granted. I love books and know what they mean to me. That you would take a risk on a book, especially one as different as this book, is a gift. I sincerely hope it has inspired and bettered you.

And now a word about sources.

In addition to the *13 Minutes to the Moon* podcast by Kevin Fong and *American Moonshot* by Douglas Brinkley, I am especially indebted to James Donovan for *Shoot for the Moon*. This would probably be my number-one, overall favorite source and the one I'd recommend if someone wanted to read only one book covering Mercury to Apollo.

Carrying the Fire by Michael Collins was extraordinary. He writes with a golden tongue. Robert Kurson's *Rocket Men* is a macro focus on Apollo 8 but sheds light on so much more. Gene Kranz, in his spectacular book *Failure Is Not an Option*, really helped me see the same missions from the perspective of mission control and upped my respect for the controllers exponentially. *Deke!* by Donald K. "Deke" Slayton was similar, as he was a grounded astronaut covering the spectrum of the whole moon shot—but since the astronauts were his brothers, it was like getting to see inside the fraternity. He was less impressed with them and more dispassionate. I enjoyed his sarcasm and jocularity immensely.

Magnificent Desolation, cowritten by Ken Abraham and Buzz Aldrin, was not only a fascinating look at Buzz's life before, during, and after the moon landing but also helped me understand how hard the success was on those involved, especially Buzz. I am grateful in particular to Ken for the time he gave me on the phone and over email, encouraging the project, and for making it possible for me to speak with Buzz Aldrin.

Brian Floca's *Moonshot* not only helped ignite Lennox's love for the Saturn V, Eagle, and Columbia but also put the cookies on the bottom shelf for me, explaining hard things in an easy-to-understand way. Reading this book made me go back to my manuscript and rethink many overly complicated paragraphs.

Norman Mailer's *Of a Fire on the Moon* was a masterpiece. No one should be allowed to be so smug and cool. He writes about himself in the third person as someone named Aquarius experiencing the moon landing. Epic.

Also, I am grateful to the many interviews with those involved with the space missions that I watched or listened to on YouTube. The NASA archives both online and in print contain more than anyone could comb through in a lifetime, but I gave it my best try. Additionally, visits to Johnson Space Center in Houston (thanks, Shane, for the BTS tour), Kennedy Space Center (which were never long enough), Space Center Houston (with the Kimbroughs—#pinchme), and the Smithsonian Air and Space Museum all helped me in my research. There are more sources I relied on but these were the main ones. I owe much to many for this book, which one friend called my magnum opus. I don't know about that, but this I do know: any errors in this book are my own.

NOTES

BEGIN HERE

xxvii **The man read from:** Buzz Aldrin, "Guideposts Classics: When Buzz Aldrin Took Communion on the Moon," *Guideposts*, October 1970, accessed September 14, 2021, https://www.guideposts.org/better-living/life-advice/finding-life-purpose/guideposts-classics-when-buzz-aldrin-took-communion-on-the-moon. The article cites the Today's English Version Bible translation of John 15:5 as the source for these words.

xxxiii **NASA plans to land:** Monica Witt and Jena Rowe, "As Artemis Moves Forward, NASA Picks SpaceX to Land Next Americans on Moon," ed. Katherine Brown, NASA, last updated April 22, 2021, https://www.nasa.gov/press-release/as-artemis-moves-forward-nasa-picks-spacex-to-land-next-americans-on-moon.

xxxiii **Aslan said:** C. S. Lewis, *The Last Battle* (1956; repr., New York: HarperTrophy, 2002), 228.

xxxvi **When rolling a pair:** Eric W. Weisstein, "Dice," Wolfram Mathworld, accessed July 18, 2021, https://mathworld.wolfram.com/Dice.html.

xxxvi **Most mammals':** Frietson Galis, "Why Do Almost All Mammals Have Seven Cervical Vertebrae? Developmental Constraints, *Hox* Genes, and Cancer," *Journal of Experimental Zoology* 285, no. 1 (April 1, 1999), https://doi.org/10.1002/(SICI)1097-010X(19990415)285:1<19::AID-JEZ3>3.0.CO;2-Z.

xxxvii **The upper limit:** Lauren Schenkman, "In the Brain, Seven Is a Magic Number," ABC News, November 27, 2009, https://abcnews.go.com/Technology/brain-memory-magic-number/story?id=9189664.

xxxvii **The Bell Telephone Company:** Malcolm Gladwell, *The Tipping Point: How Little Things Can Make a Big Difference* (New York: Little, Brown, 2006), 176.

xxxvii **George Miller wrote:** George A. Miller, "The Magical Number Seven, Plus or Minus Two: Some Limits on Our Capacity for Processing Information," *Psychological Review* 63, no. 2: 86, https://doi.apa.org/doiLanding?doi=10.1037%2Fh0043158.

xxxvii **seven hundred times:** Dolores Smith, "What Is the Biblical Significance of the Number 7?," Christianity.com, January 31, 2020, https://www.christianity.com/wiki/bible/what-is-the-biblical-significance-of-the-number-7.html.

xxxviii **The entire Jewish calendar:** Robinson Meyer, "The Ancient Math That Sets the Date of Easter and Passover," *The Atlantic*, April 19, 2019, https://www.theatlantic.com/science/archive/2019/04/why-dont-easter-and-passover-always-line/587572/.

xxxix **Buzz Aldrin expressed:** Buzz Aldrin, *Magnificent Desolation: The Long Journey Home from the Moon* (New York: Bloomsbury Publishing, 2009), 27.

CHAPTER 1: EVERYONE IS A MOON

2 **Scientists call this:** Nola Taylor Redd, "Does the Moon Rotate?," Space.com, June 17, 2021, https://www.space.com/24871-does-the-moon-rotate.html.

3 **The back side [of the moon]:** *American Experience*, "Looking at the Moon," PBS, accessed July 14, 2021, https://www.pbs.org/wgbh/americanexperience/features/moon-looking-moon-apollo-8/.

3 **The moon is essentially:** *American Experience*, "Looking at the Moon."

3 **I don't think:** Richard Hollingham, "The NASA Mission That Broadcast to a Billion People," BBC Future, December 21, 2018, https://www.bbc.com/future/article/20181220-the-nasa-mission-that-broadcast-to-a-billion-people.

3 **one-fourth the size:** Tim Sharp, "How Big Is the Moon?," Space.com, October 27, 2017, https://www.space.com/18135-how-big-is-the-moon.html.

3 **taller than the Himalayas:** Evan Gough, "Comparing Mountains on the Moon to the Earth's Peaks," Universe Today, March 5, 2020, https://www.universetoday.com/145254/comparing-mountains-on-the-moon-to-the-earths-peaks/.

4 **One known crater:** Norman Mailer, *Of a Fire on the Moon* (New York: Random House, 2014), 277.

4 **Mark Twain once said:** Mark Twain, epigraph from *Pudd'nhead Wilson's New Calendar*, in Mark Twain, *Following the Equator* (New York: Harper & Brothers Publishers, 1903), 350.

5 **Comparison is:** "Comparison Is the Thief of Joy," Quote Investigator, May 6, 2021, https://quoteinvestigator.com/2021/02/06/thief-of-joy/.

CHAPTER 2: THE CROWDED HOUR

11 **In 1927:** "Charles Lindbergh," HISTORY, last updated May 6, 2020, https://www.history.com/topics/exploration/charles-a-lindbergh.

11 **Around the clock:** Charles Lindbergh House and Museum, "Timeline," Minnesota Historical Society, accessed July 14, 2021, https://www.mnhs.org/lindbergh/learn/timeline.

11 **Six hundred and fifty million:** Brian Dunbar, "Apollo 11 Mission Overview," ed. Sarah Loff, NASA, last updated May 15, 2019, https://www.nasa.gov/mission_pages/apollo/missions/apollo11.html.

12 **They visited twenty-four:** John Uri, "50 Years Ago: Apollo 11 Astronauts Return from Around the World Goodwill Tour," ed. Kelli Mars, NASA, last updated November 5, 2019, https://www.nasa.gov/feature/50-years-ago-apollo-11-astronauts-return-from-around-the-world-goodwill-tour.

12 **The first stage:** Piers Bizony, Andrew Chaikin, and Roger D. Launius, *The NASA Archives* (Cologne, Germany: Taschen, 2019).

12 **One crowded hour:** Thomas Mordaunt, "The Call," in *The Oxford Book of English Verse: 1250–1900*, ed. Arthur Quiller-Couch (Oxford: Clarendon Press, 1912), 622.

12 **As Henry David Thoreau:** Henry David Thoreau, *The Writings of Henry David Thoreau: Journal*, ed. Bradford Torrey, vol. 4: *May 1, 1852–February 27, 1853* (New York: Houghton Mifflin, 1906), 227, https://www.walden.org/wp-content/uploads/2016/02/Journal-4-Chapter-3.pdf.

13 **propped against the wrong:** Stephen Covey, *The 7 Habits of Highly Effective People: Powerful Lessons in Personal Change* (New York: Free Press, 2004), 98.

16 **From the very beginning:** Don N. Howell Jr., *The Passion of the Servant: A Journey to the Cross* (Eugene, OR: Resource Publications, 2009), vii.

20 **C. S. Lewis said:** C. S. Lewis, *The Great Divorce*, in *The Complete C. S. Lewis Signature Classics* (New York: HarperOne, 2002), 503.

20 **In Netflix's *The Crown*:** *The Crown*, season 3, episode 7, "Moondust," directed by Jessica Hobbs, featuring Henry Pettigrew as Neil Armstrong, aired November 17, 2019, on Netflix, 00:19, https://www.netflix.com/watch/80215737.

CHAPTER 3: SO THIS IS WHAT ELEVATORS LOOK LIKE

22 **3.5 miles:** Brian Dunbar, "Vehicle Assembly Building," ed. Anna Heiney, NASA, last updated September 23, 2020, https://www.nasa.gov/content/vehicle-assembly-building.

24 **That's how you get a rocket:** Charles Fishman, "This Monster Truck Is One of the Last Pieces of Apollo-Era Space Tech Still in Use," Fast Company, July 2, 2019, https://www.fastcompany.com/90371897/this-monster-truck-is-one

-of-the-last-pieces-of-apollo-era-space-tech-still-in-use; Brian Dunbar, "Mobile Launcher," ed. Anna Heiney, NASA, last updated December 21, 2020, https://www.nasa.gov/content/mobile-launcher.

24 **thirty-six-story building:** Joel Walker, "Rocket Park: Saturn," ed. Orlando Bongat, NASA, last updated September 16, 2011, https://www.nasa.gov/centers/johnson/rocketpark/saturn_v.html.

24 **more than twenty:** Brian Dunbar, "The Crawlers," ed. Anna Heiney, NASA, last updated March 23, 2021, https://www.nasa.gov/content/the-crawlers.

25 **The flat top:** Fishman, "This Monster Truck."

26 **The real secret sauce:** Fishman.

26 **Its enormous, locomotive:** NASA, "NASA Facts: Crawler-Transporters" (Kennedy Space Center, FL: John F. Kennedy Space Center, 2021), 2, https://www.nasa.gov/sites/default/files/atoms/files/combined_crawler-transporters_fact_sheet_final.pdf.

26 **The ground this beast:** Fishman, "This Monster Truck."

26 **the star sailors:** Vocabulary.com, s.v., "astronaut," accessed July 14, 2021, https://www.vocabulary.com/dictionary/astronaut.

26 **reach a top speed:** Adam Hadhazy, "How Fast Could Humans Travel Safely Through Space?," BBC Future, August 10, 2015, https://www.bbc.com/future/article/20150809-how-fast-could-humans-travel-safely-through-space.

26 **one foot per second:** Fishman, "This Monster Truck."

28 **We three kings:** John H. Hopkins Jr., "We Three Kings of Orient Are," no. 215 in *Baptist Hymnal* (Nashville, TN: LifeWay Worship, 2008).

29 **Richard Branson's business plan**: Richard Branson, *Screw It, Let's Do It: Lessons in Life* (London: Virgin Books, 2006).

30 **Make no little plans:** "Stirred by Burnham, Democracy Champion," *Chicago Record-Herald*, October 15, 1910, in "A Chicago Tale: Why We're Happy to Erase the Asterisk from Daniel Burnham's 'Make No Little Plans,'" *Chicago Tribune*, March 6, 2019, https://www.chicagotribune.com/opinion/editorials/ct-edit-daniel-burnham-quote-20190305-story.html.

30 **Andy Andrews:** Andy Andrews (@AndyAndrews), "Think about this," Twitter, May 30, 2014, 3:00 p.m., https://twitter.com/AndyAndrews/status/472452526940631040.

CHAPTER 4: THERE IS NO MOON

35 **None will be so difficult:** John F. Kennedy, "Excerpt from an Address Before a Joint Session of Congress," 2 black-and-white 16mm film reels, 9:00, May 25, 1961, archived at the John F. Kennedy Presidential Library

and Museum website, streaming video, 3:26, accessed September 14, 2021, https://www.jfklibrary.org/learn/about-jfk/historic-speeches/address-to-joint-session-of-congress-may-25-1961.

35 **In dollars and cents:** Alex Knapp, "Apollo 11's 50th Anniversary: The Facts and Figures Behind the $152 Billion Moon Landing," *Forbes*, July 20, 2019, https://www.forbes.com/sites/alexknapp/2019/07/20/apollo-11-facts-figures-business/.

35 **most expensive thing:** Guinness World Records, s.v. "Most Expensive Man-Made Object," www.guinnessworldrecords.com/world-records/most-expensive-man-made-object.

36 **Space medicine contributed:** Douglas Brinkley, *American Moonshot: John F. Kennedy and the Great Space Race* (New York: HarperCollins, 2019), 434–35.

37 **tragically killed:** Hanneke Weitering, "OTD in Space - January 27: Apollo 1 Fire," in *On This Day in Space*, Space.com, accessed September 15, 2021, video, 00:51, https://videos.space.com/m/hfhU71e0/otd-in-space-january-27-apollo-1-fire.

37 **Three astronauts died:** Brian Dunbar, "Remembering NASA Astronauts Elliot See and Charles Bassett," NASA, February 26, 2016, https://www.nasa.gov/feature/remembering-nasa-astronauts-elliot-see-and-charles-bassett.

37 **October 1957:** Robyn Rodgers, "Sputnik and the Dawn of the Space Age," NASA History Division, accessed July 15, 2021, https://history.nasa.gov/sputnik.html.

37 **first to put a dog:** "Soviet Union Launches a Dog into Space," HISTORY, last updated November 3, 2020, https://www.history.com/this-day-in-history/the-soviet-space-dog.

38 **dubbed a Flopnik:** Brian Dunbar, "60 Years Ago: Vanguard Fails to Reach Orbit," ed. Mark Garcia, last updated January 30, 2018, https://www.nasa.gov/feature/60-years-ago-vanguard-fails-to-reach-orbit.

38 **The big problem:** John F. Kennedy, "Address at Rice University on the Nation's Space Effort," color videocassette, 19:00, September 12, 1962, John F. Kennedy Presidential Library and Museum, accessed September 14, 2021, https://www.jfklibrary.org/learn/about-jfk/historic-speeches/address-at-rice-university-on-the-nations-space-effort.

38 **As the broad strokes:** Clarence A. Robinson, "Project Apollo—Apollo 11: 50 Years," Defense Media Network, July 19, 2019, https://www.defensemedianetwork.com/stories/project-apollo-11-50-years/.

38 **This Nation has tossed:** John F. Kennedy, "Remarks at the Dedication of the Aerospace Medical Health Center, San Antonio, Texas, November 21, 1963," transcript, John F. Kennedy Presidential Library and Museum,

https://www.jfklibrary.org/archives/other-resources/john-f-kennedy-speeches/san-antonio-tx-19631121.

39 **The new ocean:** Kennedy, "Address at Rice University," 7:30.

39 **must work the works:** John 9:4.

40 **seventy-millimeter footage:** Kerri Lawrence and Sarah Garner, "National Archives Film Footage Fuels Apollo 11 Film," National Archives, March 8, 2019, https://www.archives.gov/news/articles/apollo-11-footage-debuts-in-new-documentary/.

40 **million people who camped out:** Eric Benson, "Was Apollo 11 a Beginning or an End?," *Texas Monthly,* July 2019, https://www.texasmonthly.com/being-texan/was-apollo-11-beginning-end/.

40 **three thousand reporters:** "EP-72 Log of Apollo 11," NASA History Division, accessed July 15, 2021, https://history.nasa.gov/ap11ann/apollo11_log/log.htm.

42 **Aristarchus correctly observed:** James Evans, "Aristarchus of Samos," *Encyclopedia Britannica*, February 12, 2020, https://www.britannica.com/biography/Aristarchus-of-Samos.

42 **George Washington was aided:** Bill O'Reilly and Martin Dugard, *Killing England* (New York: Henry Holt and Company, 2017), 51.

42 **fierce raids:** S. C. Gwynne, *Empire of the Summer Moon: Quanah Parker and the Rise and Fall of the Comanches, the Most Powerful Indian Tribe in American History* (New York: Scribner, 2010), 65.

42 **Nazi's rained bombs:** Erik Larson, *The Splendid and the Vile: A Saga of Churchill, Family, and Defiance During the Blitz* (New York: Crown, 2020), 5.

42 **Luftwaffe could easily see:** Larson, *Splendid and the Vile*, 265, 469.

42 **Then again at D-Day:** Jamie Carter, "What Has the Moon Got to Do with D-Day? As It Turns Out, Everything," *Forbes,* June 5, 2019, https://www.forbes.com/sites/jamiecartereurope/2019/06/05/what-has-the-moon-got-to-do-with-d-day-as-it-turns-out-everything/.

43 **It would be easier to believe:** "Was the Moon Landing Faked? Big Questions with Neil deGrasse Tyson," Penguin Books UK, November 22, 2019, YouTube video, 11:28, https://www.youtube.com/watch?v=uTChrirK-hw&t=101s.

43 **So many pages:** Rocco A. Petrone, "The Cape," in *Apollo Expeditions to the Moon*, vol. 10, ed. Edgar M. Cortright (Washington, DC: NASA, 1975), 111, https://books.google.com/books?id=szBWc8KFuXkC&newbks=1&newbks_redir=0&pg=PA111.

44 **Haynes Johnson:** Haynes Johnson, "1968 Democratic Convention," *Smithsonian Magazine*, August 2008, https://www.smithsonianmag.com/history/1968-democratic-convention-931079/.

44 **Many things led:** Caitlin Gibson, "What Happened in Chicago in 1968, and Why Is Everyone Talking About It Now?," *Washington Post*, July 18, 2016, https://www.washingtonpost.com/news/arts-and-entertainment/wp/2016/07/18/what-happened-in-chicago-in-1968-and-why-is-everyone-talking-about-it-now/.

45 **Both Senator Robert:** Johnson, "1968 Democratic Convention."

45 **President Johnson revealed:** Kenneth T. Walsh, "50 Years Ago, Walter Cronkite Changed a Nation," *US News & World Report*, February 27, 2018, https://www.usnews.com/news/ken-walshs-washington/articles/2018-02-27/50-years-ago-walter-cronkite-changed-a-nation.

45 **Major cities were rocked:** Alan Taylor, "The Riots That Followed the Assassination of Martin Luther King Jr.," *The Atlantic*, April 3, 2018, https://www.theatlantic.com/photo/2018/04/the-riots-that-followed-the-assassination-of-martin-luther-king-jr/557159/.

45 **peace protests:** Gibson, "What Happened in Chicago?"

45 **Tensions came to a fever:** Erik Larson, *The Devil in the White City: A Saga of Magic and Murder at the Fair That Changed America* (New York: Knopf Doubleday Publishing Group, 2004), 13–14.

46 **Mike Wallace:** Johnson, "1968 Democratic Convention."

46 **better angels:** Abraham Lincoln, "First Inaugural Address of Abraham Lincoln" (speech, United States Capitol, Washington, DC, March 4, 1861), The Avalon Project, Yale Law School, accessed September 17, 2021, https://avalon.law.yale.edu/19th_century/lincoln1.asp.

47 **NASA's James E. Webb:** Robert Kurson, *Rocket Men: The Daring Odyssey of Apollo 8 and the Astronauts Who Made Man's First Journey to the Moon* (New York: Random House, 2018), 46.

47 **There is a Santa:** "Apollo 8: Christmas at the Moon," NASA, last updated December 26, 2019, https://www.nasa.gov/topics/history/features/apollo_8.html.

47 **America's successful return:** Nadia Drake, "SpaceX Launches New Era of Spaceflight with Company's First Crewed Mission," *National Geographic*, May 30, 2020, https://www.nationalgeographic.com/science/article/spacex-nasa-launch-human-astronauts-crew-dragon-international-space-station-demo-2.

48 **Apollo flight controller:** Kevin Fong, "Saving 1968," June 19, 2019, in *13 Minutes to the Moon*, produced by BBC, MP3 audio, 38:45, https://www.bbc.co.uk/programmes/w3csz4dp.

48 **first foreign minister:** *American Experience*, "John Adams' Diplomatic Missions," PBS, accessed July 15, 2021, https://www.pbs.org/wgbh/americanexperience/features/adams-diplomatic-missions/.

48 **father of the United States Navy:** "John Adams I (Frigate)," Naval History and Heritage Command, July 22, 2015, https://www.history.navy.mil/research/histories/ship-histories/danfs/j/john-adams-frigate-i.html.

49 **largest library in the world:** "History of the Library of Congress," Library of Congress, accessed July 15, 2021, https://www.loc.gov/about/history-of-the-library/.

49 **he had fake teeth:** William M. Etter, "False Teeth," George Washington's Mount Vernon, accessed July 15, 2021, https://www.mountvernon.org/library/digitalhistory/digital-encyclopedia/article/false-teeth/.

49 **He conquered:** Joshua J. Mark, "Alexander the Great," World History Encyclopedia, November 14, 2013, https://www.worldhistory.org/Alexander_the_Great/.

49 **He died at age:** Mark, "Alexander the Great."

49 **He was tutored:** Mark, "Alexander the Great."

49 **after drinking a bowl:** Sarah Wolfe, "Alexander the Great Was Killed by Toxic Wine, Says Scientist," The World, January 13, 2014, https://www.pri.org/stories/2014-01-13/alexander-great-was-killed-toxic-wine-says-scientist.

49 **vat of honey:** Paul Salopek, "Honey, I'm Dead," Out of Eden Walk, *National Geographic*, May 13, 2015, https://www.nationalgeographic.org/projects/out-of-eden-walk/articles/2015-05-honey-im-dead/.

49 **He named seventy:** Nate Barksdale, "8 Surprising Facts About Alexander the Great," HISTORY, updated August 29, 2018, https://www.history.com/news/eight-surprising-facts-about-alexander-the-great.

49 **played by Colin Farrell:** *Alexander*, directed by Oliver Stone, starring Colin Farrell, Anthony Hopkins, Rosario Dawson, and Angelina Jolie (Burbank, CA: Warner Bros. Pictures, 2004), DVD.

49 **misty eyes:** Plutarch, "On Tranquillity of Mind," vol. 6 of *Moralia*, trans. W. C. Helmbold, LCL 337 (Cambridge, MA: Harvard University Press, 1939), https://penelope.uchicago.edu/Thayer/E/Roman/Texts/Plutarch/Moralia/De_tranquillitate_animi*.html.

50 **the majority of the New Testament:** Charlie Campbell, "When Was the New Testament Completed?," Always Be Ready Apologetics Ministry, accessed September 15, 2021, https://alwaysbeready.com/when-was-the-new-testament-completed/.

50 **99 percent identical:** Norman L. Geisler, "A Note on the Percent of Accuracy of the New Testament Text," Norman Geisler (website), accessed September 16, 2021, https://normangeisler.com/a-note-on-the-percent-of-accuracy-of-the-new-testament-text/.

50 **head-spinning verification:** Justin Taylor, "An Interview with Daniel B. Wallace on the New Testament Manuscripts," The Gospel Coalition,

March 22, 2012, https://www.thegospelcoalition.org/blogs/justin-taylor/an-interview-with-daniel-b-wallace-on-the-new-testament-manuscripts/.

50 **This creed of the earliest:** Kevin Porter, "Has the Bible Been Diluted by Copies of Copies of Copies? Christian Apologist Lee Strobel Answers," Christian Post, October 8, 2016, https://www.christianpost.com/news/christian-apologist-lee-strobel-bible-diluted-case-for-christ.html.

50 **drool over:** John Rodgers, quoted in Richard N. Ostling, "Who Was Jesus?," *TIME*, June 24, 2001, http://content.time.com/time/magazine/article/0,9171,149895,00.html.

51 **the literal translation:** Jeff A. Benner, "Studies in the Psalms: Psalm 1," s.v. "Blessed," Ancient Hebrew Research Center, accessed September 19, 2021, https://www.ancient-hebrew.org/psalms/studies-in-the-psalms-psalm-1.htm.

CHAPTER 4.5: ON CRUCIFIXION AND CENTRIFUGES

53 **When the different stages:** NASA, *Astronauts Answer Student Questions* (Houston: Lyndon B. Johnson Space Center), 2, accessed September 15, 2021, https://www.nasa.gov/centers/johnson/pdf/569954main_astronaut%20_FAQ.pdf.

53 **ridden the Gravitron:** Brian Dunbar, "The Pull of Hypergravity," ed. Jeanne Ryba, NASA, November 22, 2007, https://www.nasa.gov/missions/science/hyper.html.

54 **In addition to breathing:** Michael Collins, *Carrying the Fire: An Astronaut's Journey* (1974; repr., New York: First Cooper Square Press, 2001), 132–33.

54 **one revolution per second:** Bob Granath, "Gemini's First Docking Turns to Wild Ride in Orbit," NASA, March 3, 2016, https://www.nasa.gov/feature/geminis-first-docking-turns-to-wild-ride-in-orbit.

55 **Fighting off blacking out:** James Donovan, "Before the Moon Landing: How Neil Armstrong Saved NASA," *Newsweek*, March 14, 2019, https://www.newsweek.com/2019/03/22/top-flight-1359355.html.

55 **A large majority:** "America's #1 Health Problem," The American Institute of Stress, accessed July 20, 2021, https://www.stress.org/americas-1-health-problem.

56 **The word *Gethsemane*:** "Gethsemane," *Encyclopedia Britannica*, July 20, 1998, https://www.britannica.com/place/Gethsemane.

56 **An enormous stone:** Deir Samaan, "Oil Presses in the Holy Land," BibleWalks, last updated June 21, 2019, biblewalks.com/oilpresses.

57 **torment and torture:** David McClister, "The Scourging of Jesus," *Truth Magazine*, January 12, 2000, http://www.truthmagazine.com/archives/volume44/v440106010.htm.

58 **he was depleted:** Kermit Zarley, "Did Jesus Carry His Entire Cross?," *Kermit Zarley* (blog), Patheos, March 2, 2015, https://www.patheos.com/blogs/kermitzarleyblog/2015/03/did-jesus-carry-his-entire-cross/.

58 **patibulum:** Jean de Climont, *The Mysteries of the Shroud* (2013; repr., Paris: Assailly Editions, 2016), 39–40.

58 **half-mile-long:** James Stalker, *The Trial and Death of Jesus Christ: A Devotional History of Our Lord's Passion* (London: Hodder & Stoughton, 1905), Kindle locations 2572–76, Kindle Edition; Bill O'Reilly and Martin Dugard, *Killing England* (New York: Henry Holt and Company, 2017), 247, Kindle Edition.

58 **thirty thousand:** John MacArthur, *The Murder of Jesus: A Study of How Jesus Died* (Nashville, TN: Thomas Nelson, 2004), 155.

58 **The Assyrians invented:** Francois Pieter Retief and Louise Cilliers, "The History and Pathology of Crucifixion," *South African Medical Journal* 93, no. 12 (December 2003): 938–41, https://www.researchgate.net/publication/297407145_The_history_and_pathology_of_crucifixion.

59 ***Excruciating* actually means:** David Terasaka, "Medical Aspects of the Crucifixion of Jesus Christ," Blue Letter Bible, 1996, https://www.blueletterbible.org/Comm/terasaka_david/misc/crucify.cfm.

60 **Warren Wiersbe said:** Warren W. Wiersbe, *The Cross of Jesus: What His Words on Calvary Mean for Us* (La Grange, KY: 10Publishing, 2020), 67.

62 **I chose it because:** J. R. Moehringer, *The Tender Bar* (New York: Hyperion, 2006), 411.

63 **William Wallace:** *Braveheart*, directed by Mel Gibson (Hollywood: Paramount Pictures, 1995), Amazon Prime video, 2:34:05, https://www.amazon.com/gp/video/detail/amzn1.dv.gti.b2a9f746-4f5a-0622-3fe7-62100cfe478e.

63 **Drink because you are happy:** Gilbert K. Chesterton, *Heretics* (New York: John Lane Company, 1905), 103–4, https://www.google.com/books/edition/Heretics/6v5BAQAAIAAJ.

64 **Magdala was a wealthy:** Liz Curtis Higgs, *Unveiling Mary Magdalene: Discover the Truth About a Not-So-Bad Girl of the Bible* (2001; repr., Colorado Springs, CO: WaterBrook Press, 2007), 150.

65 **It means "to lift":** Strong's Concordance, s.v. "5375 nasa or nasah: to lift, carry, take," Bible Hub, accessed September 15, 2021, https://biblehub.com/hebrew/5375.htm.

69 **In 1830:** Lorraine Boissoneault, "A Brief History of Presidential Pardons," *Smithsonian*, August 2, 2017, https://www.smithsonianmag.com/history/brief-history-10-essential-presidential-pardons-arent-watergate-related-180964286/.

CHAPTER 5: LET THE PARTY CONTINUE

75 **Let the party:** N. T. Wright, *The Lord and His Prayer* (Grand Rapids, MI: Eerdmans, 1996), 25–26.

78 **The first thing:** Revelation 19:9.

78 **you *proclaim*:** Thayer's Greek Lexicon, s.v., "Strong's NT 1804: *exaggelló*," Bible Hub, accessed September 15, 2021, https://bibleapps.com/greek/1804.htm.

78 **a sore throat:** "The Death of George Washington," in *The Digital Encyclopedia of George Washington*, ed. James P. Ambuske (Mount Vernon, VA: Mount Vernon Ladies Association, 2012), https://www.mountvernon.org/library/digitalhistory/digital-encyclopedia/article/the-death-of-george-washington/.

79 **I have come home:** C. S. Lewis, *The Last Battle* (1956; repr., New York: HarperTrophy, 2002), 213.

79 **Chapter One of the Great:** Lewis, 228.

79 **Charles Spurgeon said,** Charles Spurgeon, "Many Kisses for Returning Sinners, or Prodigal Love for the Prodigal Son," sermon no. 2236, Metropolitan Tabernacle, March 29, 1891, Newington, London, The Spurgeon Archive, https://archive.spurgeon.org/sermons/2236.php.

80 **takes three days:** Valerie Stimac, "How Long Does It Take to Get to the Moon?," HowStuffWorks, March 31, 2021, https://science.howstuffworks.com/how-long-to-moon.htm.

80 **twelve manned Apollo:** Dr. David R. Williams, "The Apollo Program (1963–1972)," NASA Space Science Data Coordinated Archive, last updated September 16, 2013, https://nssdc.gsfc.nasa.gov/planetary/lunar/apollo.html.

80 **heralded as heroes:** Roger D. Launius, "Heroes in a Vacuum: The Apollo 11 Astronaut as Cultural Icon" (paper, 43rd AIAA Aerospace Sciences Meeting and Exhibit, Reno, NV, January 2005), https://doi.org/10.2514/6.2005-702.

80 **The final shuttle:** Brian Dunbar, "Space Shuttle Era," ed. Sarah Loff, NASA, last updated August 3, 2017, https://www.nasa.gov/mission_pages/shuttle/flyout/index.html.

80 **For the next decade:** Daniel Oberhaus, "The US Hitches Its Final Ride to Space from Russia—for Now," *Wired*, April 8, 2020, https://www.wired.com/story/the-us-hitches-its-final-ride-to-space-from-russia-for-now/.

81 **wasn't even broadcast:** Mark Whittington, "Apollo 13: When Flights to the Moon Stopped Being Boring," *The Hill*, April 13, 2020, https://thehill.com/opinion/technology/492447-apollo-13-when-flights-to-the-moon-were-boring-until-they-suddenly-werent.

81 **Until a tank ruptured:** Brian Dunbar, "Apollo 13," NASA, July 8, 2009, last updated January 9, 2018, https://www.nasa.gov/mission_pages/apollo/missions/apollo13.html.

81 **In a sermon:** Timothy J. Keller, "Hope for the Church," Gospel in Life, November 1, 2009, MP3, 31:50, https://gospelinlife.com/downloads/hope-for-the-church-6027/; adapted from Tony Campolo, *The Kingdom of God Is a Party: God's Radical Plan for His Family* (Dallas: Word Publishing, 1990).

CHAPTER 6: HOUSTON, WE HAVE A PROBLEM

87 **They had trained:** James Donovan, *Shoot for the Moon: The Space Race and the Extraordinary Voyage of Apollo 11* (New York: Little, Brown, 2019), 279.

87 ***Subsystem* is basically:** George M. Low, "The Spaceships," in *Apollo Expeditions to the Moon*, vol. 10, ed. Edgar M. Cortright (Washington, DC: NASA, 1975), 61.

88 **But the CSM:** Low, "Spaceships," 60–61.

88 **A scuba diver's:** Low, 61.

88 **The entire unit:** Low, 61.

88 **Uncontrollable flatulence:** Low, 65.

88 **On Gemini 7:** Donovan, *Shoot for the Moon*, 176.

89 **If an automobile:** Low, "Spaceships," 65.

89 **The spidery legs:** Anna Heiney, "Apollo's Lunar Leftovers," NASA, last updated November 22, 2007, https://www.nasa.gov/missions/solarsystem/f_leftovers.html.

89 50,000 feet: Lawrence McGlynn, "Apollo 10: 'Son of a B*tch!,'" Space Artifacts, October 2013, https://www.spaceartifactsarchive.com/2013/10/apollo-10-son-of-a-bitch-.html.

89 12, and 14–17: The Apollo Program, "Location of Apollo Lunar Modules," Smithsonian National Air and Space Museum, accessed October 10, 2021, https://airandspace.si.edu/explore-and-learn/topics/apollo/apollo-program/spacecraft/location/lm.cfm.

89 **Charts duct-taped:** Amy Shira Teitel, "Seeing Inside the Apollo Lunar Module," *Popular Science*, December 16, 2013, https://www.popsci.com/blog-network/vintage-space/seeing-inside-apollo-lunar-module/.

89 **Apparently after the space walk:** David Kerley and Samantha Spitz, "50 Years Later: The Pen That Saved Apollo 11," ABC News, July 12, 2019, https://abcnews.go.com/Politics/50-years-pen-saved-apollo-11/story?id=64228723.

89 **sister called him "Buzzer"**: James R. Hansen, *First Man: The Life of Neil A. Armstrong* (New York: Simon & Schuster, 2005), 349.

89 **to have a PhD:** Paul Ceruzzi, "Buzz Aldrin's Ph.D Thesis," Smithsonian

National Air and Space Museum, July 2, 2019, https://airandspace.si.edu/stories/editorial/buzz-aldrins-phd-thesis.

90 **take a selfie:** Bob Granath, "Gemini XII Crew Masters the Challenges of Spacewalks," NASA, November 14, 2016, last updated August 6, 2017, https://www.nasa.gov/feature/gemini-xii-crew-masters-the-challenges-of-spacewalks.

90 **In his book:** Buzz Aldrin, *Men from Earth* (New York: Bantam, 1989), 244.

92 **1986:** Elizabeth Howell, "Challenger: The Shuttle Disaster That Changed NASA," Space.com, May 1, 2019, https://www.space.com/18084-space-shuttle-challenger.html.

92 **2003:** Elizabeth Howell, "Columbia Disaster: What Happened and What NASA Learned," Space.com, February 1, 2019, https://www.space.com/19436-columbia-disaster.html.

93 **One of the oxygen tanks:** Sarah Pruitt, "What Went Wrong on Apollo 13?," HISTORY, last updated April 13, 2020, https://www.history.com/news/apollo-13-what-went-wrong.

93 **At about nine:** James A. Lovell, "'Houston, We've Had a Problem,'" in *Apollo Expeditions to the Moon*, 250.

93 **Then, at 9:08:** Pruitt, "What Went Wrong on Apollo 13?"

93 **had a problem here:** Lovell, "'Houston, We've Had a Problem,'" 249, 251.

93 **Peterson also noted:** Eugene Peterson, *Leap Over a Wall* (1997; repr., New York: HarperSanFrancisco, 1998), 187.

94 **Everyone knew space:** Tariq Malik, "NASA Honors Astronauts Lost in 3 Space Tragedies with Day of Remembrance," Space.com, January 30, 2020, https://www.space.com/nasa-day-of-remembrance-honors-astronauts-2020.html.

94 **the impossible job:** Michael Collins, *Carrying the Fire: An Astronaut's Journey* (1974; repr., New York: First Cooper Square Press, 2001), 270–71.

94 **We worried about:** Collins, 269.

95 **Spaceflight will never:** Gene Kranz, *Failure Is Not an Option: Mission Control from Mercury to Apollo 13 and Beyond* (New York: Simon & Schuster, 2000), 204.

96 **With this mindset:** Richard Hollingham, "The Fire That May Have Saved the Apollo Programme," BBC Future, January 26, 2017, https://www.bbc.com/future/article/20170125-the-fire-may-have-saved-the-apollo-programme.

96 **Apollo 2 and 3 were canceled:** Amy Shira Teitel, "What Happened to Apollos 2 and 3?," *Popular Science*, October 29, 2013, https://www.popsci.com/blog-network/vintage-space/what-happened-apollos-2-and-3/.

96 **Apollo 4, 5, and 6:** Dr. David R. Williams, "The Apollo Program

(1963–1972)," NASA Space Science Data Coordinated Archive, last updated September 16, 2013, https://nssdc.gsfc.nasa.gov/planetary/lunar/apollo.html.

96 **"We hope that if anything happens":** Kranz, *Failure Is Not an Option*, 203.

96 **Troy Aikman said:** Troy Aikman, "Session 2 - Q&A," (panel, C3 Conference 2016, February 12, 2016, Dallas, TX), https://creativepastors.com/product/c3-conference-2016/.

97 **Located on the east:** "Ephesus," HISTORY, last updated August 21, 2018, https://www.history.com/topics/ancient-greece/ephesus.

97 **The population of Ephesus:** Alfred Borcover, "'Ghosts' of Ephesus," *Chicago Tribune*, October 8, 1989, https://www.chicagotribune.com/news/ct-xpm-1989-10-08-8901200157-story.html.

97 **a 25,000-seat theater:** Smith's Bible Dictionary, s.v., "Ephesus," Bible Hub, accessed July 21, 2021, https://biblehub.com/topical/e/ephesus.htm.

97 **It was home:** Borcover, "'Ghosts' of Ephesus."

97 **center of the highway:** Henry H. Halley, *Halley's Bible Handbook with the New International Version—Deluxe Edition* (Grand Rapids, MI: Zondervan Academic, 2012).

97 **The library of Ephesus:** Joshua J. Mark, "Ephesus," World History Encyclopedia, September 2, 2009, https://www.worldhistory.org/ephesos/.

97 **temple to the goddess:** "Ephesus," *Encyclopedia Britannica*, last updated February 18, 2020, https://www.britannica.com/place/Ephesus.

97 **four times bigger:** "Ephesus," *Encyclopedia Britannica*.

97 **images of the goddess:** David Hamblin, *Unveiling the Mysteries of the Last Days* (Mustang, OK: Tate Publishing & Enterprises, 2010), 88.

97 **a figure with multiple:** F. W. Farrar, *The Life and Work of St. Paul* (New York: E. P. Dutton and Company, 1889), 14.

98 **Nicolaitans were clergy:** "Who Were the Nicolaitans?," Bibleinfo.com, accessed July 21, 2021, https://www.bibleinfo.com/en/questions/who-were-the-nicolaitans.

105 **Uzzah was the Israelite:** Peterson, *Leap Over a Wall*, 135, 138–39.

105 **Alexander Whyte said:** Alexander Whyte, *Bible Characters: Volumes 1–6*, 1901 ed. (Dallas, TX: Primedia eLaunch, 2011), 311.

106 **Watch it on YouTube:** "John Young's Lunar Salute on Apollo 16," NASA Video, May 19, 2013, YouTube video, 00:00:15, www.youtube.com/watch?v=g5aPoRtF2vw&ab_channel=TED.

106 **C. S. Lewis called:** C. S. Lewis, *Mere Christianity*, rev. ed. (New York: HarperSanFrancisco, 2001), 175.

106 **Tim Keller took:** Timothy Keller, *The Reason for God* (2008; repr., New York: Penguin Books, 2018), 222.

107 **Apollo 13 has been described:** Sean Potter and Kelly Humphries, "NASA Commemorates 50th Anniversary of Apollo 13, 'A Successful Failure,'" last updated January 4, 2021, https://www.nasa.gov/press-release/nasa-commemorates-50th-anniversary-of-apollo-13-a-successful-failure.

107 **To quote Andy Grove:** Albert Yu, *Creating the Digital Future: The Secrets of Consistent Innovation at Intel* (New York: Free Press, 1998), 93.

CHAPTER 7: FIFTEEN SECONDS TO PARADISE

110 **most complicated part:** Eric M. Jones, "The First Lunar Landing," May 10, 2018, in Apollo 11 Lunar Surface Journal, ed. Ken Glover, 102:42:35, https://www.hq.nasa.gov/alsj/a11/a11.landing.html.

110 **The third man:** Dr. Tony Phillips, "Wide Awake on the Sea of Tranquility," NASA, last updated August 7, 2017, https://www.nasa.gov/exploration/home/19jul_seaoftranquillity.html.

110 **During the planning:** Dr. James R. Hansen, "The Rendezvous That Was Almost Missed," NASA, December 1992, https://www.nasa.gov/centers/langley/news/factsheets/Rendezvous.html.

110 **The single giant booster:** Courtney G. Brooks, James M. Grimwood, and Loyd S. Swenson, *Chariots for Apollo* (Mineola, NY: Dover Publications, 2009), 62.

111 **favored by famed German:** Hansen, "Rendezvous That Was Almost Missed."

111 **Option three was:** Hansen, "Rendezvous That Was Almost Missed."

111 **three modules:** George M. Low, "The Spaceships" in *Apollo Expeditions to the Moon*, vol. 10, ed. Edgar M. Cortright (Washington, DC: NASA, 1975), 60–61.

111 **Neil described it:** Neil Armstrong, "Weird Astronaut Machines," *LIFE*, September 25, 1964, 137, https://books.google.com/books?id=rUwEAAAAMBAJ&printsec=frontcover.

111 **Michael gave insight:** Michael Collins, "I Rattled Around in My Mini-Cathedral," *LIFE*, August 22, 1969, 27, in *United States of America Congressional Record* 115, part 18 (Washington, DC: United States Government Printing Office, 1969), 24238, https://www.google.com/books/edition/Congressional_Record/mtq7mE3sEZ4C?hl=en&gbpv=0.

111 **LOR was originally:** Hansen, "Rendezvous That Was Almost Missed."

112 **He passionately believed:** Brian Dunbar, "John C. Houbolt," ed. Bob Allen, NASA, last updated August 7, 2017, https://www.nasa.gov/langley/hall-of-honor/john-c-houbolt.

112 **Despite its challenges:** Hansen, "Rendezvous That Was Almost Missed."

112 **According to NASA:** Dunbar, "John C. Houbolt."

112 **His persistence:** Dunbar.

112 **Having made:** Phillips, "Wide Awake on the Sea of Tranquility."

112 **But at a point 40,000 feet:** Neil Armstrong, quoted in *Apollo 11: Technical Crew Debriefing* (Houston: NASA, 1969), 59, https://www.hq.nasa.gov/alsj/a11/a11_tcdb.pdf.

112 **For every pound:** Thomas J. Kelly, interview with Kevin M. Rusnak, "NASA Johnson Space Center Oral History Project," September 19, 2000, NASA, https://historycollection.jsc.nasa.gov/JSCHistoryPortal/history/oral_histories/KellyTJ/KellyTJ_9-19-00.htm.

112 **As a result:** James Donovan, *Shoot for the Moon: The Space Race and the Extraordinary Voyage of Apollo 11* (New York: Little, Brown, 2019), 203–4.

112 **butter knife:** Piers Bizony, Andrew Chaikin, and Roger D. Launius, *The NASA Archives* (Cologne, Germany: Taschen, 2019), 181.

113 **Keep in mind:** Norman Mailer, *Of a Fire on the Moon* (New York: Random House, 2014), 281.

113 **Neil once said:** Neil A. Armstrong, interview by Stephen E. Ambrose and Douglas Brinkley, "NASA Johnson Space Center Oral History Project," NASA, September 19, 2001, 86, https://www.nasa.gov/sites/default/files/62281main_armstrong_oralhistory.pdf.

113 **thousand things to worry about:** Brian Dunbar, "July 20, 1969: One Giant Leap for Mankind," NASA, last updated July 20, 2021, https://www.nasa.gov/mission_pages/apollo/apollo11.html.

113 **Through dress rehearsals:** Kevin Fong, "Long Island Eagle," May 29, 2019, in *13 Minutes to the Moon*, produced by BBC, MP3 audio, 43:00, https://www.bbc.co.uk/programmes/w3csz4dl.

113 **Some simulations were done:** Christopher C. Kraft Jr., "'This Is Mission Control,'" in *Apollo Expeditions to the Moon*, 138.

113 **For this NASA created:** Gray Creech, "NASA Armstrong Recalls First Moon Landing, Preps for 'Next Giant Leap,'" ed. Yvonne Gibbs, Armstrong Flight Research Center, NASA, July 15, 2014, last updated July 9, 2018, https://www.nasa.gov/centers/armstrong/Features/armstrong_recalls_first_moon_landing.html.

113 **Its General Electric:** Creech.

114 **When he pulled:** Donovan, *Shoot for the Moon*, 269.

114 **Neil's cool, unflappable:** James R. Hansen, *First Man: The Life of Neil A. Armstrong* (New York: Simon & Schuster, 2005), 332.

114 **By the end:** Brian Dunbar, "Lunar Landing Research Vehicle," ed. Marty Curry, NASA, last updated March 3, 2008, https://www.nasa.gov/centers/dryden/about/Organizations/Technology/Facts/TF-2004-08-DFRC.html.

114 **Neil adamantly said:** Eric M. Jones, "Utility of the Lunar Landing

Training Vehicle," Apollo Lunar Surface Journal, ed. Eric M. Jones and Ken Glover, last updated June 28, 2011, https://www.hq.nasa.gov/alsj/alsj-LLTV-value.html.

114 **first portable computer:** Kevin Fong, "The Fourth Astronaut," June 12, 2019, in *13 Minutes to the Moon*, produced by BBC, MP3 audio, 47:56, https://www.bbc.co.uk/sounds/play/w3csz4dn.

114 **modern cell phone:** Tibi Puiu, "Your Smartphone Is Millions of Times More Powerful Than the Apollo 11 Guidance Computers," ZME Science, May 13, 2021, https://www.zmescience.com/science/news-science/smartphone-power-compared-to-apollo-432/.

115 **The precision required:** Robert Kurson, *Rocket Men: The Daring Odyssey of Apollo 8 and the Astronauts Who Made Man's First Journey to the Moon* (New York: Random House, 2018), 48.

115 **It gets worse:** Kurson, 48.

115 **There is no atmosphere:** Ian Sample, "'We Had 15 Seconds Left': Buzz Aldrin on the Nervy Moon Landing," *Guardian*, July 18, 2019, https://www.theguardian.com/science/2019/jul/18/we-had-15-seconds-of-fuel-left-buzz-aldrin-on-the-nervy-moon-landing.

115 **Walter Isaacson observed:** Douglas Brinkley, *American Moonshot: John F. Kennedy and the Great Space Race* (New York: HarperCollins, 2019), 395.

116 **It's a 1202:** Eric M. Jones, "The First Lunar Landing," Apollo 11 Lunar Surface Journal, ed. Eric M. Jones and Ken Glover, 102:38:30, last updated May 10, 2018, https://www.hq.nasa.gov/alsj/a11/a11.landing.html.

116 **CapCom Charlie Duke's response:** Jones, "First Lunar Landing," MP3 audio clip from the flight director's loop, 24:52, https://www.hq.nasa.gov/alsj/a11/A11_landing_FD_loop.mp3.

117 **With some urgency:** Jones, "First Lunar Landing," 102:38:42.

117 **Norman Mailer referred:** Mailer, *Of a Fire on the Moon*, 9.

117 **day-old pizza:** Frank D. Roylance, "High-Tech Can Fail, and Humans Emerge," *Baltimore Sun*, April 23, 2000, https://www.baltimoresun.com/news/bs-xpm-2000-04-23-0005080252-story.html.

117 **Steve Bales:** Jones, "First Lunar Landing," 2:38:53.

117 **eleven days before:** Gene Kranz, *Failure Is Not an Option: Mission Control from Mercury to Apollo 13 and Beyond* (New York: Simon & Schuster, 2000), 268.

118 **On his desk:** Donovan, *Shoot for the Moon*, 286, 437–38.

118 **Jack was prepared:** Donovan.

118 **Engineers at MIT:** Don Eyles, "Landing Apollo via Cambridge," MIT News, July 17, 2009, https://news.mit.edu/2009/apollo-eyles-0717.

118 **In his memoir:** Buzz Aldrin, *Magnificent Desolation: The Long Journey Home from the Moon* (New York: Bloomsbury Publishing, 2009), 34.

118 **It ended up:** Aldrin, 17.

118 **Steve let Gene know:** Jones, "First Lunar Landing," MP3, flight director's loop, 25:03.

119 **Neil replied:** Jones, "First Lunar Landing," 102:38:53.

119 **running out of fuel:** Donovan, 511.

119 **If it increased:** William Harwood, "The Inside Story of Apollo 11's Nail-Biting Descent to the Surface of the Moon," *CBS News*, July 15, 2019, https://www.cbsnews.com/news/apollo-11-moon-landing-anniversary-nail-biting-descent-to-the-surface-of-the-moon/.

119 **impossible to land:** Harwood.

119 **At about 400 feet:** Neil Armstrong, "The Moon Has Been Awaiting Us a Long Time," *LIFE*, August 22, 1969, in *United States of America Congressional Record 115*, part 18 (Washington, DC: United States Government Printing Office, 1969), 24236.

119 **an alarm tripped:** Harwood, "Inside Story."

120 **Tethers were installed:** Vida Systems, "Trip to the Moon 1966," Google Arts & Culture: NASA, accessed July 30, 2021, https://artsandculture.google.com/story/gAWhMIYZIFqczA.

120 **Aldrin: [Eagle] 540 feet:** Jones, "First Lunar Landing," 102:43:16–45:50.

120 **Buzz later said:** Buzz Aldrin, "Hear Buzz Aldrin Tell the Story of the First Moon Landing," Science Museum, July 18, 2019, YouTube video, 6:10, https://www.youtube.com/watch?v=9HvG6ZlpLrI.

123 **Those dying of crucifixion:** Mark Driscoll, "If Jesus Walked the Earth Today We'd Kill Him Too," *Real Faith*, accessed September 18, 2021, https://realfaith.com/daily-devotions/if-jesus-walked-the-earth-today-wed-kill-him-too/.

126 **In *The Lion*:** C. S. Lewis, *The Lion, the Witch and the Wardrobe* (1950; repr., New York: Harper, 2009), 150–52.

127 **place called Sheol:** Fred B. Pearson, "Sheol and Hades in Old and New Testament," *Review & Expositor* 35, no. 3 (July 1938): 304–14, https://doi.org/10.1177/003463733803500304.

127 ***Sheol* means:** Don Stewart, "What Is Sheol?," Blue Letter Bible, https://www.blueletterbible.org/faq/don_stewart/don_stewart_113.cfm. W. Edward Bedore, "Hell, Sheol, Hades, Paradise, and the Grave," Berean Bible Society, accessed July 30, 2021, https://www.bereanbiblesociety.org/hell-sheol-hades-paradise-and-the-grave/.

127 **One side was called Gehenna:** Christopher A. Pallis, "Death," *Encyclopedia Britannica*, last updated December 10, 2020, https://www.britannica.com/science/death.

127 **C. S. Lewis said:** C. S. Lewis, *The Problem of Pain* (1944; repr., New York: HarperOne, 2001), 96.

128 **I read once:** "Texas Police Officer Wraps $100 Bill in Traffic Ticket,"Officer.com, December 12, 2012, https://www.officer.com/command-hq/technology/traffic/news/10841300/plano-texas-police-officer-wraps-100-bill-inside-traffic-ticket; J. D. Miles, "Plano Police Officer Wraps $100 Bill in Traffic Ticket," CBSN Dallas–Ft. Worth, December 11, 2012, https://dfw.cbslocal.com/2012/12/11/plano-police-officer-wraps-100-bill-in-traffic-ticket/.

128 **Another compartment of Sheol:** Bedore, "Hell, Sheol, Hades, Paradise, and the Grave."

130 **As N. T. Wright:** David M. D. Lawrence, *Heaven: It's Not the End of the World* (London: Scripture Union, 1995).

130 **Heaven is important:** N. T. Wright, *Surprised by Hope: Rethinking Heaven, the Resurrection, and the Mission of the Church* (New York: HarperOne, 2008), 41.

130 **from his own lips:** Luke 23:41.

131 **A minister once faithfully:** James Ralph Grant, *The Way of the Cross* (Grand Rapids, MI: Baker Book House, 1963), 44.

CHAPTER 8: IF NOT ME, ANOTHER

133 **all the moon's light:** Elizabeth Palermo, "Why Does the Moon Shine?," Live Science, May 29, 2014, https://www.livescience.com/45979-why-does-the-moon-shine.html.

133 **the stamp should have read:** Buzz Aldrin, *Magnificent Desolation: The Long Journey Home from the Moon* (New York: Bloomsbury Publishing, 2009), 65–66.

133 **twenty minutes later:** Aldrin, 32.

134 **The case could have been made:** Aldrin, 128–30.

134 **at the time of this writing:** Alicja Zelazko, "How Many People Have Been to the Moon?," *Encyclopaedia Britannica*, accessed August 2, 2021, https://www.britannica.com/story/how-many-people-have-been-to-the-moon.

134 **Joan, said that privately:** Lily Koppel, *The Astronaut Wives Club* (2013; repr., New York: Grand Central Publishing, 2014), 230.

134 **launched an unsuccessful campaign:** James Donovan, *Shoot for the Moon: The Space Race and the Extraordinary Voyage of Apollo 11* (New York: Little, Brown, 2019), 293.

134 **He noted in his memoir:** Aldrin, *Magnificent Desolation*, 82.

134 **The kickoff was:** John Uri, "50 Years Ago: Apollo 11 Astronauts Leave Quarantine . . . ," ed. Kelli Mars, NASA, August 12, 2019, last updated August 13, 2019, https://www.nasa.gov/feature/50-years-ago-apollo-11-astronauts-leave-quarantine.

134 **Giant Leap Tour:** Ivy Donnell, "Celebrating Apollo 11 Around the World," *The Unwritten Record*, National Archives, July 30, 2019, https://unwritten-record.blogs.archives.gov/2019/07/30/celebrating-apollo-11-around-the-world/.

134 **on the president's backup plane:** Donnell, "Celebrating Apollo 11."

135 **"exhausting acclaim":** Michael Collins and Edwin E. Aldrin, "The Eagle Has Landed," in *Apollo Expeditions to the Moon*, vol. 10, ed. Edgar M. Cortright (Washington, DC: NASA, 1975), 223.

135 **invited to sleep in the White House:** Aldrin, *Magnificent Desolation*, 69–70.

135 **Over the course of the tour:** Uri, "50 Years Ago."

135 **Buzz pointed out:** Deduced from Aldrin, *Magnificent Desolation*, 82.

135 **"Fame has not worn well":** Michael Collins, *Carrying the Fire: An Astronaut's Journey* (1974; repr., New York: First Cooper Square Press, 2001), 60.

135 **Ricky Bobby is right:** *Talladega Nights*, directed by Adam McKay, starring Will Ferrell (Los Angeles: Columbia Pictures, 2006), 13:05, netflix.com/watch/70044894?source=35.

137 **He was unquestionably:** Donovan, *Shoot for the Moon*, 280.

137 **In an interview:** Neil A. Armstrong, interview with Dr. Stephen E. Ambrose and Dr. Douglas Brinkley, "NASA Johnson Space Center Oral History Project," September 19, 2001, https://historycollection.jsc.nasa.gov/JSCHistoryPortal/history/oral_histories/ArmstrongNA/ArmstrongNA_9-19-01.htm.

138 **compare him to Christopher Columbus:** Norman Mailer, *Of a Fire on the Moon* (New York: Random House, 2014), 38.

138 **At least part of Buzz's motivation:** Aldrin, *Magnificent Desolation*, 26.

139 **6.7 hours of sleep:** Norman B. Anderson et al., *Stress in America: Are Teens Adopting Adults' Stress Habits?* (Washington, DC: American Psychological Association, February 11, 2014), 14, https://www.apa.org/news/press/releases/stress/2013/stress-report.pdf.

142 **do something to prove:** Peter Scazzero, *The Emotionally Healthy Leader: How Transforming Your Inner Life Will Deeply Transform Your Church, Team, and the World* (Grand Rapids, MI: Zondervan, 2015), 39.

143 **it was a waxing crescent:** Ernie Wright, "A New Look at the Apollo 11 Landing Site," NASA's Scientific Visualization Studio, July 18, 2014, https://svs.gsfc.nasa.gov/4185.

143 **and saw a crescent earth:** Collins, *Carrying the Fire*, 437.

144 **Michael included this:** Collins, *Carrying the Fire*, 319–21.

CHAPTER 9: OUT OF THIS WORLD

149 **Neil Armstrong took his first:** Kathy Sawyer, "Armstrong's Code," *Washington Post*, July 11, 1999, https://www.washingtonpost.com/archive

/lifestyle/magazine/1999/07/11/armstrongs-code/9414e446-4611-45c4-be29-6061a6f8e854/.

151 **I resonate with:** Honoré de Balzac, *Father Goriot*, trans. Ellen Marriage (Paris: 1835), EPUB, https://www.google.com/books/edition/Father_Goriot/Eh0ZEAAAQBAJ?hl=en&gbpv=0.

153 **video explaining the backstory:** "The Coin Toss," Wimbledon, July 10, 2019, YouTube video, 1:00, youtube.com/watch?v=y_SwfbcS7Ms.

153 **The coin had been to space:** The book you are reading has too! Shane Kimbrough took it with him on a thumb drive aboard NASA's SpaceX Crew-2 Dragon that launched from the Kennedy Space Center on April 1, 2021, and it flew throughout his six-month-long mission on the International Space Station. See Brian Dunbar, "NASA's SpaceX Crew-2 Astronauts Headed to International Space Station," ed. Katherine Brown, NASA, April 23, 2021, last updated May 3, 2021, https://www.nasa.gov/press-release/nasa-s-spacex-crew-2-astronauts-headed-to-international-space-station.

153 **football field–sized:** Brian Dunbar, "International Space Station Facts and Figures," ed. Mark Garcia, NASA, last updated July 12, 2021, https://www.nasa.gov/feature/facts-and-figures.

154 **In January 1962**: James Donovan, *Shoot for the Moon: The Space Race and the Extraordinary Voyage of Apollo 11* (New York: Little, Brown, 2019), 183.

154 **In September of that same year:** Amy Shira Teitel, "Before the Moon: The Early Exploits of Neil Armstrong," BBC News, September 22, 2015, https://www.bbc.com/news/science-environment-34170799.

155 **like the X-15:** Brian Dunbar, "X-15," ed. Ruby Calzada, last updated August 7, 2017, https://www.nasa.gov/centers/dryden/multimedia/imagegallery/X-15/X-15_proj_desc.html.

155 **city in Turkey called Izmir:** Warren W. Wiersbe, *The Bible Exposition Commentary: New Testament* (Wheaton, IL: Victor Books, 1996), 2:573.

155 it was a beautiful city: Encyclopedia of the Bible, s.v. "Smyrna," BibleGateway, accessed October 10, 2021, https://www.biblegateway.com/resources/encyclopedia-of-the-bible/Smyrna.

155 ***Smyrna* means "myrrh":** John F. Walvoord, "Revelation" in *The Bible Knowledge Commentary: An Exposition of the Scriptures*, vol. 2, New Testament, ed. John F. Walvoord and Roy B. Zuck (Wheaton, IL: Victor Books, 1985), 934.

156 **center of science, medicine, and athletic competition:** Lawrence O. Richards, *The Bible Reader's Companion: Your Guide to Every Chapter of the Bible* (Wheaton, IL: Victor Books, 1991), 908.

156 **Homer, the author of *The Iliad*:** California Digital Newspaper Collection,

"Homer Birthplace Claimed by Smyrna," *San Pedro News Pilot*, September 27, 1932, UCR Center for Biographical Studies and Research, accessed October 10, 2021, https://cdnc.ucr.edu/?a=d&d=SPNP19320927.2.72&e=-------en--20--1--txt-txIN--------1.

156 **a crown to his followers:** *Encyclopedia of the Bible*, s.v. "Smyrna."

156 **the emperor, Domitian:** Wiersbe, *Bible Exposition Commentary*, 2:573.

156 **burned at the stake:** John Phillips, *Exploring Revelation: An Expository Commentary*, s.v. "Revelation 2:10a," John Phillips Commentary Series (Grand Rapids: Kregel, 2009); Chad Brand, Eric Mitchell, et al., *Holman Illustrated Bible Dictionary*, rev. and expanded ed., s.v. "Smyrna," (Nashville, TN: Holman Bible Publishers, 2003), 1512.

156 **added to their suffering:** Wiersbe, *Bible Exposition Commentary*, 2:573.

157 **the blood of Christians is seed:** P. A. Hartog, "Martyr," in The Lexham Bible Dictionary, ed. J. D. Barry et al. (Bellingham, WA: Lexham Press, 2016).

160 **one-sixth gravity:** "Here Are 13 Nuggets of Lunar Knowledge," *National Geographic*, July 16, 2004, https://www.nationalgeographic.com/science/article/moon-facts.

160 **"Our cause is never more in danger":** C. S. Lewis, *The Screwtape Letters* (1942; repr., New York: HarperOne, 1996), 39.

CHAPTER 10: NINE AND A HALF FINGERS

163 **one of the stories:** W. David Woods, Kenneth D. MacTaggart, and Frank O'Brien, "Apollo 11—Day 2, Part 1: Midcourse Correction," Apollo 11 Flight Journal, last updated May 1, 2021, 023:14:23, https://history.nasa.gov/afj/ap11fj/05day2-mcc.html.

170 **Middle eastern sheep pens:** "Ancient Sheep Fold," Bible History, accessed August 4, 2021, https://www.bible-history.com/sketches/ancient/sheep-fold.html.

171 **"The planet just hung there":** Robert Kurson, *Rocket Men: The Daring Odyssey of Apollo 8 and the Astronauts Who Made Man's First Journey to the Moon* (New York: Random House, 2018), 248.

171 **All Mercury flights:** Brian Dunbar, "What Was Project Mercury?," ed. Sandra May, NASA, last updated August 7, 2017, https://www.nasa.gov/audience/forstudents/5-8/features/nasa-knows/what-was-project-mercury-58.html.

171 **Cape Canaveral was renamed:** Cliff Lethbridge, "History of Cape Canaveral Chapter 3," Spaceline, accessed August 3, 2021, https://www.spaceline.org/history-cape-canaveral/history-cape-canaveral-chapter-3/.

171 **the first word:** Eric M. Jones, "The First Lunar Landing," Apollo 11 Lunar

Surface Journal, ed. Eric M. Jones and Ken Glover, 102:45:58, last updated May 10, 2018, https://www.hq.nasa.gov/alsj/a11/a11.landing.html.

171 **NASA named a computer building:** Sam McDonald, "Computational Facility Named After Langley 'Human Computer' Katherine Johnson," last updated August 6, 2017, https://www.nasa.gov/feature/langley /computational-facility-named-in-tribute-to-nasa-langley-math-master -katherine-johnson.

171 **Imagine firing a bullet:** Piers Bizony, Andrew Chaikin, and Roger D. Launius, *The NASA Archives* (Cologne, Germany: Taschen, 2019).

172 **That is the accuracy it took:** Kurson, *Rocket Men*, 50–51.

172 **the nickname made sense:** Alaina, "Snoopy, Charlie Brown and Apollo 10," *The Payload Blog*, Kennedy Space Center Visitor Complex, NASA, May 16, 2019, https://www.kennedyspacecenter.com/blog/snoopy-charlie-brown -and-apollo-10.

172 **Snoopy had become the symbol:** Jeremy Hsu, "Snoopy Celebrates 40th Anniversary of His Moon Flight," Space.com, January 28, 2010, https:// www.space.com/6700-snoopy-celebrates-40th-anniversary-moon-flight .html.

172 **"Snoopy cap":** Brian Dunbar, "NASA Astronaut Andrew Morgan Wears a Communications Cap," ed. Mark Garcia, NASA, April 3, 2020, https:// www.nasa.gov/image-feature/nasa-astronaut-andrew-morgan-wears-a -communications-cap.

CHAPTER 11: RITE OF PASSAGE

175 **we know more about outer space:** Dan Stillman, "Oceans: The Great Unknown," NASA, October 8, 2009, https://www.nasa.gov/audience /forstudents/5-8/features/oceans-the-great-unknown-58.html.

176 **the Malcolm Gladwell concept:** Malcolm Gladwell, "The 10,000-Hour Rule," chap. 2 in *Outliers: The Story of Success* (New York: Little, Brown, 2008).

177 **"On the individual level":** Stephen M. R. Covey, *The Speed of Trust: The One Thing That Changes Everything* (New York: Free Press, 2018), 105.

177 ***Harvard Business Review* article:** Chip Heath and Dan Heath, "The Curse of Knowledge," *Harvard Business Review*, December 2006, https://hbr.org /2006/12/the-curse-of-knowledge.

179 **General George Patton had it right:** General George S. Patton Jr., *War as I Knew It* (Boston: Houghton Mifflin, 1975), 354.

179 **"People would rather follow a leader":** *Craig Groeschel Leadership Podcast*, Life.church, accessed September 18, 2021, https://www.life.church /leadershippodcast/.

181 **You should watch it:** Ron Fritz, "An Exploration of Coming of Age Rituals & Rites of Passage in a Modern Era," TEDx Talks, July 6, 2017, YouTube video, 18:30, https://www.youtube.com/watch?v=Obta5WPfse4.

182 **In his book:** Matthew McConaughey, *Greenlights* (New York: Crown, 2020), 85.

187 **General Grant interrupted:** Horace Porter, *Campaigning with Grant* (New York: The Century Co., 1897), 357, https://www.google.com/books/edition/Campaigning_with_Grant/mC8OAAAAIAAJ.

188 **acknowledged this tension:** C. S. Lewis, *On Stories and Other Essays on Literature* (1966; repr., New York: HarperCollins, 1982), 50.

190 **In the case of the Saturn V:** Norman Mailer, *Of a Fire on the Moon* (New York: Random House, 2014), 56.

190 **a fraction of that:** Michael Collins, *Carrying the Fire: An Astronaut's Journey* (1974; repr., New York: First Cooper Square Press, 2001), 362.

190 **leaves earth tipping the scales:** Collins, 439.

190 **Incredibly, it loses:** Mailer, *Of a Fire on the Moon*, 65.

190 **command module in the Smithsonian:** "Apollo 11 Command Module *Columbia*," Smithsonian National Air and Space Museum, accessed August 5, 2021, https://airandspace.si.edu/collection-objects/command-module-apollo-11/nasm_A19700102000.

CHAPTER 12: BURN, BABY! BURN!

193 **Saturn V was burning as much oxygen:** Norman Mailer, *Of a Fire on the Moon* (New York: Random House, 2014), 82.

193 **five nozzles each so large:** Brian Dunbar, "The F-1 Engine Powered Apollo into History, Blazes Path for Space Launch System Advanced Propulsion," NASA, last updated August 7, 2017, https://www.nasa.gov/topics/history/features/f1_engine.html.

193 **to control the force:** Nexus_2006, "How Is a Rocket Stabilized During the Initial, Slow Speed, Portion of Launch?," Space Exploration, Stack Exchange, September 15, 2015, https://space.stackexchange.com/questions/12093/how-is-a-rocket-stabilized-during-the-initial-slow-speed-portion-of-launch.

193 **swiveled intelligently:** Wernher von Braun, "An All-Up Test for the First Flight," in *Apollo Expeditions to the Moon*, vol. 10, ed. Edgar M. Cortright (Washington, DC: NASA, 1975), 52.

194 **"pogo" vibrations:** James Donovan, *Shoot for the Moon: The Space Race and the Extraordinary Voyage of Apollo 11* (New York: Little, Brown, 2019), chap. 11.

194 **the computer code showed up online:** Avianne Tan, "Apollo 11's Source Code Has Tons of Easter Eggs, Including an Ignition File Titled 'Burn Baby

Burn,'" ABC News, July 12, 2016, https://abcnews.go.com/Technology/apollo-11s-source-code-tons-easter-eggs-including/story?id=40515222.

195 **Computer programmers combed:** Jack D'Isidoro and Eliza Lambert, "Hidden Messages Found in Computer Code of Apollo 11 Mission," *The Takeaway*, WNYC Studios, July 11, 2016, https://www.wnycstudios.org/podcasts/takeaway/segments/hidden-messages-discovered-apollo-moon-mission-computer-code.

195 **There were references to:** Tan, "Apollo 11's Source Code."

195 **Peter put it this way:** Peter Adler, "Apollo 11 Program Alarms," Apollo 11 Lunar Surface Journal, ed. Eric M. Jones and Ken Glover, last updated March 26, 2019, https://www.hq.nasa.gov/alsj/a11/a11.1201-pa.html.

196 **became a rallying cry:** Magnificent Montague with Bob Baker, *Burn, Baby! BURN! The Autobiography of Magnificent Montague* (Urbana, IL: University of Illinois Press, 2003), 1.

196 **The average age of the engineers:** Joe P. Hasler, "Is America's Space Administration Over the Hill? Next-Gen NASA," *Popular Mechanics*, May 26, 2009, https://www.popularmechanics.com/space/a4288/4318625/.

197 **The city of Pergamum:** Corinne K. Hoexter, "The Ancient Heights of Pergamon," *New York Times*, October 1, 1995, https://www.nytimes.com/1995/10/01/travel/the-ancient-heights-of-pergamon.html.

197 **nothing that fails like success:** G. K. Chesterton, *Heretics and Orthodoxy: Two Volumes in One* (1905; repr., Bellingham, WA: Lexham Press, 2017), 8.

198 **As Ronald Reagan said:** Ronald Reagan, "January 5, 1967: Inaugural Address (Public Ceremony)," Ronald Reagan Presidential Library & Museum, accessed September 19, 2021, https://www.reaganlibrary.gov/archives/speech/january-5-1967-inaugural-address-public-ceremony.

199 **LeBron James spends over $1 million:** Darren Heitner, "The $1.5 Million Expended by LeBron James Every Offseason Is Money Well Spent," *Forbes*, May 20, 2018, https://www.forbes.com/sites/darrenheitner/2018/05/20/the-1-5-million-expended-by-lebron-james-every-offseason-is-money-well-spent/.

201 **According to Eugene Peterson:** Eugene Peterson, *As Kingfishers Catch Fire: A Conversation on the Ways of God Formed by the Words of God* (Colorado Springs, CO: WaterBrook Press, 2017), 47–48.

CHAPTER 13: GO, NO GO

203 **the command and service module separated:** W. David Woods, Kenneth D. MacTaggart, and Frank O'Brien, "Apollo 11—Day 1, Part 3: Transposition, Docking and Extraction," Apollo 11 Flight Journal, last updated March 1, 2021, https://history.nasa.gov/afj/ap11fj/03tde.html.

203 **They gradually slowed down:** Dave Roos, "Apollo 11 Moon Landing Timeline: From Liftoff to Splashdown," HISTORY, last updated December 8, 2020, https://www.history.com/news/apollo-11-moon-landing-timeline.

204 **One hour and twenty minutes later:** W. David Woods, Kenneth D. MacTaggart, and Frank O'Brien, "Apollo 11—Day 5, Part 2: Undocking and the Descent Orbit," Apollo 11 Flight Journal, last updated February 10, 2017, https://history.nasa.gov/afj/ap11fj/15day5-undock-doi.html.

204 **Norman Mailer called this:** Norman Mailer, *Of a Fire on the Moon* (New York: Random House, 2014), 109.

204 **Here is what they heard:** Eric M. Jones, "The First Lunar Landing," Apollo 11 Lunar Surface Journal, ed. Eric M. Jones and Ken Glover, 102:28:34, last updated May 10, 2018, https://www.hq.nasa.gov/alsj/a11/a11.landing.html.

204 **"Flight Director Gene Kranz":** Jones, 102:28:34.

204 **These White Team men:** "The White Team," Honeysuckle Creek Tracking Station, accessed August 1, 2021, https://honeysucklecreek.net/msfn _missions/Apollo_11_mission/a11_White_Team.html.

204 **"It was, for all of us:** Gene Kranz, *Failure Is Not an Option: Mission Control from Mercury to Apollo 13 and Beyond* (New York: Simon & Schuster, 2000), 152.

205 **The go/no go sounded like this:** Jones, "First Lunar Landing," 102:28:34.

205 **In the audio:** Jones, "First Lunar Landing," MP3 audio clip from the network controller's loop, 14:02, 23:15, https://www.hq.nasa.gov/alsj/a11 /A11_landing_NC.mp3.

205 **"We're *go*":** Jones, "First Lunar Landing," MP3, network controller's loop, 14:12.

206 **When deployed:** Steven Siceloff, "Launch Aborts Challenge Rocket Engineers," NASA, January 30, 2012, https://www.nasa.gov/exploration /commercial/crew/LASdevelopment.html.

207 **stay, no stay:** Jones, "First Lunar Landing," MP3, network controller's loop, 33:04, 37:40.

207 **it would take two hours:** Jay Bennett, "The Enduring Legacy of Michael Collins, Astronaut and Chronicler of Apollo 11," *National Geographic*, April 28, 2021, https://www.nationalgeographic.com/science/article /the-literary-legacy-of-michael-collins-the-forgotten-astronaut-of-apollo-11.

207 **They had less than two minutes:** Jones, "First Lunar Landing," MP3, network controller's loop, 32:10, 33:04.

207 **Thirty lives have been lost:** Jake Parks, "How Many People Have Died in Outer Space?," *Discover*, October 8, 2019, https://www.discovermagazine .com/the-sciences/how-many-people-have-died-in-outer-space.

208 **Spurgeon said they were:** Charles Haddon Spurgeon, "The Saddest Cry from the Cross," sermon, Metropolitan Tabernacle, January 7, 1877,

Newington, London, The Spurgeon Center, https://www.spurgeon.org/resource-library/sermons/the-saddest-cry-from-the-cross/#flipbook/.

208 **One author said:** James Stalker, *The Trial and Death of Jesus Christ: A Devotional History of Our Lord's Passion* (New York: American Tract Society, 1894), 227.

208 **crucifixion within the crucifixion:** Dr. Clovis G. Chappell, *The Seven Words* (Whitefish, MT: Literary Licensing, 2011), 39.

210 **It's been called the Great Exchange:** Wyatt Hout, "Martin Luther and The Great Exchange," PostBarthian, March 28, 2012, https://postbarthian.com/2012/03/28/martin-luther-and-the-great-exchange/.

211 **Two thousand years before this happened:** Matthew Henry, *Matthew Henry's Concise Commentary on the Whole Bible* (Nashville, TN: Thomas Nelson, 2003), 38.

212 **John Stott wrote:** John R. W. Stott, *The Cross of Christ*, 20th anniversary ed. (Downers Grove, IL: InterVarsity Press, 2006), 81.

212 **And Erwin Lutzer added:** Erwin W. Lutzer, *Cries from the Cross: A Journey into the Heart of Jesus* (Chicago: Moody Publishers, 2002), 98.

212 **Author James Stalker said:** Stalker, *Trial and Death of Jesus Christ*, 238.

213 **The actual Greek word for *forsaken*:** ἐγκαταλείπω *enkataleipō*

213 **ten thousand charms:** Dr. Hawn, "History of Hymns: 'Come, Ye Sinners, Poor and Needy,'" Discipleship Ministries, June 20, 2013, https://www.umcdiscipleship.org/resources/history-of-hymns-come-ye-sinners-poor-and-needy.

215 **When he doesn't answer in silver:** Paraphrased from Charles Spurgeon, "When God Doesn't Give That for Which You Asked," in *Morning and Evening* (1869), reposted on Reasons for Hope, November 16, 2020, https://reasonsforhopejesus.com/when-god-doesnt-give-that-for-which-you-asked-by-charles-spurgeon/.

220 **I read in the news:** Emily Shapiro, "Wyoming Woman Stunned by Gift from Husband Who Died Last Year," ABC News, February 17, 2015, https://abcnews.go.com/US/wyoming-woman-stunned-gift-husband-died-year/story?id=29017007.

CHAPTER 14: HAKUNA MATATA

222 **Neil Armstrong attended the ribbon cutting:** Stacy Conradt, "15 Out-of-This-World Facts About Space Mountain," Mental Floss, April 6, 2016, https://www.mentalfloss.com/article/77347/15-out-world-facts-about-space-mountain.

222 **Jules Vernes's 1865 book:** M. Bogaert, "Remembering Space Mountain, from the Earth to the Moon," Designing Disney, accessed October 10,

2021, https://www.designingdisney.com/parks/disneyland-paris/disneyland-park/discoveryland/remembering-space-mountain-earth-moon/.

222 **help from NASA advisers:** "NASA Lends Reality to Disney Thrill Ride," NASA, October 7, 2003, https://www.nasa.gov/vision/earth/everydaylife/mission_space.html.

222 **Grant, a prolific author:** Adam Grant, "Why So Many Ideas Are Pitched as 'Uber for X,'" *The Atlantic*, February 4, 2016, https://www.theatlantic.com/business/archive/2016/02/adam-grant-originals-uber-for-x/459321/.

222 **It ended up as:** Grant.

222 **"*Bambi* in Africa":** Grant.

223 **"To be or not to be":** Grant.

223 **Alan Shepard's first flight:** Brian Dunbar, "60 Years Ago: Alan Shepard Becomes the First American in Space," ed. Kelli Mars, NASA, May 5, 2021, https://www.nasa.gov/image-feature/60-years-ago-alan-shepard-becomes-the-first-american-in-space.

224 **The Camelot era:** Tierney McAfee and Liz McNeil, "How Jackie Kennedy Invented Camelot Just One Week After JFK's Assassination," *People*, November 22, 2017, https://people.com/politics/jackie-kennedy-invented-camelot-jfk-assassination/.

224 **He died on the same day:** Simon Usborne, "C. S. Lewis: In the Shadow of JFK's Death . . . ," *Independent*, November 22, 2013, https://www.independent.co.uk/life-style/health-and-families/features/cs-lewis-shadow-jfk-s-death-8955470.html.

224 **he had been an atheist:** Peter Schakel, "C. S. Lewis," *Encyclopaedia Britannica*, December 17, 2020, https://www.britannica.com/biography/C-S-Lewis.

224 **He called it Joy:** C. S. Lewis, *Surprised by Joy* (1955; repr., San Francisco: HarperOne, 2017), 274.

224 **Tolkien and Lewis were talking:** C. S. Lewis, quoted in Jerram Barrs, "Echoes of Eden," CSLewis.com, December 23, 2016, https://www.cslewis.com/echoes-of-eden-2/.

225 **he wrote his friend:** C. S. Lewis, quoted in James M. Houston, "The Prayer-Life of C. S. Lewis," *Knowing & Doing* (Summer 2006), 9, https://www.cslewisinstitute.org/The_Prayer_Life_of_CS_Lewis_FullArticle.

225 **Lewis believed:** C. S. Lewis, *The Weight of Glory* (1949; repr., New York: HarperCollins, 2001), 43.

226 **forty miles southeast of Pergamum:** Steve Tucker, "Pergamum: The Seven Churches of Revelation," https://sermons.faithlife.com/sermons/4176-pergamum-the-seven-churches-of-revelation.

226 **it existed to guard:** "Akhisar," *Encyclopaedia Britannica*, April 7, 2014, https://www.britannica.com/place/Akhisar.

226 **because of an army base:** Mark Wilson, "The Social and Geographical World of Thyatira" in *Lexham Geographic Commentary on Acts Through Revelation*, ed. Barry Beitzel (Bellingham, WA: Lexham Press, 2019), 657.

226 **the smallest of the seven:** Crickett Keeth, "Thyatira—The Tolerant Church," Bible.org, January 25, 2008, https://bible.org/seriespage/5-thyatira-tolerant-church.

227 **It was a city known for:** Wilson, "Social and Geographical World of Thyatira," 659.

229 **touched down in the Pacific Ocean:** "A Historic Moment in World History," USS Hornet, accessed August 12, 2021, https://uss-hornet.org/history/splashdown#1607460826038-ef0f9c03-e5f5.

229 **an unknown citizen had left:** Douglas Brinkley, *American Moonshot: John F. Kennedy and the Great Space Race* (New York: Harper, 2019), Apple Books locations 1297–98.

229 **As Gene Kranz put it:** Gene Kranz, *Failure Is Not an Option: Mission Control from Mercury to Apollo 13 and Beyond* (New York: Simon & Schuster, 2009), 93.

CHAPTER 15: GOOD LUCK AND GODSPEED

231 **Apollo 11 flew up:** "Good Luck and Godspeed," *Scientific American*, July 19, 1999, https://www.scientificamerican.com/article/good-luck-and-godspeed/.

231 **As this enormous bull:** "Good Luck and Godspeed."

232 **Scott Carpenter had quipped:** "John Glenn Was Everything Right About America," *USA Today*, December 8, 2016, https://eu.usatoday.com/story/news/2016/12/08/short-list-thursday/95136358/.

234 **Here is one author's description:** Robert Kurson, *Rocket Men: The Daring Odyssey of Apollo 8 and the Astronauts Who Made Man's First Journey to the Moon* (New York: Random House, 2018), 76.

234 **According to Michael Collins:** Michael Collins, *Carrying the Fire: An Astronaut's Journey* (1974; repr., New York: First Cooper Square Press, 2001), 329.

234 **Michael said his time:** Collins, 350.

234 **buoyant and got goose pimples:** Michael Collins and Edwin E. Aldrin Jr., "The Eagle Has Landed: A Yellow Caution Light," in Apollo Expeditions to the Moon, ed. Edgar M. Cortright (Washington, DC: National Aeronautics and Space Administration, 1975), https://history.nasa.gov/SP-350/ch-11-4.html.

235 **the audio recording of the communication:** Eric M. Jones, "The First Lunar Landing" in Apollo 11 Lunar Surface Journal, ed. Eric M. Jones and Ken Glover, MP3 audio clip from the flight director's loop, 31:55, last updated May 10, 2018, https://www.hq.nasa.gov/alsj/a11/A11_landing_FD_loop.mp3.

235 **That may have seemed like:** Eric M. Jones, "Post-Landing Activities," March 18, 2018, in Apollo 11 Lunar Surface Journal, ed. Eric M. Jones and Ken Glover, MP3 audio clip from the public affairs loop, 7:30, last updated March 18, 2018, https://www.hq.nasa.gov/alsj/a11/a11.postland.html.

236 **What Walter Cronkite called:** "'Man on the Moon': The 50th Anniversary of the Apollo 11 Landing," CBS News, July 16, 2019, https://www.cbsnews.com/news/man-on-the-moon-50th-anniversary-of-the-apollo-11-landing-cbs-news-special/.

236 **President Richard Nixon was so excited:** Frank Gannon, "24 July 1969: Home from the Moon," Richard Nixon Foundation, July 23, 2008, https://www.nixonfoundation.org/2008/07/24-july-1969-home-from-the-moon/.

236 **Omega Speedmaster watches:** "What Watch Do Astronauts Wear? (Watches Worn on Moon & in Space)," Watch Ranker, accessed August 14, 2021, https://watchranker.com/astronauts-watches-space/.

236 **a trip around the earth:** "Month," *Encyclopaedia Britannica*, September 21, 2011, https://www.britannica.com/science/month#ref225844.

237 **what we call a day:** Molly Wasser, "What Do You Wonder?," NASA Science: Earth's Moon, accessed August 14, 2021, https://moon.nasa.gov/inside-and-out/top-moon-questions/.

237 **Since the moon keeps one side**: BI India Bureau, "What Is a Lunar Day and a Lunar Night—All You Need to Know," *Business Insider*, September 21, 2019, https://www.businessinsider.in/what-is-a-lunar-day-and-a-lunar-night/articleshow/71234498.cms.

237 **GET—Ground Elapsed Time:** Richard Orloff, ed., "Introduction," in Apollo by the Numbers, rev. ed. (Washington, DC: NASA History Division, 2004), https://history.nasa.gov/SP-4029/Apollo_00c_Introduction.htm.

237 **The Eagle landed on the moon:** "Apollo 11 (AS-506)," Smithsonian National Air and Space Museum, accessed August 14, 2021, https://airandspace.si.edu/explore-and-learn/topics/apollo/apollo-program/landing-missions/apollo11.cfm.

242 **Like Aslan in The Chronicles:** C. S. Lewis, *The Lion, the Witch and the Wardrobe* (1950; repr., New York: Harper, 2009), 182.

242 **the Becoming One:** David Guzik, "Exodus 3—Moses and the Burning Bush," Enduring Word (website), 2018, https://enduringword.com/bible-commentary/exodus-3/.

245 **Neil deGrasse Tyson noted:** Neil deGrasse Tyson, May 26, 2021, *The Joe Rogan Experience*, podcast, Spotify, 43:50, https://open.spotify.com/episode/032MLx3jJ2ZNg0sQsuAueb.

CHAPTER 16: EIGHT COMES BEFORE ELEVEN

247 **missiles built by a former Nazi:** Piers Bizony, Andrew Chaikin, Roger D. Launius, *The NASA Archives* (Cologne, Germany: Taschen, 2019), 9.

247 **Going from 0 to 17,500:** National Aeronautics and Space Administration, *Astronauts Answer Student Questions*.

247 **ten times faster than a bullet:** Bizony, Chaikin, and Launius, *NASA Archives*.

248 **begun at the outset of World War I:** Elizabeth Suckow, "NACA: Overview," NASA, April 23, 2009, https://history.nasa.gov/naca/overview.html.

248 **a new civilian organization:** Brian Dunbar, "What Is NASA?," ed. Sandra May, NASA, last updated February 6, 2020, https://www.nasa.gov/audience/forstudents/5-8/features/nasa-knows/what-is-nasa-58.html.

248 **selling NASA shirts:** Samantha Masunaga and Hailey Mensik, "The NASA Logo Is Having a Moment," *Los Angeles Times*, July 19, 2019, https://www.latimes.com/business/story/2019-07-19/nasa-logo-shirts-swimsuits-everything.

248 **translunar injection (TLI):** "Translunaar Injection," WestEastSpace, February 20, 2020, https://westeastspace.com/encyclopedia/translunar-injection/.

248 **early days of the Third Reich:** James Donovan, *Shoot for the Moon: The Space Race and the Extraordinary Voyage of Apollo 11* (New York: Little, Brown, 2019), 16.

248 **was his team's creation:** Donovan, *Shoot for the Moon*, 19–20.

248 **But his true passion:** Donovan, 17–18.

248 **Wernher sought a way out:** Donovan, 22–23.

249 **managed to stage a daring escape:** Donovan, 23–24.

249 **They fled to the Harz Mountains:** Howard Benedict, "German Scientists Recall Era of Revolution in Space," *Los Angeles Times*, January 20, 1985, https://www.latimes.com/archives/la-xpm-1985-01-20-mn-10432-story.html.

249 **a program called Operation Paperclip:** Donovan, *Shoot for the Moon*, 25.

249 **a lot of latitude:** Donovan, 25.

249 **Wernher was moved:** Donovan, 25–26.

249 **opportunities at Disney:** Donovan, 27.

249 **A popular comedian suggested:** John Crosby, "Our Sense of Humor Is Back," *Lowell Sun*, April 24, 1961, 5, in Donovan, 33–34.

249 **He worked on:** Brian Dunbar, "Biography of Wernher von Braun," NASA, August 3, 2017, https://www.nasa.gov/centers/marshall/history/vonbraun/bio.html.

250 **250-mile mark:** "Low Earth Orbit," *Science Direct*, https://www.sciencedirect.com/topics/engineering/low-earth-orbit.

250 **They set a speed record:** Cliff Lethbridge, "Apollo 8 Fact Sheet," Spaceline, accessed August 15, 2021, https://www.spaceline.org/united-states-manned-space-flight/apollo-mission-program-facts-sheet-index/apollo-8-fact-sheet/.

250 **achieving a top speed:** Kelli Mars, ed., "50 Years Ago: Apollo 8, You Are Go for TLI!" NASA, December 21, 2018, https://www.nasa.gov/feature/50-years-ago-apollo-8-you-are-go-for-tli.

250 **An unprecedented billion people:** Kevin Wilcox, "This Month in NASA History: Apollo 8 Changes the World's Perspective," APPEL Knowledge Service, NASA, December 18, 2018, https://appel.nasa.gov/2018/12/18/this-month-in-nasa-history-apollo-8-changes-the-worlds-perspective/.

250 **it is a figure eight:** Amy Shira Teitel, "Why Apollo Flew in a Figure 8," *Discover*, April 21, 2018, https://www.discovermagazine.com/the-sciences/why-apollo-flew-in-a-figure-8.

250 **After lifting off:** Mars, "50 Years Ago."

250 **completing the figure eight:** W. David Woods, *How Apollo Flew to the Moon*, 2nd ed. (New York: Springer–Praxis, 2011), 123–29.

250 **This also explains the mission patch:** Brian Dunbar, "Apollo 8," NASA, last updated July 9, 2018, https://www.nasa.gov/mission_pages/apollo/missions/apollo8.html.

250 **In August 1968:** Robert Kurson, *Rocket Men: The Daring Odyssey of Apollo 8 and the Astronauts Who Made Man's First Journey to the Moon* (New York: Random House, 2018), chap. 1.

251 **the crazy idea:** Joel Achenbach, "Apollo 8: NASA's First Moonshot Was a Bold and Terrifying Improvisation," *Washington Post*, December 21, 2018, https://www.washingtonpost.com/history/2018/12/20/apollo-nasas-first-moonshot-was-bold-terrifying-improvisation/.

251 **had only four months**: Kevin Fong, "Saving 1968," June 23, 2019, in *13 Minutes to the Moon*, produced by BBC, MP3 audio, 12:00, https://www.bbc.co.uk/programmes/w3csz4dp.

251 **The photo they took:** NASA's Scientific Visualization Studio, "The Story Behind Apollo 8's Famous Earthrise Photo," NASA Science, December 21, 2018, https://solarsystem.nasa.gov/resources/2234/the-story-behind-apollo-8s-famous-earthrise-photo/.

251 **on Christmas Eve:** "Apollo 8 Departs for Moon's Orbit," HISTORY, last updated December 21, 2020, https://www.history.com/this-day-in-history/apollo-8-departs-for-moons-orbit.

251 **said Teasel Muir-Harmony:** Richard Hollingham, "The NASA Mission That Broadcast to a Billion People," BBC Future, December 21, 2018, https://www.bbc.com/future/article/20181220-the-nasa-mission-that-broadcast-to-a-billion-people.

252 **Here is a transcript:** W. David Woods and Frank O'Brien, "Apollo 8—Day 4: Lunar Orbit 9," February 27, 2021, in Apollo 8 Flight Journal, 086:06:40, https://history.nasa.gov/afj/ap08fj/21day4_orbit9.html.

252 **Check it out on YouTube:** Frank Borman, James Lovell, and William Anders, "Apollo 8's Christmas Eve 1968 Message," NASA Video, May 19, 2013, YouTube video, 2:01, https://www.youtube.com/watch?v =ToHhQUhdyBY.

253 **hair on his neck to stand up:** Fong, "Saving 1968," 36:18.

253 **because of the controversy:** Joe Carter, "9 Things You Should Know About the Communion Service on the Moon," The Gospel Coalition, July 17, 2019, https://www.thegospelcoalition.org/article/9-things-you-should-know -about-the-communion-service-on-the-moon/.

253 **a Saturn 1B rocket:** David S. F. Portree, "A Forgotten Rocket: The Saturn 1B," *Wired*, September 16, 2013, https://www.wired.com/2013/09/a -forgotten-rocket-the-saturn-ib/.

253 **180 million horsepower:** Wernher von Braun, "Saturn the Giant," in *Apollo Expeditions to the Moon*, vol. 10, ed. Edgar M. Cortright (Washington, DC: NASA, 1975), 52.

253 **eighty-five Hoover Dams:** Joel Walker, "Rocket Park: Saturn," ed. Orlando Bongat, NASA, last updated September 16, 2011, https://www.nasa.gov /centers/johnson/rocketpark/saturn_v.html.

253 **for over an hour:** "Saturn V Is the Biggest Engine Ever Built," *Popular Mechanics*, December 7, 2004, https://www.popularmechanics.com/science /a227/1280801/.

253 **tallest and heaviest rocket ever:** Anna Versai, "10 Most Powerful Rockets Ever Built," Technowize, January 8, 2021, https://www.technowize.com /10-most-powerful-rockets-ever-built/.

253 **0 to 17,500 miles per hour:** Bizony, Chaikin, and Launius, *The NASA Archives*.

253 **around the world eight hundred times:** Walker, "Rocket Park: Saturn V."

254 **"rain from ceilings":** von Braun, "Saturn the Giant," 52.

254 **The closest distance:** Norman Mailer, *Of a Fire on the Moon* (New York: Random House, 2014), 56.

254 **six seconds before they can hear:** Mailer, 97.

254 **rattling the fillings:** *The Saturn V Story,* directed by Elliot Weaver and Zander Weaver, featuring Dr. Nigel Bannister, Dr. David Baker, and Professor Mike Cruise (Elliander Pictures/ Free Spirit Film and TV, 2014), 52:00.

254 **The rocket has three stages:** "Saturn V Launch Vehicle," Smithsonian National Air and Space Museum, accessed August 15, 2021, https:// airandspace.si.edu/explore-and-learn/topics/apollo/apollo-program /spacecraft/saturn_v.cfm.

254 **The third and final stage:** Kurson, *Rocket Men*, 163.

254 **It was believed:** Kurson, 43.

254 **shoot through the instrument panel:** Francis French and Colin Burgess, *In the Shadow of the Moon* (Lincoln, NE: University of Nebraska Press, 2007), 304.

254 **everybody's a rookie:** *American Experience*, "Apollo 8 Insider Stories," PBS, https://www.pbs.org/wgbh/americanexperience/features/moon-apollo-8-insider-stories/.

255 **Apollo 8 has 5,600,000:** Michael Collins, *Carrying the Fire: An Astronaut's Journey* (1974; repr., New York: First Cooper Square Press, 2001), 304.

257 **more than three hundred:** "351 Old Testament Prophecies Fulfilled in Jesus Christ," New Testament Christians, accessed August 15, 2021, https://www.newtestamentchristians.com/bible-study-resources/351-old-testament-prophecies-fulfilled-in-jesus-christ/.

257 **Some of the smartest people:** Famed rocket scientist Wernher von Braun himself became a Christian when a neighbor invited him to church and he accepted. The experience shattered his expectations because he thought it would be a country club, but there he encountered vibrant community and he made a faith decision. Darrin J. Rodgers, "This Week in AG History—June 26, 1966," Assemblies of God, June 23, 2016, https://news.ag.org/Features/This-Week-in-AG-History-June-26-1966.

260 **Jim Lovell saw his mission:** Thom Patterson, "It's Been 50 Years Since Apollo 8 United a Fractured World," CNN, December 22, 2018, https://www.cnn.com/2018/05/18/us/apollo-8-anniversary-1968/index.html.

260 **He resonated with:** Neil Armstrong, Jim Lovell, and Gene Cernan, "Column: Is Obama Grounding JFK's Space Legacy?," *USA TODAY*, May 24, 2011, https://usatoday30.usatoday.com/news/opinion/forum/2011-05-24-Obama-grounding-JFK-space-legacy_n.htm.

260 **their westward journey:** Jay H. Buckley, "Lewis and Clark Expedition," *Encyclopaedia Britannica*, May 6, 2021, https://www.britannica.com/event/Lewis-and-Clark-Expedition.

261 **Matthew Henry said:** Matthew Henry, *Matthew Henry's Commentary on the Whole Bible: Volume VI–III—Titus–Revelation*, ed. Anthony Uyl (Ontario, Canada: Devoted Publishing, 2018), 153.

263 **He remarked how:** "Apollo 8 Astronaut Remembers Looking Down at Earth," Smithsonian National Air and Space Museum, December 21, 2018, https://airandspace.si.edu/stories/editorial/apollo-8-astronaut-remembers-looking-down-earth.

263 **He referred to earth as:** Woods and O'Brien, "Apollo 8—Day 4: Lunar Orbit 9," 085:46:23.

263 **Michael Collins would later:** T. S. Eliot, "Little Gidding," in *Four Quartets* (New York: Houghton Mifflin Harcourt, 2014), quoted in Collins, *Carrying the Fire*, 453.

263 **the first heart transplant surgery:** Allen Silbergleit, "Norman E. Shumway and the Early Heart Transplants," *Texas Heart Institute Journal* 33, no. 2 (2006): 274–75, https://www.ncbi.nlm.nih.gov/pmc/articles/PMC1524691/.

CHAPTER 17: ROCK[S] OF AGES

266 **The original plan:** Wernher von Braun, "Saturn the Giant," in *Apollo Expeditions to the Moon*, vol. 10, ed. Edgar M. Cortright (Washington, DC: NASA, 1975), 50–52.

266 **Neil's pressure suit:** "Pressure Suit, A7-L, Armstrong, Apollo 11, Flown," Smithsonian National Air and Space Museum, accessed August 13, 2021, https://airandspace.si.edu/collection-objects/pressure-suit-a7-l-armstrong-apollo-11-flown/nasm_A19730040000; "Apollo 11 Command Module *Columbia*," Smithsonian National Air and Space Museum, accessed August 26, 2021, https://airandspace.si.edu/collection-objects/command-module-apollo-11/nasm_A19700102000.

266 **Apollo 11 was the first:** Meghan Bartels, "Apollo 11 Astronauts Spent 3 Weeks in Quarantine, Just in Case of Moon Plague," Space.com, July 24, 2019, https://www.space.com/apollo-11-astronauts-quarantined-after-splashdown.html.

267 **The measures in place**: Johannes Kemppanen, "Apollo Lunar Quarantine: A 50th Anniversary View," Apollo Flight Journal, NASA History Division, last modified July 22, 2019, https://history.nasa.gov/afj/lrl/apollo-quarantine.html.

267 **who had been flustered:** Michael Collins, *Carrying the Fire: An Astronaut's Journey* (1974; repr., New York: First Cooper Square Press, 2001), 349.

267 **Lord God, our Heavenly Father:** Richard Nixon, "Remarks to Apollo 11 Astronauts Aboard the U.S.S. Hornet Following Completion of Their Lunar Mission," The American Presidency Project, UC Santa Barbara, accessed August 26, 2021, https://www.presidency.ucsb.edu/node/239653.

268 **sent out a piece of moon rock:** Ann Morgan, "Operation Moon Rock: The Hunt for Lost Lunar Samples," PBS SoCal, July 11, 2019, https://www.pbssocal.org/science/in-1970-president-richard-nixon-gave-apollo-11-lunar-samples-to-135-friendly-countries-and-u-s-states-now-many-of-those-samples-are-unaccounted-for.

268 **In 2002:** "The Case of the Stolen Moon Rocks: Last of Three NASA Interns Sentenced for Grievous Theft," Federal Bureau of Investigation,

November 18, 2003, https://archives.fbi.gov/archives/news/stories/2003/november/apollo_111803.

268 **A piece of the moon that Nixon gave Honduras:** Chris Lefkow, "What Happened to the Apollo Goodwill Moon Rocks?," Phys.org, June 16, 2019, https://phys.org/news/2019-06-apollo-goodwill-moon.html.

268 **NASA lore has it:** Jim Beckerman, "The Strange Afterlife of the Apollo 11 Moon Rocks," Northjersey.com, July 19, 2019, https://www.northjersey.com/story/entertainment/2019/07/19/apollo-11-moon-rocks-had-strange-afterlife/1563939001/.

269 **Neil stiffly replied:** Norman Mailer, *Of a Fire on the Moon* (New York: Random House Trade Paperbacks, 2014), 38–39.

269 **Spoke in computerese:** Mailer, 39.

269 **In 1969:** Dora Jane Hamblin, "He Could Fly Before He Could Drive," *LIFE*, August 11, 1969, https://books.google.com/books?id=oEwEAAAAMBAJ&pg=PT36.

269 **also said that silence:** Hamblin.

269 **In 2009:** Toby Sterling, "'Moon Rock' in Museum Is Just Petrified Wood," NBC News, August 27, 2009, https://www.nbcnews.com/id/wbna32581790.

270 **Founded in 1400 BC:** R. L. Drouhard, "Sardis," in *The Lexham Bible Dictionary*, ed. John D. Barry (Bellingham, WA: Lexham Press, 2016).

271 **Their necropolis was:** Rick Barbare, "Sardis: The Church Warm to the Eyes but Cold to the Touch, Part 1," accessed August 29, 2021, https://rickbarbarebiblestudies.blog/2019/07/21/sardis-a-church-warm-to-the-eyes-but-cold-to-the-touch-part-1/.

271 **It's called back-burning:** *Merriam-Webster.com Dictionary*, s.v. "back-burn," accessed August 26, 2021, https://www.merriam-webster.com/dictionary/back-burn.

273 **Astronomers tell us:** John Phillips, *Exploring Revelation: An Expository Commentary* (Grand Rapids, MI: Kregel, 1974), 61.

277 **Rock of Ages, cleft for me:** Augustus Toplady, "Rock of Ages, Cleft for Me," in *The United Methodist Hymnal* (Nashville, TN: United Methodist Publishing House, 1989), 361, https://hymnary.org/text/rock_of_ages_cleft_for_me_let_me_hide.

CHAPTER 18: ON SHUTTLES AND ROADS

279 **The plan was to transport them:** Kevin Kelly, *What Technology Wants* (2010; repr., New York: Penguin Books, 2011), 180.

280 **Lincoln made the final decision:** George W. Hilton, "A History of Track Gauge," *Trains*, May 1, 2006, https://www.trains.com/trn/train-basics/abcs-of-railroading/a-history-of-track-gauge/.

280 **These paths were all precisely:** Adam Morgan and Mark Barden, *A Beautiful Constraint: How to Transform Your Limitations into Advantages, and Why It's Everyone's Business* (Hoboken, NJ: John Wiley & Sons, 2015), 36.

281 **what happened to Apollo 12**: "Apollo 12 Lifts Off," HISTORY, last updated November 12, 2020, https://www.history.com/this-day-in-history/apollo-12-lifts-off.

281 **The word for *way*:** "Hodos," Bill Mounce (website), accessed August 26, 2021, https://www.billmounce.com/greek-dictionary/hodos.

283 **Max Lucado said:** Max Lucado, *Traveling Light: Releasing the Burdens You Were Never Intended to Bear* (Nashville, TN: Thomas Nelson, 2001), 240.

284 **Though he would not live:** "Transcontinental Railroad Completed, Unifying United States," HISTORY, last updated May 6, 2021, https://www.history.com/this-day-in-history/transcontinental-railroad-completed.

284 **When the final golden spike:** Stephen E. Ambrose, *Nothing Like It in the World: The Men Who Built the Transcontinental Railroad, 1863–1869* (2000; repr., New York: Simon & Schuster Paperbacks, 2005), 181.

285 **Humans went from traveling:** Ambrose, 42, 57.

285 **Communication also became:** Ambrose, 369.

285 **They needed a unified system:** Ambrose, 20.

CHAPTER 19: STEAK AND EGGS

287 ***The Four-Hour Body:*** Timothy Ferriss, *The Four-Hour Body* (New York: Crown Archetype, 2010).

288 **In Tim's words:** Ferriss, *Four-Hour Body*, 88.

289 **Buzz, Neil, and Michael had steak and eggs:** Michael Collins, *Carrying the Fire: An Astronaut's Journey* (1974; repr., New York: First Cooper Square Press, 2001), 355–56.

289 **Alan Shepard had eaten**: Tom Wolfe, *The Right Stuff* (New York: Farrar, Straus and Giroux, 1979), 192.

289 **He flew to the moon:** Lawrence K. Altman, "A Tube Implant Corrected Shepard's Ear Disease," *New York Times*, February 2, 1971, https://www.nytimes.com/1971/02/02/archives/a-tube-implant-corrected-shepards-ear-disease.html.

289 **Shepard also has the distinction:** "Facts About Spacesuits and Spacewalking," NASA, last updated July 5, 2018, https://www.nasa.gov/audience/foreducators/spacesuits/facts/index.html.

289 **It was chosen:** Robert Kurson, *Rocket Men: The Daring Odyssey of Apollo 8 and the Astronauts Who Made Man's First Journey to the Moon* (New York: Random House, 2018), 142.

290 **high-protein, low-residue makeup:** Kurson, 142, 308.

290 **the Battle of Midway:** "The One World War II Battle Where America Crushed Japan (and They Never Recovered)," *The Buzz* (blog), *National Interest*, September 23, 2017, https://nationalinterest.org/blog/the-buzz/the-one-world-war-ii-battle-where-america-crushed-japan-they-22409?page=0%2C1.

290 **Every time they served:** *Greatest Events of WWII in Colour*, season 1, episode 4, "Battle of Midway," directed by Sam Taplin, featuring Derek Jacobi, aired November 8, 2019, https://www.netflix.com/title/80989924.

290 **"live in infamy":** Franklin D. Roosevelt, "Address to the Congress Asking That a State of War Be Declared Between the United States and Japan" (speech, New York, December 8, 1941), Library of Congress, http://hdl.loc.gov/loc.afc/afc1986022.ms2201.

291 **The battle had more twists:** *Greatest Events of WWII in Colour*, "Battle of Midway."

291 **At 10:25 a.m.:** Barton Biggs, *Wealth, War and Wisdom* (Hoboken, NJ: John Wiley & Sons, 2008), 150.

292 **"the greatest single word ever uttered":** James Stalker, *The Trial and Death of Jesus Christ: A Devotional History of Our Lord's Passion* (New York, 1894), 254.

292 **In James Stalker's book:** Stalker, 254.

293 **Josephus says:** Flavius Josephus, *The Works of Flavius Josephus*, vol. 2, trans. William Whiston (London, 1845), 465.

293 **When you finally paid off:** Charles R. Swindoll, *Swindoll's New Testament Insights: Insights on John* (Grand Rapids, MI: Zondervan, 2010), 334.

294 **I told the whole story:** Levi Lusko, introduction to *Swipe Right: The Life-and-Death Power of Sex and Romance* (Nashville, TN: W Publishing Group, 2017).

298 **JFK's May 25, 1961, message to Congress:** John F. Kennedy, "Address to Joint Session of Congress" (speech, United States Capitol, Washington, DC, May 25, 1961), Space.com, https://www.space.com/11772-president-kennedy-historic-speech-moon-space.html.

298 **on the NASA screen:** NASA's Johnson Space Center (@NASA_Johnson), "50 years ago, #Apollo11 astronauts splashed down in the Pacific Ocean," Twitter, July 24, 2019, https://twitter.com/nasa_johnson/status/1154029751268282368.

CHAPTER 20: TATTOOED SOUL

301 **The last man to stand:** Elizabeth Howell, "Eugene Cernan: Last Man on the Moon," Space.com, January 16, 2017, https://www.space.com/20790-eugene-cernan-astronaut-biography.html.

301 **in December 2020:** Yaron Steinbuch, "China Plants Its Flag on Moon Before Return Trip to Earth," *New York Post*, December 4, 2020, https://nypost.com/2020/12/04/china-plants-its-flag-on-moon-before-return-trip-to-earth/.

302 **Harrison Schmitt was a geologist:** Robert Sherrod, "Men for the Moon," in *Apollo Expeditions to the Moon*, vol. 10, ed. Edgar M. Cortright (Washington, DC: NASA, 1975), 147.

302 **packed like a jigsaw puzzle:** Amy Shira Teitel, "The Lunar Rover: Designing and Unpacking a Car on the Moon," *Popular Science*, September 8, 2014, https://www.popsci.com/blog-network/vintage-space/lunar-rover-designing-and-unpacking-car-moon/.

302 **They drove about twenty-one miles:** Elizabeth Howell, "Apollo 17: The Last Men on the Moon," Space.com, October 3, 2018, https://www.space.com/17287-apollo-17-last-moon-landing.html.

302 **they spent three days:** Brian Dunbar, ed., "Remembering Gene Cernan," NASA, last updated October 20, 2017, https://www.nasa.gov/astronautprofiles/cernan.

302 **we know a hundred times more:** Kevin Fong, et al., "Live from Houston," July 20, 2019, in *13 Minutes to the Moon*, produced by BBC, MP3 audio, 28:20, https://www.bbc.co.uk/sounds/play/w3csz534.

303 **one of the highest murder rates:** "Philadelphia Has Highest Murder Rate Per Capita Among Country's 10 Largest Cities," CBS Philly, July 23, 2021, https://philadelphia.cbslocal.com/2021/07/23/philadelphia-highest-murder-rate-per-capita-countrys-10-largest-cities/.

303 **has also been given:** Jane Recker, "DC's Murder Rate Reaches Its Highest Number in 15 Years," *Washingtonian*, December 3, 2020, https://www.washingtonian.com/2020/12/03/dcs-murder-rate-reaches-its-highest-number-in-15-years/.

303 **it's also jokingly referred to:** Ashwin Verghese, "With This Name Game, a City of Brotherly Love Is Blamed," *Temple News*, May 1, 2007, https://temple-news.com/with-this-name-game-a-city-of-brotherly-love-is-blamed/.

303 **It was the youngest:** "The Letters to the Seven Early Christian Churches (Revelation 2:1–3:22)," Bible Blender, accessed August 26, 2021, https://www.bibleblender.com/2016/bible-stories/new-testament/revelation/letters-to-seven-early-christian-churches-revelation-2-1-3-22.

304 **a center of Greek culture:** Paul Himes, *Where Is Your Allegiance? The Message to the Seven Churches* (Gonzalez, FL: Energion Publications, 2017), 106.

304 **"Gateway to the East":** Himes, 106.

304 **They were known for their vineyards:** Himes, 106.

304 **Eusebius recorded:** Eusebius, *Church History* 5.17.2–4, available at "Church

History (Book V)," New Advent, accessed August 26, 2021, https://www.newadvent.org/fathers/250105.htm.

304 **The Greek word is *hagios*:** *Strong's Concordance*, s.v. "40. hagios," Bible Hub, accessed August 26, 2021, https://biblehub.com/greek/40.htm.

306 **As I wrote in my first book:** Levi Lusko, *Through the Eyes of a Lion: Facing Impossible Pain, Finding Incredible Power* (Nashville, TN: W Publishing Group, 2015), 63, 96.

308 **Listen to this excerpt:** C. S. Lewis, *The Last Battle* (1956; repr., New York: HarperTrophy, 2000), 156, 159, 161.

309 **"There is no such thing":** G. K. Chesterton, *Heretics*, vol. 1 in *The Collected Works of G. K. Chesterton: Heretics, Orthodoxy, The Blatchford Controversies*, ed. David Dooley (San Francisco: Ignatius, 1986), 54.

309 **Bill Gates once said:** Dax Shepard and Bill Gates, "Bill Gates," August 20, 2020, in *Armchair Expert*, produced by Dax Shepard, Monica Padman, and Rob Holysz, podcast, MP3 audio, 19:10–20:50, https://armchairexpertpod.com/pods/bill-gates.

310 **Ryan Holiday pointed out:** Ryan Holiday, *The Obstacle Is the Way: The Timeless Art of Turning Trials into Triumph* (New York: Penguin Group, 2014), 27.

311 **At 150 miles:** Holiday, 27–28.

311 **Holiday quoted Publilius Syrus:** Holiday, 27.

311 **The first time the transit authority:** James Donovan, *Shoot for the Moon: The Space Race and the Extraordinary Voyage of Apollo 11* (New York: Little, Brown, 2019), 110.

312 **The poet John Greenleaf Whittier said:** John G. Whittier, *Maud Muller* (Boston, 1870), 12.

312 **In 2021 a giant vessel:** Katie Hunt, "The Full 'Worm Moon' Helped Break the Logjam in the Suez Canal," CNN, March 30, 2021, https://www.cnn.com/2021/03/30/world/worm-moon-spring-tide-suez-canal-container-ship-scn/index.html.

312 **Speaking of regret:** R. Morlock, "Tattoo Prevalence, Perception and Regret in US Adults: A 2017 Cross-Sectional Study," *Value in Health* 22, supp. 3 (November 2019), https://doi.org/10.1016/j.jval.2019.09.1998.

313 **What's in your wallet:** "What is 'What's in Your Wallet?,'" Reference, last updated May 27, 2020, https://www.reference.com/business-finance/s-wallet-19c2cb5d428be010.

CHAPTER 21: NASA MEETS NAPA

316 **The Apollo space suits:** Norman Mailer, *Moonfire: The Epic Journey of Apollo 11,* 50th anniversary ed. (Cologne, Germany: Taschen, 2019), 174.

316 **Or having your tears boil:** Col. Chris Hadfield, *An Astronaut's Guide to Life on Earth* (New York: Little, Brown, 2013), 88.

316 **With a price tag:** Mailer, *Moonfire*, 173.

316 **Each Apollo mission:** "Facts About Spacesuits and Spacewalking," NASA, last modified July 5, 2018, https://www.nasa.gov/audience/foreducators/spacesuits/facts/index.html.

316 **A vintage space suit on earth:** "Facts About Spacesuits and Spacewalking," NASA.

317 **Gene Cernan said:** Eugene Cernan and Don Davis, *The Last Man on the Moon: One Man's Part in Mankind's Greatest Adventure* (New York: St. Martin's Griffin, 1999), 134.

317 **Once the astronaut is loaded up:** Hadfield, *Astronaut's Guide to Life on Earth*, 88.

317 **You can snag it:** Kevin Fong, et al., "Live from Houston," June 20, 2019, in *13 Minutes to the Moon*, produced by BBC, MP3 audio, 30:00, https://www.bbc.co.uk/programmes/w3csz534.

317 **"Spam in a can":** Tom Wolfe, *The Right Stuff* (New York: Farrar, Straus and Giroux, 1979), 60.

318 **break the sound barrier:** Andy Pasztor, "Chuck Yeager, Pioneer of Supersonic Flight, Dies at Age 97," *Wall Street Journal*, December 8, 2020, https://www.wsj.com/articles/chuck-yeager-pioneer-of-supersonic-flight-dies-at-age-97-11607404925.

318 **a job that a chimp could do:** Wolfe, *Right Stuff*, 99–100.

318 **on June 3, 1965:** Tim Childers, "Ed White: The First American to Walk in Space," Space.com, July 9, 2019, https://www.space.com/ed-white.html.

318 **This is the saddest moment:** James Donovan, *Shoot for the Moon: The Space Race and the Extraordinary Voyage of Apollo 11* (New York: Little, Brown, 2019), 250.

319 **In his book:** Hadfield, *Astronaut's Guide to Life on Earth*, 90–96.

323 **Buzz Aldrin acknowledged this:** Buzz Aldrin, *Magnificent Desolation: The Long Journey Home from the Moon* (New York: Bloomsbury Publishing, 2009), 26.

324 **said one *LIFE* magazine article:** Loudon Wainwright, "The View from Here: All Systems Are Ho-Hum," *LIFE*, December 2, 1966, 30–31, quoted in Donovan, *Shoot for the Moon*, 192.

324 **NASA has a word:** Donovan, 106.

325 **Below the verse:** Erin Blakemore, "Buzz Aldrin Took Holy Communion on the Moon. NASA Kept It Quiet," HISTORY, September 6, 2018, https://www.history.com/news/buzz-aldrin-communion-apollo-11-nasa.

CHAPTER 22: RADIO SILENCE

327 **the only astronaut:** Norman Mailer, *Of a Fire on the Moon* (New York: Random House Trade Paperbacks, 2014), 307.

328 **Intelligence without genius:** Robert Sherrod, “Men for the Moon,” in *Apollo Expeditions to the Moon*, vol. 10, ed. Edgar M. Cortright (Washington, DC: NASA, 1975), 146.

329 **the height and age limitations:** James Donovan, *Shoot for the Moon: The Space Race and the Extraordinary Voyage of Apollo 11* (New York: Little, Brown, 2019), 39.

329 **testing of every sort:** Sherrod, “Men for the Moon,” 146. See also Donovan, *Shoot for the Moon*, 39–43.

329 **There would be another six waves:** Sherrod, “Men for the Moon,” 146.

329 **“the New Nine”:** Donovan, *Shoot for the Moon*, 151.

329 **“the Final Fourteen”:** Donovan, 175.

329 **no one can hear you scream:** *Alien*, directed by Ridley Scott (Brandywine Productions, 1979).

330 **They all three privately:** Kona N. Smith, “This Is What Michael Collins Did During the Apollo 11 Moon Landing,” *Forbes*, July 21, 2018, https://www.forbes.com/sites/kionasmith/2018/07/21/this-is-what-michael-collins-did-during-the-apollo-11-moon-landing/?sh=665c9c3363ca. See also Michael Collins and Edwin E. Aldrin, “The Eagle Has Landed,” in *Apollo Expeditions*, 204.

331 **Fate has ordained:** Bill Andrews, “If the Apollo 11 Astronauts Died, Here’s the Speech Nixon Would Have Read,” *Discover*, July 17, 2019, https://www.discovermagazine.com/the-sciences/if-the-apollo-11-astronauts-died-heres-the-speech-nixon-would-have-read.

331 **Michael orbited above:** Smith, “This Is What Michael Collins Did.”

332 **he put it this way:** Michael Collins, *Carrying the Fire: An Astronaut’s Journey* (1974; repr., New York: First Cooper Square Press, 2001), 402–4.

332 **In the days leading up:** Marcia Dunn, “Apollo 11 at 50: Celebrating First Steps on Another World,” AP News, July 13, 2019, https://apnews.com/article/neil-armstrong-us-news-ap-top-news-michael-collins-tx-state-wire-b8980cbe0b544cffb19398f5d3e99bfa.

333 **As Michael orbited, one author said:** Robin McKie, “How Michael Collins Became the Forgotten Astronaut of Apollo 11,” *The Guardian*, July 18, 2009, https://www.theguardian.com/science/2009/jul/19/michael-collins-astronaut-apollo11.

333 **Collins later wrote:** Jane Lavender, “Horrifying Preparations to Leave Neil Armstrong on Moon to Die if Mission Went Wrong,” *Mirror*, July 20, 2020, https://www.mirror.co.uk/science/horrifying-preparations-leave-neil-armstrong-22373457.

333 **He has been called:** Brian Floca, acknowledgments for *Moonshot: The Flight of Apollo 11*, rev. ed. (New York: Simon & Schuster, 2019).

339 **I wrote in my first book:** Levi Lusko, *Through the Eyes of a Lion: Facing Impossible Pain, Finding Incredible Power* (Nashville, TN: W Publishing Group, 2015), 68.

341 **from whence he will return:** "Apostles' Creed," Loyola Press, accessed August 13, 2021, https://www.loyolapress.com/catholic-resources/prayer/traditional-catholic-prayers/prayers-every-catholic-should-know/apostles-creed/.

342 **7.5 million pounds of thrust:** Karl Tate, "NASA's Mighty Saturn V Moon Rocket Explained (Infographic)," Space.com, November 9, 2012, https://www.space.com/18422-apollo-saturn-v-moon-rocket-nasa-infographic.html.

343 **In the photo is:** Michael Collins, "Earth's Moon—Apollo 11," Catalog of Spaceborne Imaging, accessed September 16, 2021, https://nssdc.gsfc.nasa.gov/imgcat/html/object_page/a11_h_44_6642.html.

343 **The first one through is Buzz:** Collins, *Carrying the Fire*, 419.

344 **"Let's do it":** Dan Wieden, "The History of Nike's Just Do It Slogan," Creative Review, accessed August 27, 2021, https://www.creativereview.co.uk/just-do-it-slogan/.

344 **A design team called Wieden+Kennedy:** Mark Barden and Adam Morgan, *A Beautiful Constraint: How to Transform Your Limitations into Advantages, and Why It's Everyone's Business* (Hoboken, NJ: John Wiley & Sons, 2015), 26.

344 **In a 2015 interview:** Marcus Fairs, "Nike's 'Just Do It' Slogan Is Based on a Murderer's Last Words, Says Dan Wieden," *Dezeen*, March 14, 2015, https://www.dezeen.com/2015/03/14/nike-just-do-it-slogan-last-words-murderer-gary-gilmore-dan-wieden-kennedy/.

344 **In a recent interview:** Simon Sinek, "Live Conversation with Simon Sinek and Scott Harrison," published by charitywater, April 18, 2020, YouTube video, 14:45–16:30, https://www.youtube.com/watch?v=21P0NENWEnw.

345 **what the president foretold:** John F. Kennedy, "Address to Joint Session of Congress" (speech, United States Capitol, Washington, DC, May 25, 1961), Space.com, accessed September 19, 2021, https://www.space.com/11772-president-kennedy-historic-speech-moon-space.html.

CHAPTER 23: MASKS AND THERMOMETERS

347 **There was a pandemic:** Eric Spitznagel, "Why American Life Went on as Normal during the Killer Pandemic of 1969," *New York Post* online, May 16, 2020, https://nypost.com/2020/05/16/why-life-went-on-as-normal-during-the-killer-pandemic-of-1969/.

347 **On July 5, 1969:** "Lunar Lander to Bear Name 'Eagle'; Command Ship Named 'Columbia,'" *Orlando Sentinel*, July 6, 1969, 3A, https://www.newspapers.com/clip/30951306/apollo-11-news-conference/.

348 **enclosed in a three-sided plastic tent:** "Lunar Lander to Bear Name 'Eagle,'" 3A.

348 **President Lyndon Johnson spread:** Michael Beschloss (@BeschlossDC), "LBJ had Apollo 8 astronauts to White House 12 days before Dec. 21 flight," Twitter, May 30, 2020, https://twitter.com/beschlossdc/status/1266842351177211904?lang=en; see also Joanna Brenner, "Rewind: On Christmas Eve 1968, Apollo 8 Orbited the Moon," *Newsweek*, December 24, 2016, https://www.newsweek.com/rewind-christmas-eve-1968-apollo-8-orbited-moon-535844.

348 **NASA's self-important flight surgeon:** James Donovan, *Shoot for the Moon: The Space Race and the Extraordinary Voyage of Apollo 11* (New York: Little, Brown, 2019), 426.

348 **the worst, he said, he had felt in his life:** "Hong Kong Flu," Biomedical Scientist, Institute of Biomedical Science, October 2, 2020, https://thebiomedicalscientist.net/resources/hong-kong-flu.

348 **Frank Borman became sick:** Robert Kurson, *Rocket Men: The Daring Odyssey of Apollo 8 and the Astronauts Who Made Man's First Journey to the Moon* (New York: Random House, 2018), 199–201.

349 **all three astronauts got head colds:** Kurson, 81–82.

349 **refusing to wear their helmets:** Roger Simmons, "Apollo 11 Astonauts Took a Lot of Drugs for Trip Around Dark Side of the Moon," *Orlando Sentinel*, April 25, 2019, https://www.orlandosentinel.com/business/space/apollo-11-anniversary/os-ne-apollo-11-moon-astronauts-drugs-20190426-story.html.

349 **NASA's Christopher Columbus Kraft determined:** Kurson, *Rocket Men*, 83.

349 **only crew to not suffer any divorce:** Kurson, 336.

351 **Laodicea was widely known:** Warren W. Wiersbe, *Be Victorious* (Colorado Springs, CO: David C. Cook, 1985), 44; see also Grant R. Osborne, "Letters to the Seven Churches (2:1–3:22)" in *Revelation* (Grand Rapids, MI: Baker Academic, 2002), 201.

352 **John Stott said:** John Stott, *What Christ Thinks of the Church: Preaching from Revelation 1 to 3*, rev. ed. (Carlisle, UK: Langham, 2019), 89.

353 **In March 2019:** "About: envihab," NASA, accessed August 28, 2021, https://www.nasa.gov/analogs/envihab/about.

353 **Astronauts on the International Space Station:** Loren Grush, "How Do Astronauts Exercise in Space?," The Verge, updated December 23, 2019, https://www.theverge.com/2017/8/29/16217348/nasa-iss-how-do-astronauts-exercise-in-space.

353 **If you're not intentionally moving forward:** Mikhail Gorbachev, "Overcoming Nuclear Danger" (lecture, Harvard, Cambridge, MA, December 4, 2007), quoted in "If You Don't Move Forward—You Begin to

Move Backward," *Harvard Magazine*, December 5, 2007, https://www.harvardmagazine.com/breaking-news/if-you-dont-move-forward-you-begin-move-backward.

356 **Martyn Lloyd-Jones put it this way:** David Martyn Lloyd-Jones, *Studies in the Sermon on the Mount*, vol. 2 (Grand Rapids, MI: Eerdmans, 1959), 248.

357 **the scene in *The Avengers*:** *The Avengers*, directed by Joss Whedon, featuring Tom Hiddleston and Mark Ruffalo (Burbank, CA: Marvel Studios, 2012), DVD, 2:00:10.

357 **stars in our galaxy:** Fraser Cain, "How Many Stars Are There in the Universe?," Universe Today, June 3, 2013, https://www.universetoday.com/102630/how-many-stars-are-there-in-the-universe/.

357 **more bacteria microbes in your stomach:** Andrew Holmes and Carly Rosewarne, "Gut Bacteria: The Inside Story," Australian Academy of Science, April 4, 2016, https://www.science.org.au/curious/people-medicine/gut-bacteria.

CHAPTER 24: BLOOD, SWEAT, AND TEARS

360 **like how young Indy:** *Indiana Jones and the Last Crusade*, directed by Steven Spielberg, scene featuring River Phoenix (Hollywood: Paramount, 1989), DVD, 8:25.

367 **"aliens from *Toy Story 2*" voice:** *Toy Story 2*, directed by John Lasseter, scenes featuring Jeff Pidgeon (Burbank, CA: Walt Disney Pictures, 1999), DVD, 1:13:51.

367 **it's sort of like brimstone:** Chris Hadfield, "Astronaut Chris Hadfield Debunks Common Space Myths," *Wired* video, April 23, 2018, 1:58–2:32, https://www.wired.com/video/watch/space-myths.

367 **like the smell of gunpowder:** Megan Garber, "What Space Smells Like," *The Atlantic*, July 19, 2012, https://www.theatlantic.com/technology/archive/2012/07/what-space-smells-like/259903/.

367 **one of his more famous speeches:** Winston Churchill, "Blood, Toil, Tears and Sweat" (speech, Westminster, London, May 13, 1940), shared by the International Churchill Society, https://winstonchurchill.org/resources/speeches/1940-the-finest-hour/blood-toil-tears-and-sweat-2/.

367 **Smithsonian National Air and Space Museum:** "Apollo 11 Command Module *Columbia*," Smithsonian National Air and Space Museum, accessed August 26, 2021, https://airandspace.si.edu/collection-objects/command-module-apollo-11/nasm_A19700102000.

368 **The Smithsonian also boasts:** "1903 Wright Flyer," Smithsonian National Air and Space Museum, accessed August 13, 2021, https://airandspace.si.edu/collection-objects/1903-wright-flyer/nasm_A19610048000.

368 **brought a piece:** David McCullough, *The Wright Brothers* (New York: Simon & Schuster Paperbacks, 2015), 262.

368 **Orville and Wilbur flew:** "1903 Wright Flyer."

369 **Six and a half hours:** "Apollo 11 (AS-506)," Smithsonian National Air and Space Museum, accessed August 27, 2021, https://airandspace.si.edu/explore-and-learn/topics/apollo/apollo-program/landing-missions/apollo11.cfm.

369 **four hundred thousand people at NASA:** Richard Hollingham, "Apollo in 50 Numbers: The Workers," BBC Future, June 19, 2019, https://www.bbc.com/future/article/20190617-apollo-in-50-numbers-the-workers.

369 **The mission patch for Apollo 11:** Cat Baldwin, "The Making of the Apollo 11 Mission Patch," NASA, last updated August 6, 2017, https://www.nasa.gov/feature/the-making-of-the-apollo-11-mission-patch.

369 **Breaking the trend:** Hannah Baker, "Apollo Mission Patches," National Space Centre, January 16, 2019, https://spacecentre.co.uk/blog-post/apollo-mission-patches/.

370 **for All Mankind:** Brian Dunbar, "We Came in Peace for All Mankind," NASA, last updated July 20, 2020, https://www.nasa.gov/image-feature/we-came-in-peace-for-all-mankind.

370 **Old Glory is still on the moon:** James Fincannon, "Six Flags on the Moon: What Is Their Current Condition?," in Apollo Lunar Surface Journal, last updated April 12, 2012, https://www.hq.nasa.gov/alsj/ApolloFlags-Condition.html.

370 **ran an apology:** Tom Kuntz, "150th Anniversary: 1851–2001; The Facts That Got Away," *New York Times*, November 14, 2001, https://www.nytimes.com/2001/11/14/news/150th-anniversary-1851-2001-the-facts-that-got-away.html.

370 **father of modern rocketry:** Brian Dunbar, "Dr. Robert H. Goddard, American Rocketry Pioneer," ed. Rob Garner, NASA, last updated August 3, 2017, https://www.nasa.gov/centers/goddard/about/history/dr_goddard.html.

370 **And he theorized:** Dunbar.

370 **In 1920:** Elaine M. Marconi, "Robert Goddard: A Man and His Rocket," NASA, March 9, 2004, https://www.nasa.gov/missions/research/f_goddard.html.

371 **had actual gold:** "Gold Coating," NASA Technology Transfer Program, 1997, https://spinoff.nasa.gov/spinoff1997/hm2.html.

END HERE

373 **President Nixon called them:** Michael Collins and Edwin E. Aldrin, "The Eagle Has Landed," in *Apollo Expeditions to the Moon*, vol. 10, ed. Edgar M. Cortright (Washington, DC: NASA, 1975), 216.

374 **In the most famous photo:** "Moonman," Out of This World (exhibit), Smithsonian National Air and Space Museum, accessed August 14, 2021, https://airandspace.si.edu/exhibitions/out-of-this-world/online/1969/MoonmanPhoto.cfm.

374 **Love God:** Bulleted quotes are from Matt. 22:37; Matt. 22:39; Mark 16:15; Mark 16:15 MSG.

375 **It's a beautiful day:** James Donovan, *Shoot for the Moon: The Space Race and the Extraordinary Voyage of Apollo 11* (New York: Little, Brown, 2019), 3; Jennifer Lu, "Deke Slayton's Moon Shot: How the Man Who Picked the Apollo 11 Crew Finally Got to Fly," *La Crosse Tribune*, July 15, 2019, https://lacrossetribune.com/news/local/deke-slaytons-moon-shot-how-the-man-who-picked-the-apollo-11-crew-finally-got/article_b283eb03-1f16-5da4-9949-1a066febd445.html.

ACKNOWLEDGMENTS

377 Gene Kranz, *Failure Is Not an Option: Mission Control from Mercury to Apollo 13 and Beyond* (New York: Simon & Schuster, 2000), 295.

ABOUT THE AUTHOR

LEVI LUSKO IS THE FOUNDER AND LEAD PASTOR OF FRESH Life Church located in Montana, Wyoming, Oregon, Utah, and everywhere online. He is the bestselling author of *Through the Eyes of a Lion*, *Swipe Right*, *I Declare War*, *Take Back Your Life*, and *Roar Like a Lion*. Levi also travels the world speaking about Jesus. He and his wife, Jennie, have one son, Lennox, and four daughters: Alivia, Daisy, Clover, and Lenya, who is in heaven.